Advances in Tuna Aquaculture

From Hatchery to Market

Advances in Tuna Aquaculture

From Hatchery to Market

Edited by

Daniel D. Benetti

Gavin J. Partridge

Alejandro Buentello

ELSEVIER

AMSTERDAM • BOSTON • HEIDELBERG • LONDON
NEW YORK • OXFORD • PARIS • SAN DIEGO
SAN FRANCISCO • SINGAPORE • SYDNEY • TOKYO
Academic Press is an imprint of Elsevier

Academic Press is an imprint of Elsevier
The Boulevard, Langford Lane, Kidlington, Oxford OX5 1GB
225 Wyman Street, Waltham MA 02451

Library of Congress Cataloging-in-Publication Data
A catalog record for this book is available from the Library of Congress

British Library Cataloguing-in-Publication Data
A catalogue record for this book is available from the British Library

ISBN: 978-0-12-411459-3

Publisher: Nikki Levy
Acquisition Editor: Patricia Osborn
Editorial Project Managers: Jaclyn Truesdell and Karen Miller
Production Project Manager: Melissa Read
Designer: Maria Ines Cruz

Contents

List of Contributors

Yasuo Agawa Fisheries Laboratories, Kindai University, Kushimoto, Wakayama, Japan

Jimena Balli Institute of Marine and Antarctic Studies, University of Tasmania, Launceston, TAS, Australia

William H. Bayliff Inter-American Tropical Tuna Commission, La Jolla, CA, USA

Daniel D. Benetti University of Miami, Rosenstiel School of Marine and Atmospheric Science, Miami, FL, USA

Alejandro Buentello Ichthus Unlimited, LLC., West Des Moines, IA, USA

Ben Nan Chen Clean Seas Tuna Ltd., Port Lincoln, SA, Australia

Aldo Corriero Department of Emergency and Organ Transplantation, Section of Veterinary Medicine and Animal Production, University of Bari Aldo Moro, Bari, Italy

Fernando de la Gándara Centro Oceanografico de Murcia, Spanish Institute of Oceanography, Murcia, Spain

David J. Die Department of Marine Ecosystems and Society, University of Miami, Coral Gables, FL, USA

David Ellis David Ellis and Associates, Port Lincoln, SA, Australia

Jose Fernandez-Polanco Department of Business Management, University of Cantabria, Cantabria, Spain

Craig Foster Clean Seas Tuna Ltd., Port Lincoln, SA, Australia

Wayne Hutchinson South Australian Research and Development Institute, Urrbrae, SA, Australia

F. Saadet Karakulak Faculty of Fisheries, Istanbul University, Laleli, Istanbul, Turkey

Keitaro Kato Fisheries Laboratory of Kinki University, Kinki University, Wakayama, Japan

Ilse Kiessling David Ellis and Associates, Port Lincoln, SA, Australia

Ignacio Llorente Department of Business Management, University of Cantabria, Cantabria, Spain

Daniel Margulies Inter-American Tropical Tuna Commission, La Jolla, CA, USA

Ivona Mladineo Laboratory of Aquaculture, Institute of Oceanography & Fisheries, Croatia

C.C. Mylonas Institute of Marine Biology, Biotechnology and Aquaculture, Hellenic Center for Marine Research, Aqualabs, Iraklion, Crete, Greece

Barbara F. Nowak Institute of Marine and Antarctic Studies, University of Tasmania, Launceston, TAS, Australia

Aurelio Ortega Centro Oceanografico de Murcia, Spanish Institute of Oceanography, Murcia, Spain

Bayram Öztürk Faculty of Fisheries, Istanbul University, Laleli, Istanbul, Turkey

Gavin J. Partridge Australian Centre for Applied Aquaculture Research, Challenger Institute of Technology, Fremantle, WA, Australia; Freshwater Fish Group & Fish Health Unit, School of Veterinary & Life Sciences, Murdoch University, Murdoch, WA, Australia

H. Rosenfeld National Center for Mariculture, Israel Oceanographic and Limnological Research Institute, Eilat, Israel

Yoshifumi Sawada Fisheries Laboratories, Kindai University, Kushimoto, Wakayama, Japan

Vernon P. Scholey Laboratorio Achotines, Comision Interamericana del Atun Tropical, Correo General, Las Tablas, Provincia de Los Santos, Republica de Panama

Manabu Seoka Skretting ARC/Japan, Fukuoka, Japan

Sho Shirakashi Fisheries Laboratory, Kinki University, Wakayama, Japan

Maria S. Stein Inter-American Tropical Tuna Commission, La Jolla, CA, USA

John Stieglitz University of Miami, Rosenstiel School of Marine and Atmospheric Science, Miami, FL, USA

Jorge Suarez Rosenstiel School of Marine and Atmospheric Science, University of Miami, Miami, FL, USA

Jeanne B. Wexler Inter-American Tropical Tuna Commission, La Jolla, CA, USA

Taner Yıldız Faculty of Fisheries, Istanbul University, Laleli, Istanbul, Turkey

Y. Zohar Department of Marine Biotechnology, Institute of Marine and Environmental Technology, University of Maryland Baltimore County, Baltimore, MD, USA

About the Editors

Dr Daniel D. Benetti

PhD Marine Biology and Fisheries, Rosenstiel School of Marine and Atmospheric Science, University of Miami, FL, USA
MS Aquaculture and Fisheries Management, University of Stirling, Scotland, UK
BSc Biological Oceanography, University of Rio Grande, Brazil

Dr Daniel Benetti is a Professor and Director of Aquaculture at the University of Miami's Rosenstiel School of Marine and Atmospheric Science. He has over 30 years of experience in aquaculture worldwide. Besides his academic and research responsibilities, he carries out scientific and R&D projects on technology development, production optimization, and environmental issues related to aquaculture. He specializes in advanced hatchery, land-based and open ocean growout technologies of marine finfish such as tuna, cobia, Seriola, mahi, snapper, grouper, pompanos and flounder and others.

He has published over 100 articles in aquaculture technology and production, has extensive experience with the industry, and has been a consultant for the private and government sectors in Latin America, the USA, Europe, Asia, Caribbean, Africa, Australia, and the Middle East. He collaborates on tuna aquaculture with researchers and institutions the world over and is internationally recognized for his contributions to modern aquaculture. His work is centered on innovative approaches to ensure that seafood production through mariculture is wholesome, environmentally sustainable, and economically viable.

Dr Gavin J. Partridge

PhD Veterinary and Biomedical Sciences, Murdoch University, WA, Australia
BSc (Hons I) Zoology, James Cook University, QLD, Australia
BSc Physical and Inorganic Chemistry, University of Western Australia, WA, Australia

Dr Gavin Partridge is the Principal Research Scientist at the Australian Centre for Applied Aquaculture Research and an Adjunct Associate Professor at the School of Veterinary and Life Sciences at Murdoch University. He has been working in the field of marine fish aquaculture for

20 years, during which time he has worked on all aspects of production science including hatchery production and growout in both land-based recirculating systems and sea cages. He has an applied focus and works closely with industry to overcome production challenges and bottlenecks. He has worked throughout the world on a wide range of marine fish species including tunas and has published widely on a variety of topics relating to marine fish aquaculture including hatchery production, nutrition, health, and environmental issues.

Dr Alejandro Buentello
PhD Nutrition and Biochemistry, Texas A&M University College Station, TX, USA
MS Fish Nutrition, Texas A&M University College Station, TX, USA
BS Biochemical Engineering and Aquatic Resources Management, Monterrey Institute of Technology & Higher Education Sonora, Mexico

Dr Alejandro Buentello is a Fish Nutritionist and Physiologist and presently he is the Director of Animal Nutrition Research at Archer Daniels Midland Co. (ADM), in Decatur, IL. He is also the founder and President of Ichthus Unlimited, LLC, an aquaculture consulting and research and development service company in West Des Moines, IA.

Dr Buentello has over 25 years of experience in applied research in aquaculture nutrition and feed technology and has coordinated numerous extramurally funded research programs in collaboration with leading nutritionists worldwide. He has also conducted extensive research on various aspects of tuna nutrition and physiology for more than 15 years, often serving as scientific consultant for tuna ranching operations in Mexico and South America.

Before joining ADM, Dr Buentello was Vice President of Research at Navita Premium Feed Ingredients, serving as the company's point-person before global aquafeed manufacturers. He also served as Senior Associate Scientist and Lecturer at Texas A&M University and as Assistant Professor at the Northwest Center for Biological Research in Baja California Sur, Mexico, conducting both research and teaching activities. Presently, Dr Buentello holds adjunct appointments at the University of Maryland Baltimore County, Iowa State University, and Texas A&M University.

Chapter 1

Overview on Status and Technological Advances in Tuna Aquaculture Around the World

Daniel D. Benetti[1], Gavin J. Partridge[2,3] and John Stieglitz[1]
[1]University of Miami, Rosenstiel School of Marine and Atmospheric Science, Miami, FL, USA, [2]Australian Centre for Applied Aquaculture Research, Challenger Institute of Technology, Fremantle, WA, Australia, [3]Freshwater Fish Group & Fish Health Unit, School of Veterinary & Life Sciences, Murdoch University, Murdoch, WA, Australia

1.1 INTRODUCTION

Advances in Tuna Aquaculture is the first book that encompasses all aspects related to this industry and it merges them into a state-of-the-art compendium that points the reader in the right direction, whether a science student, a researcher, a fisherman, or a farmer. It presents developments in tuna aquaculture throughout the world, from ranching wild juvenile fish to closed-cycle cultivation of a variety of tuna species with a focus on the high-value bluefin species. Reputed experts in their fields provide detailed accounts of the various disciplines directly or indirectly associated with tuna aquaculture. This introductory chapter summarizes the content of the book and provides an outlook for the future of the industry.

Tuna are some of the best-known and highly regarded species of fish. While "tuna" refers to a large number of scombrid species including skipjacks (SJT), bonitos, bullets, and frigates, most people associate the term with the large, high-value species of the genus *Thunnus* such as bluefin, yellowfin, and bigeye tuna that roam the world's oceans. Serving important ecological roles as both predator and prey, depending on the life stage of the fish, these species are now some of the most sought after marine fish for the global seafood market. The high market value of tuna stocks has led to intensified fishing pressure that, in turn, resulted in drastic population reductions in every ocean where these fish are found. High prices sustained by strong market demand also create opportunities for tuna ranching, which is

D.D. Benetti, G.J. Partridge, A. Buentello (Eds): Advances in Tuna Aquaculture.
DOI: http://dx.doi.org/10.1016/B978-0-12-411459-3.00001-1

arguably the most profitable form of fish farming in the world. Today, the practice of tuna ranching occurs at the intersection between the aquaculture and fisheries sectors which makes it difficult to consider them independently from each other. Complicating the issue is the fact that ranched tuna are typically fattened using wild-caught small pelagic fish species such as sardines and mackerel, as opposed to balanced diets, which is the case for the grow-out of other marine fish. With aquaculture now providing over half of the seafood consumed in the world, and future forecasts predicting consistent growth in global fish farming, it comes as no surprise that efforts to mass-produce tuna are following this trend.

Tuna industry stakeholders, from fishermen, farmers, and scientists to seafood business professionals and consumers, face enormous challenges and uncertainties when considering the sustainability of this industry. In an effort to better understand these issues and effectively manage the world's tuna stocks, five regional intergovernmental fisheries organizations—the Inter-American Tropical Tuna Commission (IATTC), the Western and Central Pacific Fisheries Commission (WCPFC), the Commission for the Conservation of Southern Bluefin Tuna (CCSBT), the Indian Ocean Tuna Commission (IOTC), and the International Commission for the Conservation of the Atlantic Tunas (ICCAT)—study tuna populations and the biotic, abiotic, and anthropogenic factors that influence these fisheries. In the present book, renowned scientists from both IATTC and ICCAT discuss important management challenges, as well as those brought about by the shift from traditional tuna fisheries to fattening (i.e., ranching) and farming, in Chapters 2 and 3.

While tuna fishing has been practiced for several millennia, tuna aquaculture is a relatively new industry. Although early tuna aquaculture efforts date back to the late 1960s, it was not until the early 1990s that industrial-scale ranching and farming developed into the modern industry that it is today. Similarly, research into closed-cycle aquaculture production of tuna began in the 1970s. However, due to the many challenges associated with hatchery production, the rewards of this protracted effort are only now being realized, with significant numbers of hatchery produced juveniles currently being put to sea, predominantly in Japan and to a smaller extent in Europe. The accomplishments by Japanese scientists, which represent an important milestone, have led to advances in hatchery technology throughout the world, and have made closed-cycle tuna production a reality.

It is difficult to accurately report on total global catches and farmed tuna production. Even data reported from the most credible and reputable sources are incomplete, contrasting, and often conflicting. This is reflected, to a great extent, throughout the different chapters of this book. According to the latest fisheries statistics available, catches of tuna and tuna-like species continue to increase and set a new record at more than 7 million metric tons in 2012 (FAO, 2014). However, the majority of this catch comprised low-value skipjack, primarily sold to the canning industry. Seven species have consistently accounted for about 90% of the total tuna catch since 2000. Catches of small tuna (such as skipjack, frigate, and bullet tuna), seer fishes (*Scomberomorus* spp.), and albacore have grown significantly. In 2012, catches of yellowfin tuna exceeded the 2000 level, after fluctuations in prior years, while bigeye tuna had the only decreasing trend with catches down by 5% (FAO, 2014).

The total volume of the higher value, fresh or frozen product destined primarily for the sashimi market, is approximately one-third of the total. The actual quantity of tuna produced via aquaculture is difficult to accurately quantify, and there is great disparity among the figures quoted by the various reporting agencies and governments. The reasons for these discrepancies are described in detail in a recent paper by Metian et al. (2014). The data from the FAO in Table 1.1 suggests that the annual global aquaculture production of bluefin tuna in the years between 2011 and 2013 ranged from ca. 9400 to 23,500 metric tons, however, this dataset omitted the production from several countries and underreported production from others. For example no production was included for Japan in 2011, the Australian production was underreported and the production of ABFT was listed as only ca. 3000 to 4000 metric tons, despite a TAC of ca. 13,000 metric tons during this period. We believe a more accurate estimate of the current total bluefin aquaculture production is that provided by Tveteras et al. (2015) of

TABLE 1.1 Aquaculture Production (metric tons) of Pacific, Southern and Atlantic Bluefin Tuna as Reported by the FAO

Country	Species	2011	2012	2013
Japan	PBFT		9639	10,396
Mexico	PBFT	3557	1784	6228
Australia	SBFT	1987	2486	3482
Croatia	ABFT	1610	1125	915
Spain	ABFT	575	555	305
Malta	ABFT	960	530	985
Turkey	ABFT	100	395	470
Tunisia	ABFT	70	220	630
Italy	ABFT	435	85	85
Greece	ABFT	95	30	55
Total		9389	16,849	23,551

Source: FAO (FishStat).

TABLE 1.2 Aquaculture Production (metric tons) of Pacific, Southern and Atlantic Bluefin Tuna in 2014 as Reported by Tveteras et al. (2015)

Region	Species	2014
Japan	PBFT	9000
Mexico	PBFT	4500
Australia	SBFT	8350
Mediterranean	ABFT	14,500
Sum		36,350

ca. 36,000 metric tons (Table 1.2). Tuna aquaculture production currently comprises only the three highly regarded bluefin species, namely, Pacific bluefin tuna (PBFT, *Thunnus orientalis*), Atlantic bluefin tuna (ABFT, *Thunnus thynnus*), and southern bluefin tuna (SBFT, *Thunnus maccoyii*). PBFT is cultured in Japan and Mexico, ABFT in several countries bordering the Mediterranean Sea, and SBFT only in Australia. While yellowfin tuna (YFT, *Thunnus albacares*) has previously been cultured in Mexico and Oman (see Chapter 8), there is currently no aquaculture production of this species. However, this is changing rapidly, as we are currently witnessing breakthroughs with YFT hatchery technology (Chapter 5).

Production volumes of bluefin tuna vary considerably between species and the regions in which they are grown, and also over time as catch quotas change. For recent years, indicative figures are summarized in Tables 1.1 and 1.2. While Japan is now the largest producer, this is a recent occurrence, and the production of Japanese farmed PBFT only exceeded that of ranched SBFT and ABFT in recent years (Tada, 2010).

As described throughout this book, the vast majority of tuna aquaculture production is still reliant upon the capture of wild-caught juveniles. This production is referred to as ranching, fattening, or farming, and these terms are often used interchangeably. In the true sense, however, ranching and fattening differ from farming. The former two terms are more accurately applied to the short-term practice of holding subadult or adult tuna primarily for the purposes of increasing condition and fat content, thus improving their market value. However, significant increases in biomass are often also achieved. This is the practice employed in Australia and Mexico, and by the majority of Mediterranean producers. The ranching or fattening period differs between regions but typically only lasts up to several months. The use of such fattening practices has grown exponentially in the last decade, to the point where today, a large proportion of all bluefin tuna caught in many parts of the world are stocked into cages for increasing their size and fat content prior to being sold. In spite of the inherent issues of the operation (they are fed sardines and other forage fish which are ecologically very important), this practice could be considered a "value-added" process, since the individual value of the tuna kept in the cages increases considerably in what can be viewed as a more energetically and ecologically efficient way than would occur nature. Farming involves the capture of juveniles which are then held for extended periods of up to several years. Farming of wild-caught juvenile tuna occurs only in Japan and Croatia, although preadults are also caught and grown in Mexico as they migrate across the Pacific Ocean. The more recent practice of growing hatchery reared juveniles is also termed "farming." While we have attempted to apply these terms consistently throughout this book, the term "farming" is often applied as a generic term to all of these practices.

The boom in the tuna farming industry over the last 25 years has been predominantly driven by the high prices achieved in Japan, where some 400,000–500,000 metric tons of tuna are consumed annually, primarily as sashimi and sushi (OPRT, 2015) (Figure 1.1).

FIGURE 1.1 (A) Toro, the fattest, finest, and most expensive bluefin tuna sashimi in the world; (B) Maguro, also greatly appreciated in the sashimi market worldwide, especially in Japan; (C) Increased demand for high-quality tuna for sashimi and sushi markets in Japan and elsewhere is driving the exploitation of their fishery stocks and efforts to develop close-cycle aquaculture. *Photos: Daniel Benetti.*

Exorbitant and headline-making prices are often paid for bluefin at the first auction of the season at the Tsukiji, Nagoya, and Osaka markets. However, average prices are significantly less. While the current record price for a bluefin stands at US$1.763 million for a 222 kg PBFT (equating to nearly 8000 USD/kg), the average prices for bluefin range widely from approximately 7–100 USD/kg, with high-grade tuna currently ranging from 40 to 100 USD/kg and low-grade from 10 to 20 USD/kg. Fernandez-Polanco and Llorente describe in Chapter 14 that the market in Japan is currently stagnating or slightly decreasing, yet the explosion in the popularity of Japanese cuisine throughout the rest of the world has continued to sustain demand. The history and current status of the farming and hatchery production of the main tuna species are summarized below.

1.2 PACIFIC BLUEFIN TUNA

Japan's farming history is described in Chapter 7. While it dates back to the 1970s, these early attempts were aimed at capturing immature fish and growing them to maturity to be used as broodstock. Commercial-scale farming did not begin until the early 1990s, with the first harvest of some 900 metric tons occurring in 1993 (Tada, 2010). As previously described, PBFT in Japan are farmed rather than ranched, and the methods therefore differ from the ranching of other species and regions. The main difference lies in the fact that early juvenile PBFT weighing only 100–500 grams are stocked and grown to a market size of 30–50 kg over a period of 2–3 years (Masuma et al., 2008). Farmed tuna production in the late 1990s and the early years of the 2000s ranged from 2000 to 7000 metric tons/annum (Tada, 2010). It was not until 2010 that the Japanese Ministry of Agriculture, Forestry and Fisheries (MAFF) introduced mandatory reporting requirements for Japanese tuna farmers (Koya, 2010), and these more recent production figures, which now exceed 9000 metric tons/annum, are detailed in Chapter 8. In 2012, MAFF reported a production of 9592 metric tons (similar to that reported by the FAO in Table 1.1), of which 244 metric tons were derived from hatchery-reared juveniles. Presumably these were the juveniles harvested from the first large-scale commercial stocking of ca. 45,000 hatchery reared juveniles in 2009 (Tada, 2010).

The evolution and advances in hatchery technology for PBFT since 1970 are also described in Chapter 7. These developments have been driven primarily by Kinki University (now known as Kindai University) with the federal government (Fisheries Research Agency), prefectural governments, and private companies also playing important roles. Egg production from PBFT in Japan is derived almost entirely from naturally spawning fish held in sea cages, with the first land-based broodstock system only recently commissioned in Nagasaki in 2013. The development of hatchery technology has been a very challenging road, with significant mortality

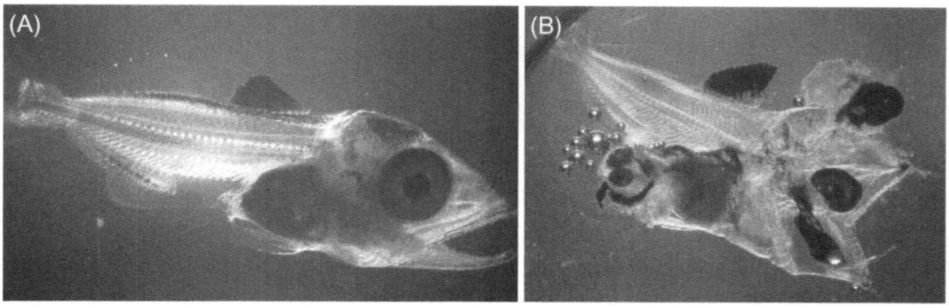

FIGURE 1.2 (A) During the third week post-hatch, yellowfin tuna larvae have large mouths, well-developed eyes and teeth, and begin their piscivorous stage, becoming extremely aggressive and cannibalistic. (B) Gut analysis of a 15-day post hatch (15 DPH) yellowfin tuna larvae showing another larvae of the same batch in the gut. All tuna species are highly cannibalistic. *(A) Photo: Craig Purcell, (B) Photo: Zack Daugherty.*

events occurring during every life stage. These can be summarized as (i) floating death and sinking death during the first 8 days of life (Ishibashi, 2010), (ii) cannibalism and collision death during the post-flexion stages, and (iii) transfer-related mortalities after they are moved from land-based hatcheries to sea cages. Mortality rates during each period can approach 90% and the overall average survival of tuna larvae ranges from 0.01% to 4.5% (Masuma et al., 2008). Feeding tuna larvae also represents a challenge not experienced in the culture of other marine fish, as they must be fed on the yolk-sac larvae of other marine fish, with feeding on *Artemia* alone resulting in "growth failure" (Seoka et al., 2007). Arguably, cannibalism remains the most serious problem causing low survival rates of all species of tuna larvae (Figure 1.2).

As is described throughout this book, these same challenges have also been experienced in all other tuna species reared to date. The life cycle of PBFT was closed by Kinki University in 2002, with the first successful spawning of first generation broodstock (Sawada et al., 2005). Between 2002 and 2007, approximately 10,000 juveniles per year were produced, and in 2009 this increased to 45,000 (Tada, 2010). The outlook from recent MAFF reports for the increasing contribution of hatchery reared tuna to the total tuna production is positive, as they report that in 2012, 56% of the 474,000 juvenile tuna stocked in cages were hatchery produced. Large commercial companies such as Maruha Nichiro and Nippon Formula Feed Manufacturing Company, among others, have invested in tuna hatcheries and are now producing commercial quantities of juveniles (Masuma et al., 2011).

Progress in PBFT hatchery production has also been reported from South Korea with several thousand PBFT juveniles having been produced at the Future Farming Research Center of the National Fisheries Research and Development Institute (NFRDI), the Jeju Ocean and Fishery Institute, and Gyeongsangnam-do Fisheries Resources Research Institute. The juveniles are being grown in SeaStation submersible sea cages in high-energy sites off Jeju Island, where the occurrence of typhoons is a significant challenge to the success of the grow out industry.

Mexico produces the second largest volume of ranched PBFT after Japan and this industry is also described in Chapter 8. Mexican tuna farms exploit the same stock of PBFT as the Japanese, as the fish migrate across the Pacific Ocean. Ranching in Mexico follows similar methods to those used in Australia and the Mediterranean. There have been no attempts in Mexico at PBFT hatchery production.

1.3 ATLANTIC BLUEFIN TUNA

Farming of the eastern stock of ABFT occurs in several countries which border the Mediterranean Sea and the Atlantic coast of Portugal, and has occurred on an industrial-scale since the mid-1990s after beginning in Spain in 1979 (Ottolenghi, 2008). The tenuous state of wild ABFT stock in the eastern Atlantic has been the subject of much controversy and quota restrictions and is well documented throughout this book.

ABFT for ranching are targeted by purse-seiners and traditional trap fishermen during their spawning migrations, and typically range in size from 40 to 400 kg (Figure 1.3). Fish targeted for farming in Croatia range from 8 to 30 kg. The fattening period in the Mediterranean ranges from 3–7 months, and the farming period in Croatia lasts up to 2 years (Mylonas et al., 2010). As with all ranching operations, production is closely linked to the quotas placed on the wild stocks, and this is discussed in detail in Chapters 3 and 6.

In an effort to reduce the reliance on wild-caught juveniles and enable consistent supply, there has been an intensive effort on closing the life cycle of ABFT in Europe since the early 2000s. The European Union has invested in excess of €10 million in major R&D consortiums such as

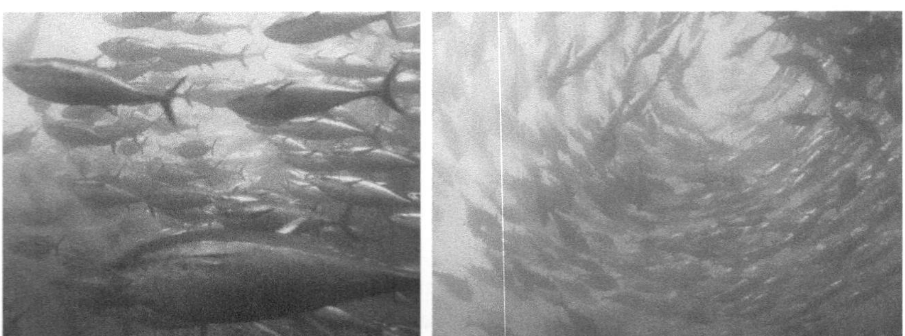

FIGURE 1.3 Atlantic bluefin tuna cultured in cages in the Mediterranean. *Photos: Valerio Vitalini and Daniel Benetti.*

the REPRODOTT and SELFDOTT, with considerable funding from other European governments and companies. There has been a strong collaborative focus during these projects, which has yielded excellent outcomes in terms of both pure science and the important applied outcome of juveniles put to sea. These outcomes and achievements are described in detail by Karakulak et al. and de la Gandara et al. in Chapters 4 and 6, respectively. Egg production has again been heavily reliant on broodstock held in cages, both from natural spawning and hormone induction techniques using a variety of specialized sustained release hormone implants (primarily GnRH-a). Spawning induction using this practice has led to remarkable improvements in the quantity of eggs collected and in extending the captive-spawning season to over 2 months instead of only a few weeks. These techniques are described in detail in Chapter 7. The largest land-based bluefin tuna broodstock facility in the world has recently been completed at the Spanish Institute of Oceanography (IEO) in Mazarron, Spain, and is described in Chapter 6. This facility will allow ABFT to be spawned in captivity under controlled conditions, and represents a significant advancement in the closed-cycle production of this species. In 2011, the IEO put some 3000 juvenile ABFT to sea and these are now approaching maturity.

Great progress is also being reported from Cyprus and Turkey: Kiliç Seafood Co., the largest producer of sea bream and sea bass in Europe, has recently begun developing hatchery technology and grow out of ABFT in the Aegean Sea. In 2014–2015, the company successfully produced 15,000 juveniles from eggs collected from its own broodstock cages. Juveniles weighing 5–10 grams were transferred to nursery cages and as of the time of this publication, a total of 1000 fish weighing approximately 10 kg at 420 days post hatch remain.

Europeans are therefore on the verge of closing the life cycle of ABFT as has been achieved with PBFT in Japan.

1.4 SOUTHERN BLUEFIN TUNA

The ranching of SBFT is described in detail by Ellis and Kiessling in Chapter 9. South Australia is the only place in the world where SBFT are farmed, despite several other countries also owning quota. Total global quota of SBFT has been diminishing since the introduction of quotas in 1989. In 2014, however, CCSBT's Scientific Committee reported that the wild stocks of SBFT have been recovering, and the global TAC was increased from the 9449 metric tons set in 2010 to 12,449 metric tons in 2014 and to 14,647 metric tons for the period 2015–2017. Australia's share of the global TAC between 2015 and 2017 will be 5665 metric tons, the highest since 1989.

Ranching of SBFT in Australia involves the purse-seining of subadult fish in the Great Australian Bight, which are then towed into the ranching sites in the waters surrounding Port Lincoln. At this stage in their migration, the fish are approximately 2 years old and weigh around 15 kg. They are fattened for 6–9 months on local and imported baitfish, during which time they increase in size to 30–40 kg. Approximately 20% of the harvested product is flown fresh to Japan, with the remainder shipped frozen.

Efforts to close the life cycle of SBFT have been driven by one Australian company, Clean Seas Tuna (CST), with funding support from the Australian Federal Government. These efforts are described by Chen et al. in Chapter 10. In summary, CST's tuna propagation program began in 1999 with approximately 500 of their farmed stock being retained in sea cages for on-growing as future broodstock (Stehr, 2010). CST built the world's first land-based broodstock system for any bluefin species in 2006, and some of the retained fish were transferred from the sea cages to the broodstock tank via helicopter (Partridge, 2013; Stehr, 2010). Closing the lifecycle of SBFT presents additional challenges compared to similar efforts with ABFT and PBFT. While the latter two species mature at 3–5 years and ca. 50–60 kg (Masuma et al., 2008; Ottolenghi, 2008), SBFT mature somewhere between 8 and 14 years at sizes in excess of 100 kg. Despites this challenge, CST first achieved successful spawning from hormone-induced broodstock in 2008 (Thomson et al., 2010). In 2009, approximately 50 million eggs were produced (Stehr, 2010). Juveniles were successfully reared on land to an age of 238 days and *Time Magazine* named CST's "Tank-Bred Tuna" the second best invention of the year. The first successful transfer of 149 juvenile SBFT to the sea was achieved in 2011. However, after growing to a size of ca. 500 grams, they succumbed to South Australia's cold winter water temperatures (CST, 2012a). It was planned to advance the spawning season in subsequent years in order to enable larger and more robust fish which could tolerate the low winter water temperatures to be put to sea. While some advanced spawning events were achieved in the controlled environment broodstock tank, the results were poor in terms of the number of fertilized eggs produced (Chapter 10). In 2012, CST announced the suspension of its SBFT propagation program in order to focus their efforts and resources on their yellowtail kingfish operation (CST, 2012b). The broodstock have been retained in the hope of restarting the program at some point in the future (CST, 2015).

1.5 YELLOWFIN TUNA AND OTHER TUNA SPECIES

Although the value of YFT is less than that of bluefin tuna, there is still considerable interest in the aquaculture of this species. It is in Mexico that the majority of attempts to ranch YFT have been made, but as described in Chapter 8, problems have arisen primarily due to availability of suitably sized wild stock, rather than issues regarding their biological or technical suitability for ranching. Hatchery production of YFT would overcome these issues and as described in detail in Chapter 5, great progress is being made toward this goal.

YFT grow rapidly, mature at an earlier age, smaller in size than bluefin, and spawn readily in captivity nearly year-round and without hormone induction. Research into hatchery production of YFT has been undertaken primarily at the Inter-American Tropical Tuna Commission's (IATTC's) laboratory in Panama, and, to a lesser extent, at the Gondol Research Institute for Mariculture (GRIM) in Indonesia. Both land-based broodstock facilities were constructed in collaboration with the Japanese Overseas Fishery Cooperation Agency (OFCF). While the focus of the work at the IATTC has been primarily to gain a better understanding of the ecology of this species in the wild, many of their findings have direct relevance to commercial hatchery production of both YFT and the bluefin species. Recent research collaboration between the IATTC, Kinki (Kindai) University, and the Autoridad de los Recursos Acuáticos de Panamá (ARAP) has advanced the science of both YFT and PBFT through comparative studies of these two species. As part of this research, many notable achievements have been made including the recent transfer of the first juvenile YFT to an experimental floating sea cage for grow out off the Pacific Coast of the Republic of Panama. Worldwide, this represents the first successful transfer of hatchery-produced YFT juveniles to a sea cage. The juveniles stocked in the sea cage ranged in length from 9 to 13 cm, and were raised from fertilized eggs spawned by captive YFT broodstock kept at the ARAP/IATTC Achotines Laboratory in Panama. This achievement marks an important milestone toward closed-cycle production of this species and increased understanding of the later life stages of YFT in captivity.

Unlike the IATTC facility, GRIM in Indonesia has a focus on commercial aquaculture production. Since beginning operations in 2003 it has, however, received intermittent funding, which has hampered its ability to continuously retain broodstock and therefore sustain spawning. The occurrence of the egg protozoan parasite described in Chapter 11 and by Hutapea and Permana (2007) has also hampered larval rearing efforts, and the oldest juvenile produced at GRIM has been 53 days of age. Aside from the challenges faced by the tuna research program at GRIM, much progress has been made in the development of capture, transport, and acclimation of YFT to the land-based holding facilities. The GRIM facilities and their achievements are described in detail by Partridge (2013), Hutchinson et al. (2011), and Hutapea et al. (2009).

A number of other research and development projects have occurred in the past few decades examining the feasibility of closed-cycle production of other tuna and tuna-like species such as the bigeye tuna (*Thunnus obesus*),

FIGURE 1.4 Blackfin tuna (*Thunnus atlanticus*) metabolism and swimming energetics studies are providing the basic scientific information required to develop sound aquaculture technologies for this species in the near future. *Photo: John Stieglitz.*

blackfin tuna (*Thunnus atlanticus*), and bonito species (Atlantic: *Sarda sarda*; Pacific: *Sarda chiliensis*). These species, while not as well-known as bluefin and yellowfin tuna, are nonetheless valuable to the fishing industry. Given the smaller size of blackfin tuna and bonito, these species are easier to transport and maintain in smaller land-based facilities compared to the infrastructure involved in maintaining bluefin or yellowfin tuna broodstock in land-based tanks. In the case of the bonito species, the life cycle has been closed on an experimental scale (McFarlane et al., 2000; Ortega et al., 2013).

Blackfin tuna is the smallest of the true tuna species of the genus *Thunnus*, reaching 1 m and 20 kg within a relatively short life span of 5 years. While it is unlikely that blackfin tuna will ever compete in the same global marketplace of its larger counterparts, as a true *Thunnus* species it deserves attention for commercial aquaculture development, targeting another segment of the market—a smaller size, yet high-quality, sashimi-grade tuna. Attempts to spawn blackfin tuna in land-based tanks have not been successful thus far (Benetti et al., 2009). Despite ongoing research efforts investigating their bioenergetics (Figure 1.4), spawning this species of tuna in captivity remains challenging.

1.6 HEALTH, NUTRITION, AND GENETICS

While Balli and colleagues describe in Chapter 11 a wide range of diseases (and particularly parasites) that have been found in tuna, it is also clear from this chapter that tuna are remarkably robust and have not suffered a great deal from acute or chronic mortalities associated with disease outbreaks. Indeed, many of the parasites hosted on the wild-caught fish introduced to ranching are lost once they are held in cages. These authors also effectively point out, however, that more diseases are likely to be encountered as farming intensifies and once the life cycle of more species are closed and naïve fish are put to sea.

With consistent growth in the global demand for high-value tuna comes a need to develop cost-effective and environmentally sustainable methods to satisfy this demand. Given that feeds comprise, on average, over 60% of operating costs for tuna farming operations, research into tuna nutrition offers the potential to help offset the high costs involved in producing farm-raised tuna for the global market. The small pelagic fish (erroneously termed "trash fish"), still primarily being used to feed tuna in cages, are less expensive than formulated diets but have a significant ecological footprint. Consequently, its use is highly controversial and constantly challenged by some influential scientists and powerful environmental groups. Even though many believe that feeding small pelagics to grow tuna to market size is "value-added fisheries" (i.e., it is more ecologically efficient than the efficiencies that exist in the food chain in the wild), this ongoing practice is generally recognized as ecologically unsustainable, costly, and unpractical given the logistics required to obtain, transport, and handle the small pelagics from the fishing source to the feeding of the tuna in the cages. The obvious route will be to develop feed formulations that are more economically and ecologically efficient, and much progress is being made toward this goal, as reviewed by Buentello and colleagues in Chapter 12. There are many inherent qualities in tuna that make them challenging candidates for robust nutritional studies, yet as tuna culture expands globally, more nutrition research is being conducted on these high-value animals. Given the high-performance metabolism of tuna, and of scombrids in general, it is intuitive that high-energy diets have been found to be positively correlated with growth, survival, and food conversion ratios in such species. However, different species and different life stages of tuna possess specific nutritional demands, and one of the keys to unlocking the full aquaculture potential of these animals lies in gaining a better understanding of their physiological and nutritional requirements.

Finally, recent efforts to develop techniques for genetic improvement of PBFT through a novel breeding program at Kindai University in Japan are described by Sawada and Agawa in Chapter 13. The breeding program is targeting improved reproduction, increased survival and growth rates, disease resistance meat quality and yield. Ultimately, the program aims at the domestication of PBFT while maintaining genetic diversity to preserve the gene pool of natural populations in case of escapements. Although the breeding program has just begun, it is developing fast and requires the simultaneous implementation of a strategy for intellectual property protection in anticipation of potential conflicting issues in the future development of full-cycle tuna aquaculture technologies. Sawada and Agawa discuss these important topics in Chapter 13.

1.7 MOVING TO COMMERCIAL-SCALE HATCHERY PRODUCTION

The vast majority of tuna industry stakeholders now widely accept the view that aquaculture will play a key role in the future of tuna fishery stocks. However, to provide the economic stimulus for moving from wild-caught to hatchery-produced juveniles, the price of producing the latter should be cheaper than the cost of catching the former. While there appears to be no data on the current costs of producing juvenile PBFT in Japan, they are likely to be very high given the low survival rate and the high cost of maintaining tuna broodstock. Alternative technologies under investigation in Japan for overcoming the high costs of maintaining broodstock include the development of true surrogates in which smaller, highly fecund species such as mackerel could deliver viable tuna gametes, as has been conducted with other species (Masuma, 2010; Okutsu et al., 2007; Partridge, 2013). Other methods with potential include obtaining viable gametes from commercially farmed PBFT as they are being harvested for market and inducing early maturation (Elizur et al., 2009; Masuma, 2010; Endo et al., In Press).

While the number of hatchery-reared juvenile PBFT produced in Japan has increased rapidly in recent years, much of this increase can be attributed to the increases in the number of eggs used rather than dramatic improvements in larval survival. Figure 1.5 highlights the relationship between the number of eggs used and the number of juveniles produced. Although this graph shows different numbers of juveniles produced compared to those reported elsewhere in this book, these inconsistencies are likely the result of the high post-transfer mortality and therefore, the stage at which the assessment of the number of juveniles produced is taken.

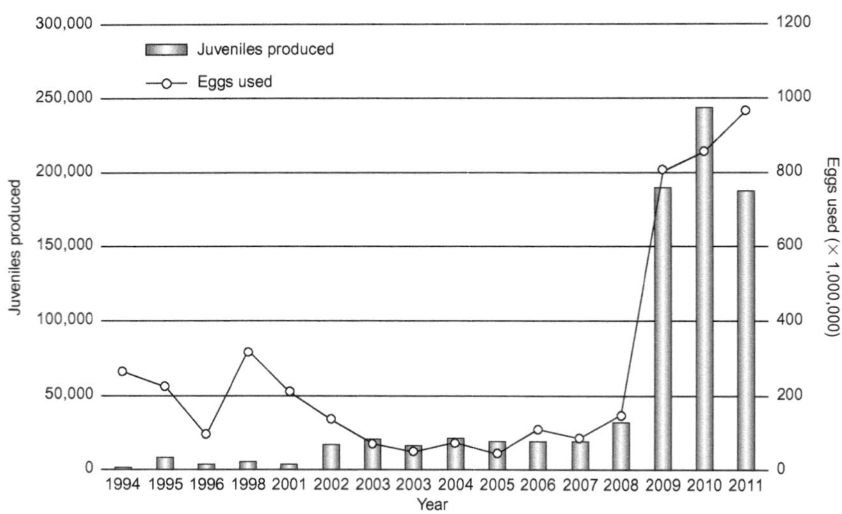

FIGURE 1.5 Relationship between number of eggs and juveniles production of Pacific bluefin tuna in Japan.

The reliance on increasing egg numbers to increase juvenile supply is a strategy that may be viable for early-maturing species such as PBFT and YFT, which may begin spawning during the normal farming period and which spawn in cages or relatively simple land-based systems. For species such as SBFT this strategy will not be as straightforward, as they mature at a much older age and do not undergo final maturation in the cool waters where they are currently ranched (Bubner, 2012).

A number of land-based bluefin broodstock facilities have recently been commissioned to control and extend the natural breeding season. While this will be an effective strategy to supplement the eggs collected from caged broodstock, land-based facilities may not provide sufficient eggs to produce commercial quantities of juveniles in their own right, unless concomitant and significant increases in hatchery and nursery survival rates also occur. For example, Figure 1.5 demonstrates that the production of ca. 250,000 juvenile tuna in Japan in 2011 was derived from an egg collection during that period of nearly 1 billion eggs. Obtaining such quantities of eggs only from land-based facilities may be cost-prohibitive. However, because of such high stakes and of the success of natural year-round spawns of YFT in captivity at IATTC's Achotines Laboratory in Panama, major efforts and investments have been made in Japan, Spain, Australia, and South Korea to build sophisticated land-based facilities to spawn bluefin tuna in captivity with or without the use of hormone implants. Results of these efforts remain uncertain to date, and egg collection from both PBFT and ABFT continue to be conducted successfully from natural and induced spawnings in cages (Figure 1.6).

FIGURE 1.6 Collecting Atlantic bluefin tuna naturally spawned eggs in a cage in Calabria, Southern Italy. Similar practices are carried out extensively wherever tuna are spawned naturally or through the use of hormone induction, including in Japan, Spain, Malta, and Croatia. *Photo: Daniel Benetti.*

To reduce the production costs of juvenile tuna, significant increases in survival rates will also be required, thereby leading to more stable and consistent production. A great deal of research is therefore still required to overcome the various bottlenecks that cause significant mortalities during each life stage which are described throughout this book. The development of effective early weaning diets will also greatly assist in overcoming the many disadvantages of feeding yolk-sac larvae, including the costs of maintaining separate broodstock populations and the risks of disease transmission (Masuma et al., 2011).

1.8 CONCLUSIONS

While fisheries management strategies will continue to play an important role in preserving wild tuna stocks at sustainable levels, it is clear that tuna aquaculture will be a crucial component of global tuna conservation. Closing the life cycles of the primary high-value tuna species will allow for more sustainable production to satisfy the world's insatiable demand. The fisheries trends for these species will likely follow that of the seafood industry as a whole, where aquaculture now produces more than half of the seafood for human consumption worldwide, and has recently surpassed global beef production in terms of annual production tonnage. In fact, aquaculture development is entirely reshaping the seafood industry and fisheries management strategies worldwide. It is reasonable to assume that tuna production will follow this trend. At the current pace, fattening practices combined with closed-cycle tuna aquaculture clearly have the potential to supersede wild capture production of bluefin tuna in the near future as well. While critics question the ecological sustainability of raising carnivorous pelagic species such as tuna, ingenuity, and economic drivers ensure that the industry will develop in the right direction. This argument has become increasingly important amidst strict regulations aimed at curbing the harvest of wild tuna, as seen in recent bans on yellowfin tuna fishing and calls for bans on international trade of bluefin tuna and for their inclusion in the red lists of influential environmental groups and NGOs that shape public opinion and government policy.

However, such progress will not come easily given that some aspects of tuna aquaculture remain controversial. For instance, when it comes to capture-based aquaculture, which comprises the majority of current tuna farming efforts, serious problems arise in the quantification of such production as the size of the tuna is only estimated when they are initially stocked in the cages. This practice has the potential to cause significant uncertainty in the fishery statistics. Previously, there was no documentation for the trade of live tuna, allowing for the proliferation of IUU (illegal, unregulated, and unreported) fishing (Miyake, 2003). Other criticisms include those normally associated with traditional aquaculture operations, such as, excess feeds and waste entering the water.

Furthermore, the demand for small pelagics to feed tuna in the cages continues to grow and the majority of these fish come from fishery stocks outside the tuna on-growing regions. Such practices add further ecological and economic costs to the environmental footprint of the operations, and legitimize criticism from opposing NGOs and other stakeholder groups. However, one might argue that even under the most adverse circumstances, such as capturing juveniles for stocking cages and feeding them with small pelagics, the practice of tuna fattening could still be considered a value-added fishery. This is because captive tuna convert feeds more efficiently than their wild counterparts, largely as a result of the lower metabolic cost of a captive lifestyle, and the added fat increases the quality and value of the final product. Additionally, this practice allows tuna to be harvested in response to market demand, while also providing better usage of the biomass captured within strict tuna quotas since cumulative biomass is increased without a concurrent increase in mortality and fishing effort. As previously pointed out, we reiterate that tuna fattening blurs the line between traditional fisheries and aquaculture operations, and this intersection provides ample opportunity for future research into the economical and ecological impacts of the industry. This is further discussed in Chapter 3.

Nonetheless, most stakeholders, including this book's editors and authors, concur that closing the life cycle of high-value tuna species in captivity to provide juveniles for stocking grow out cages combined with the development of practical diets to reduce reliance on small pelagics for feeding the fish are necessary steps to ensure the sustainable expansion of the industry. These goals have been recognized by governmental agencies throughout the world and a great deal of effort is being carried out in numerous countries throughout Europe, the Americas, Australia, and Asia to develop hatchery technology and aquafeeds for tuna. This book details many of these efforts, while providing a roadmap for the future development of the tuna aquaculture industry. As the ink dries on the pages within, considerable progress in this industry is being made throughout the world. Undoubtedly, by the time this book goes to press, additional important breakthroughs will have happened. Tuna aquaculture is a rapidly developing field, and it is predicted that advanced technologies from hatchery to market will dramatically improve fisheries and aquaculture practices to ensure the future of tuna as a sustainable resource in the decades to come.

REFERENCES

Benetti, D.D., Stieglitz, J.D., Hoenig, R.H., Welch, A.W., Brown, P.B., Sardenberg, B. Miralao, S. 2009. Developments in blackfin tuna (*Thunnus atlanticus*) aquaculture. In: Proc. 2nd Global Center of Excellence Program Tuna Symposium of Kinki University. Adelaide, Australia, Dec. 1–3, 2009, pp. 12–14.

Bubner, E., 2012. Assessment of reproductive maturation of southern bluefin tuna (*Thunnus maccoyii*) in captivity. Aquaculture 364–365, 82–95.

CST. 2012a. Company announcement: Early commencement of 2012 Southern bluefin tuna spawning season. http://www.cleanseas.com.au/uploads/pdfs/investor-information/ASX%20announcement%20-%20Clean%20Seas%20spawning%20update%20Jan2012.pdf.

CST. 2012b. Company announcement: Clean Seas Tuna Ltd market update and change to operations. http://www.cleanseas.com.au/uploads/pdfs/investor-information/ASX%20announcement%20Market%20Update%2021%20Dec%202012.pdf.

CST. 2015. Company announcement: Clean Seas Tuna Ltd half year financial results. http://www.cleanseas.com.au/uploads/pdfs/investor-information/150217FfinalAsxCombinedH1FY15Results.pdf.

Elizur, A., Diechmann, M., Wise, M., Zohar, Y., Nocillado, J., Lee, Y., Abraham, L., et al. 2009. Strategies to control reproduction in southern bluefin tuna (*Thunnus maccoyii*) in South Australia. In: Allan, G., Booth, M., Mair, G., Clarke, S., Biswas, A. (Eds.), Proceedings of the 2nd Global COE Program Symposium of Kinki University. Kinki University Press, SARDI, Adelaide, South Australia, pp. 31–33.

Endo, T., Ishida, M. Yazawa, R. Takeuchi, Y. Kumakura, N., Hara, T. Adachi, S. In Press. Mass production of fertilized eggs by artificial insemination 1 from captive-reared Pacific bluefin tuna (*Thunnus orientalis*). Aquaculture, in press.

FAO, 2014. Food and Agriculture Organization of the United Nations. The State of World Fisheries and Aquaculture (SOFIA), Italy: 223 pp. http://www.fao.org/fishery/sofia/en.

Hutapea, J.H. Permana, I.G.N. 2007. Life cycle of endoparasite, *Ichthyodinium chabelardi* which infect marine fish eggs, 3rd Marine and Fisheries National Seminar Proceedings, Surabaya, Indonesia, pp. 68–72. (In Indonesian.)

Hutapea, J.H., Permana, I.G.N., Giri, I.N.A. 2009. Achievements and bottlenecks for yellowfin tuna, *Thunnus albacares*, propagation at the Gondol Research Institute for Mariculture, Bali, Indonesia. In: Allan, G., Booth, M., Mair, G., Clarke, S., Biswas, A. (Eds.), The 2nd Global COE Program Symposium of Kinki University. Kinki University Press, Adelaide, Australia, pp. 34–37.

Hutchinson, W., Partridge, G.J., Hutapea, J. 2011. Achieving consistent spawning of captive yellowfin tuna (*Thunnus albacares*) broodstock at Gondol Research Institute for Mariculture, Bali, Indonesia. Australian Centre for International Agricultural Research, Canberra, Australia, p. 33.

Ishibashi, Y. 2010. Seedling Production of the Pacific Bluefin Tuna, *Thunnus orientalis* at Kinki University. In: Miyashita, S., Takii, K., Sakamoto, W., Biswas, A. (Eds.), Joint International Symposium of Kinki University and Setouchi Town on The 40th Anniversary of Pacific Bluefin Tuna Aquaculture. Kinki University Press, Setouchi Town, Japan, pp. 64–70.

Koya, T. 2010. Actions to be introduced by government of Japan toward effective conservation and management for Pacific bluefin tuna. In: Miyashita, S., Takii, K., Sakamoto, W., Biswas, A. (Eds.), Joint International Symposium of Kinki University and Setouchi Town on The 40th Anniversary of Pacific Bluefin Tuna Aquaculture. Kinki University Press, Setouchi Town, Japan, pp. 12–15.

Masuma, S. 2010. Seedling Production Of Pacific Bluefin Tuna–1: National Activities. In: Miyashita, S., Takii, K., Sakamoto, W., Biswas, A. (Eds.), Joint International Symposium of Kinki University and Setouchi Town on The 40th Anniversary of Pacific Bluefin Tuna Aquaculture. Kinki University Press, Setouchi Town, Japan, pp. 64–70.

Masuma, S., Miyashita, S., Yamamoto, H., Kumai, H., 2008. Status of Bluefin Tuna Farming, Broodstock Management, Breeding and Fingerling Production in Japan. Rev. Fish. Sci. 16, 385–390.

Masuma, S., Takebe, T., Sakakura, Y., 2011. A review of the broodstock management and larviculture of the Pacific northern bluefin tuna in Japan. Aquaculture 315, 2–8.

McFarlane, M.B., Cripe, D.J., Thompson, S.H., 2000. Larval growth and development of cultured Pacific bonito. J. Fish. Biol. 57, 134–144.

Metian, M., Pouil, S., Boustany, A., Troell, M., 2014. Farming of bluefin tuna–reconsidering global estimates and sustainability concerns. Rev. Fish. Sci. Aquaculture 22, 184–192.

Miyake, P.M., De la Serna, J.M., Di Natale, A., Farrugia, A., Katavic, I., Miyabe, N., et al. 2003. General review of bluefin tuna farming in the Mediterranean area. Col. Vol. Sci. Pap. ICCAT 55 (1), 114–124.

Mylonas, C.C., de la Gándara, F., Corriero, A., Belmonte Rios, A., 2010. Atlantic Bluefin Tuna (*Thunnus thynnus*) Farming and Fattening in the Mediterranean Sea. Rev. Fish. Sci. 18, 266–280.

Okutsu, T., Shikina, S., Kanno, M., Takeuchi, Y., Yoshizaki, G., 2007. Production of trout offspring from triploid salmon parents. Science 317, 1517.

OPRT. 2015. Global consumption of tuna and sashimi tuna. http://oprt.or.jp/eng/data/global-tuna-supply-sashimi-tuna/.

Ortega, A., Viguri, F.J., de la Gandara, F. 2013. Cierre del ciclo biologico en cautividiad del bonito Atlantico *Sarda sarda* (Bloch, 1793). Actas del XIV Congreso Nacional de Acuicultura, Gijon, Spain, pp. 286–287.

Ottolenghi, F., 2008. Capture-based aquaculture of bluefin tuna. In: Lovatelli, A., Holthus, P.F. (Eds.), Capture-Based Aquaculture. Global Overview. FAO, Rome (Italy), pp. 169–182.

Partridge, G.J., 2013. Closed-cycle hatchery production of tuna. In: Allan, G., Burnell, G. (Eds.), Advances in Aquaculture Hatchery Technology. Woodhead Publishing Limited, Cambridge, England, p. 626.

Sawada, Y., Okada, T., Miyashita, S., Murata, O., Kumai, H., 2005. Completion of the Pacific bluefin tuna *Thunnus orientalis* (Temminck and Schlegel) life cycle. Aquac. Res. 36, 413–421.

Seoka, M., Kurata, M., Kumai, H., 2007. Effect of docosahexaenoic acid enrichment in *Artemia* on growth of Pacific bluefin tuna *Thunnus orientalis* larvae. Aquaculture 270, 193–199.

Stehr, H. 2010. Building a Sustainable Southern Bluefin Tuna Aquaculture Industry in South Australia – A Progress Report. In: Miyashita, S., Takii, K., Sakamoto, W., Biswas, A. (Eds.), Joint International Symposium of Kinki University and Setouchi Town on The 40th Anniversary of Pacific Bluefin Tuna Aquaculture. Kinki University Press, Setouchi Town, Japan, pp. 30–35.

Tada, M. 2010. Challenges and opportunities for the full cycle farmed tuna in Japan. In: Miyashita, S., Takii, K., Sakamoto, W., Biswas, A. (Eds.), Joint International Symposium of Kinki University and Setouchi Town on The 40th Anniversary of Pacific Bluefin Tuna Aquaculture. Kinki University Press, Setouchi Town, Japan, pp. 40–44.

Thomson, M., Deichmann, M., Cyypionka, K., Czypionka, A., Crawford, J., Miller, A., et al. 2010. Recent Developments in Southern Bluefin Tuna Larval and Juvenile Rearing. In: Miyashita, S., Takii, K., Sakamoto, W., Biswas, A. (Eds.), Joint International Symposium of Kinki University and Setouchi Town on The 40th Anniversary of Pacific Bluefin Tuna Aquaculture. Kinki University Press, Setouchi Town, Japan, pp. 53–58.

Tveteras, R., Nystoyl, R., Jory, D. 2015. Aquaculture production forecast. Global Aquaculture Alliance GOAL 2015 meeting, Vancouver, BC, Canada, October 2015.

Chapter 2

The Fisheries for Tunas in the Eastern Pacific Ocean

William H. Bayliff
Inter-American Tropical Tuna Commission, La Jolla, CA, USA

2.1 INTRODUCTION

The eastern Pacific Ocean (EPO) is the area bounded by the west coast of the Americas, 50°N, 150°W, and 50°S (Figure 2.1). The principal species of tunas taken by the fisheries of the EPO are yellowfin, *Thunnus albacares*, bigeye, *T. obesus*, skipjack, *Katsuwonus pelamis*, Pacific bluefin, *Thunnus orientalis*, and albacore, *T. alalunga*. The first three are tropical species, while the other two inhabit temperate waters, but they spawn in tropical waters, and their progeny spend some time there before migrating to temperate waters as subadults. Yellowfin and bigeye are relatively large fish, reaching maximum weights of roughly 175 to 200 kg and averaging about 30 to 60 kg in the subsurface fishery of the EPO. Skipjack are considerably smaller, reaching a maximum weight of about 9 kg in the surface fisheries of the EPO. Pacific bluefin are the largest of the tunas occurring in the EPO, with a maximum weight exceeding 450 kg (Foreman and Ishizuka, 1990). Albacore are intermediate in size, reaching a maximum weight of roughly 40 kg and an average of about 15 kg in the subsurface fishery of the EPO.

Data on the catches of yellowfin, skipjack, bigeye, bluefin, and North and South Pacific albacore tuna during the 2001–2012 period are shown in Tables 2.1–2.6. The nations in which the vessels with the greatest catches during 2001–2012, by species, gear types, and nations in which they were registered are shown in Table 2.7.

Most of the catch of bluefin in the EPO is taken by Mexican-flag vessels, and nearly all of this catch is transferred to floating pens for fattening and eventual slaughter and transportation, by air, to Far Eastern nations, where it brings high prices. Pacific bluefin spawn only in the western Pacific Ocean (Bayliff, 1994). The fish in the pens, if not caught in the EPO, would have

D.D. Benetti, G.J. Partridge, A. Buentello (Eds): Advances in Tuna Aquaculture.
DOI: http://dx.doi.org/10.1016/B978-0-12-411459-3.00003-5

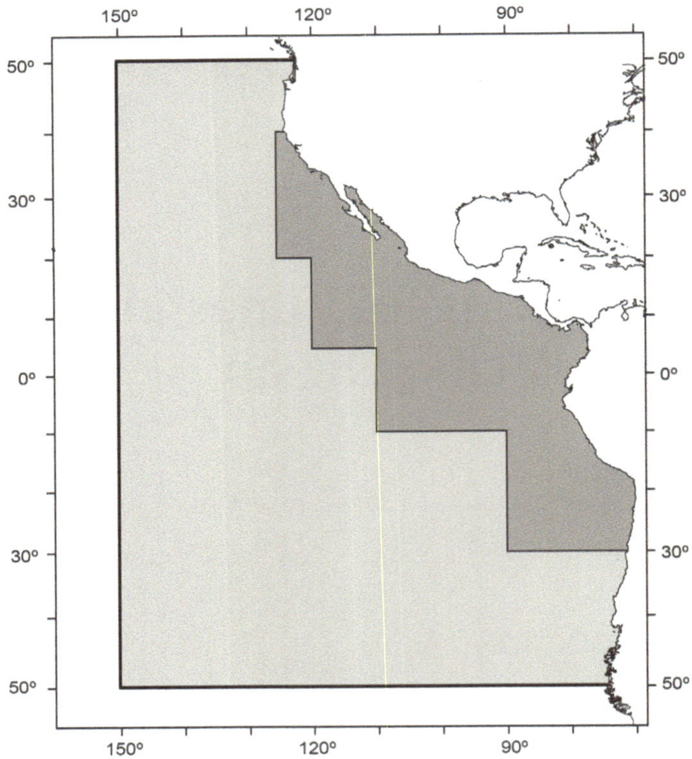

FIGURE 2.1 The Commission's Yellowfin Regulatory Area (heavy shading) and the area of jurisdiction of the IATTC defined in the Antigua Convention, effective since August 27, 2010 (light shading).

TABLE 2.1 Estimated Catches of Yellowfin Tuna by Gear Type, in Metric Tons, in the EPO

Year	Purse Seine	Baitboat	Longline	Other + Unknown	Total
2001	390,964	3916	29,608	520	425,008
2002	416,426	950	25,531	551	443,458
2003	389,144	470	25,174	1145	415,933
2004	275,557	1884	18,779	627	296,847
2005	270,872	1822	12,118	1852	286,664
2006	168,165	686	9316	1458	179,625
2007	171,741	894	8067	1439	182,141
2008	185,753	814	9812	941	197,320
2009	238,034	709	10,459	994	250,196
2010	252,040	460	8329	958	261,787
2011	207,266	276	8071	806	216,419
2012	198,468	400	9956	528	209,352

TABLE 2.2 Estimated Catches of Skipjack Tuna by Gear Type, in Metric Tons, in the EPO

Year	Purse Seine	Baitboat	Longline	Other + Unknown	Total
2001	155,980	448	1214	34	157,676
2002	166,052	616	261	119	167,048
2003	296,421	638	634	2777	300,470
2004	214,902	528	713	1106	217,249
2005	280,144	1299	231	1779	283,453
2006	307,445	435	224	986	309,090
2007	214,745	276	238	1065	216,324
2008	304,852	499	1185	1163	307,699
2009	236,587	151	1584	1112	239,434
2010	149,961	47	1815	1270	153,093
2011	281,250	24	1384	851	283,509
2012	269,726	303	162	1040	271,231

TABLE 2.3 Estimated Catches of Bigeye Tuna by Gear Type, in Metric Tons, in the EPO

Year	Purse Seine	Baitboat	Longline	Other + Unknown	Total
2001	61,772	–	68,755	19	130,546
2002	58,370	–	74,424	12	132,806
2003	55,378	–	59,776	21	115,175
2004	67,045	–	43,483	194	110,722
2005	69,795	–	41,377	25	111,197
2006	85,518	–	33,802	40	119,360
2007	64,340	–	29,855	44	94,239
2008	77,114	–	26,148	28	103,290
2009	77,818	–	31,520	15	109,353
2010	58,316	–	37,029	2	95,347
2011	57,143	–	32,630	–	89,773
2012	66,493	–	31,722	8	98,223

TABLE 2.4 Estimated Catches of Bluefin Tuna by Gear Type, in Metric Tons, in the EPO

Year	Purse Seine	Baitboat	Longline	Other + Unknown	Total
2001	1158	1	148	401	1708
2002	1759	3	71	653	2486
2003	3233	3	87	372	3695
2004	8899	–	16	62	8977
2005	4758	–	–	85	4843
2006	9928	–	–	101	10,029
2007	4189	–	–	16	4205
2008	4406	15	–	104	4525
2009	3452	0	0	181	3633
2010	7746	0	3	123	7872
2011	2833	–	1	518	3352
2012	6705	–	–	615	7320

TABLE 2.5 Estimated Catches of North Pacific Albacore Tuna by Gear Type, in Metric Tons, in the EPO

Year	Troll	Baitboat	Longline	Other + Unknown	Total
2001	11,570	157	1822	1638	15,187
2002	11,905	381	1227	2388	15,901
2003	17,749	59	1129	2260	21,197
2004	20,162	126	854	1623	22,765
2005	13,752	66	643	1741	16,202
2006	18,514	1	3482	408	22,405
2007	17,927	21	2520	1415	21,883
2008	16,135	1050	1085	308	18,578
2009	16,268	2218	39	736	19,261
2010	19,161	–	1602	753	21,516
2011	16,382	–	2395	466	19,243
2012	16,634		5,294	916	22,844

TABLE 2.6 Estimated Catches of South Pacific Albacore Tuna by Gear Type, in Metric Tons, in the EPO

Year	Troll	Longline	Other + Unknown	Total
2001	516	17,588	5	18,109
2002	131	14,062	40	14,233
2003	419	23,772	3	24,194
2004	331	17,590	0	17,921
2005	181	10,754	7	10,942
2006	48	10,246	119	10,413
2007	19	8511	87	8617
2008	0	7878	159	8037
2009	0	12,148	213	12,361
2010	0	12,286	246	12,532
2011	0	17,558	222	17,780
2012	0	17,231	210	17,441

TABLE 2.7 Nations in Which the Vessels with the Greatest Catches during 2001–2012, by Species and Gear Type, were Registered

Species	Gear	Percentages of Catches by Nation			
Yellowfin	purse seine	Mexico (41%)	Venezuela (17%)	Ecuador (12%)	Panama (9%)
Yellowfin	longline	Japan (40%)	Taiwan (14%)	Korea (12%)	Costa Rica (8%)
Skipjack	purse seine	Ecuador (51%)	Panama (11%)	Venezuela (8%)	Mexico (7%)
Bigeye	purse seine	(Ecuador (51%)	Panama (13%)	Venezuela (7%)	Spain (5%)
Bigeye	longline	Japan (47%)	Korea (20%)	Taiwan (17%)	China (9%)
North Pacific albacore	longline	Taiwan (37%)	China (26%)	Vanuatu (24%)	Belize (7%)
South Pacific albacore	longline	Taiwan (34%)	Vanuatu (22%)	French Poly-nesia (18%)	Japan (14%)

soon begun a return trip to the western Pacific Ocean.) In addition, there are, or have been, at least six yellowfin "ranches" on the Pacific coast of Mexico (Alejandro Buentello, personal communication).

Additional information about these and the other species of tunas can be found in Collette and Nauen (1983) and Joseph et al. (1988).

2.2 THE PRINCIPAL TYPES OF FISHING

2.2.1 Baitboat Fishing

Baitboat fishing was first employed by Japanese fishermen off Baja California, Mexico, in about 1918 (Estes, 1977). Baitboats employed in the EPO during the 1930s are described in detail by Godsil (1938). Baitboats were the principal type of gear employed to catch tropical tunas from the early twentieth century to about 1960, when technical advances (to be described later) made purse seining a more efficient way to catch tunas. Briefly, after catching or purchasing a supply of live bait (virtually always sardines or anchovies) in inshore waters, the vessel proceeds to the fishing grounds and begins to search for signs of fish, for example, tunas breaking the surface of the water, birds that prey on small fish upon which tunas may also be preying, or flotsam, which often attracts tunas. When fish are located, live bait is thrown into the water to attract the fish to the vessel. Meanwhile, the fishermen station themselves in steel racks mounted just above the surface of the water and cast feathered jigs, attached with short lines to fiberglass poles, into the water. When a fish is hooked it is quickly brought aboard the vessel, and then the jig is immediately returned to the water. When the fish stop biting the fishermen leave the racks, store the fish in the wells of the vessel, where they are frozen, and resume their search for fish. When the supply of bait is exhausted, the vessel returns to inshore waters to catch or purchase more bait. As can be seen in Tables 2.1–2.4, the baitboat catches of the tropical tunas and Pacific bluefin in the EPO during 2001–2012 were far exceeded by those of purse seiners, and fewer than 10 baitboats have been fishing in the EPO in recent years.

2.2.2 Trolling

Trolling gear, which is used mainly to catch albacore, is described in detail by Scofield (1956) and Sainsbury (1996). Lines with artificial lures are towed behind the vessel, and when a fish strikes one of the lures it is brought aboard the boat with the assistance of a hydraulically powered gurdy. Most trolling vessels also fish for Pacific salmon, *Oncorhynchus* spp. Most of the catch of albacore in the northern EPO is taken by trolling (Table 2.5), whereas in the southern EPO most of the catch is taken by longline gear (Table 2.6).

2.2.3 Purse Seining

Purse seines have been used to catch sardines and some other species of fish off California since the late nineteenth century (Scofield, 1951). They were first used successfully to catch bluefin tuna in 1918 (Whitehead, 1931). Since about 1960, most of the tunas caught in the EPO have been taken with purse seines. Two technical developments at about that time, the invention of the "power block," which makes retrieval of the net easier and safer, and the availability of synthetic netting, which does not rot under tropical conditions, made purse seining for tropical tunas feasible for the first time (Felando and Medina, 2011). Briefly, a purse seine is a wall of synthetic netting roughly 2,000 m long and 200 m deep. One of the longer edges has plastic corks attached to it, and the other has a steel chain and steel rings attached to it. A steel cable, the "purse line," with one end attached to the winch of the vessel, passes through the rings. When a school of tunas is sighted the vessel approaches the school. Then the net, with the "skiff" (a workboat about 8 m long with a powerful diesel engine) attached to one end of it, is released off the stern of the vessel in a circle around the fish. The corks cause one edge of the net to float, while the chain and rings cause the other edge to sink, forming a circular wall around the fish at the port side of the vessel. The bottom of the net is then closed by pulling the two ends of the purse line aboard the vessel with its winch. This is called "pursing" because it is similar to pulling the drawstring on an old-fashioned purse. The net is then pulled aboard the vessel with the hydraulic power block attached to the end of the boom. Crewmen, standing under the power block, stack the net on the stern of the vessel in such a way that it will come smoothly off the stern at the beginning of the next set. When most of the net has been retrieved the fish are removed from it with a large dipnet, the "brailer," and stored in the wells of the vessel, where they are frozen. The rest of the net and the skiff are then retrieved and the gear is prepared for the next set. Further information on purse seining is given by Ben-Yami (1994) and Sainsbury (1996).

Purse-seine sets are made on three types of schools of tuna in the EPO. The fish are caught in association with dolphins, in association with floating objects (either flotsam or man-made "fish-associating devices" (FADs)), or in association only with other fish ("unassociated schools"). Only Class-6 vessels (vessels with fish-carrying capacities greater than 363 MT) are permitted to make sets on tunas associated with dolphins. Virtually all of the fish caught in association with dolphins are yellowfin tuna. A great majority of the sets made on fish associated with floating objects are made by Class-6 vessels. Roughly half the sets made on unassociated schools of tuna are made by vessels with fish-carrying capacities not greater than 363 MT. The yellowfin caught in association with floating objects and in unassociated schools are nearly always considerably smaller than those caught in association with dolphins.

When a set is made on tunas associated with dolphins, the dolphins must be released unharmed. This is accomplished mainly by "backing down," which temporarily submerges part of the corkline, encouraging the dolphins to swim out of the net. Some of the dolphins could get their snouts entangled in the net during this process, so a fine-meshed "safety panel" is substituted for the regular netting in the "backdown channel" (Felando and Medina, 2011).

The vessels that make sets on tunas associated with floating objects have a wide variety of electronic equipment aboard, which increases the ability of the captain to determine whether to set the net around a school of fish (Fuller and Schaefer, 2014).

2.2.4 Longlining

Longlines, which catch fish well below the surface of the ocean, have been used in the EPO since the 1950s, first by Japanese fishermen, and later by fishermen of Taiwan, Korea, China, and other nations, to catch tunas and billfishes. The gear is made up of "baskets," each of which consists of a main horizontal line about 250 to 800 m long with 4 to 15 branch lines, each with a wire leader and a baited hook. A typical set consists of about 200 or more baskets connected together, with a buoy at each connection, and a total of about 3,000 hooks. The hooks fish at depths typically between 100 and 300 m. Squid and various species of fish are used for bait. Longline-caught fish, which are considerably larger than baitboat- or purse seine-caught fish, are usually handled with great care and frozen at extremely low temperatures to maintain their high quality. Those that have been handled this way are eventually sold to fresh-fish markets for high prices. Longlines catch billfishes, in addition to tunas, and these are handled like the tunas and eventually sold at high prices. Skipjack are seldom caught by longline gear, because the longline hooks are larger than those used aboard baitboats to catch skipjack, and possibly also because skipjack are more reluctant to try to consume dead bait than are other species (Miyake, 1968). Longline vessels and gear are described by Bjordal and Løkkeborg (1996), and Sainsbury (1996).

2.3 DEVELOPMENT OF THE FISHERIES

2.3.1 The Period Prior to World War II

Fishing for tunas probably began in the EPO more than 5,000 years ago. Crockford (1997) presents evidence that bluefin tuna were fished by American Indians off what is now Washington and off Vancouver Island, British Columbia. She stated that "the Nuu-Chah-Nulth people, in particular, were especially adept at using this resource, and their material culture included large sea-going canoes, detachable harpoon heads, braided ropes, and floats required for the successful hunting of both whales and large tuna."

She also stated that "recent oral accounts ... confirm that large bluefin tuna were being harvested in the northeastern Pacific until the late 19th century." (Bluefin do not occur in the northeastern Pacific every year, which Crockford realized, as she cited papers by Neave (1959) and Radovich (1961) that documented the occurrence of bluefin far to the north of their normal range.)

Smith (2012) describes the recreational fishery for large bluefin tuna at Santa Catalina Island, California, in detail. "Charles Holder, an East Coast naturalist, first visited Santa Catalina Island ... in 1886. ... There were plenty of fishermen on the island, and they caught plenty of tuna, but they did so using thick hand-lines with multiple hooks ... tied to boats or piers." Holder was interested in catching bluefin tuna with recreational gear, and finally, in 1896, after acquiring heavier gear, he succeeded in landing a 183-pound (83-kg) bluefin. Shortly thereafter, Holder and some of his friends founded the Tuna Club of Avalon (Avalon being the largest community on the island). Recreational fishing for bluefin at Santa Catalina Island became quite popular for a few years thereafter, but eventually, for unknown reasons, large bluefin virtually disappeared from the vicinity of Santa Catalina Island.

Richardson (1981), Joseph (1989), Zolezzi and Bradley (2008), Gaffey (2010), Smith (2012), and Felando and Medina (2012) describe the development of commercial fishing for tunas off California and Baja California. Canning of sardines, *Sardinops caeruleus*, was carried out in Southern California during the late nineteenth and early twentieth centuries, but when they became scarce in 1902 the owners of the canneries had to look for some other kinds of seafood to can. Albert P. Halfhill of California Fish Company and others, after considerable effort with various species of fish and other marine animals, found something that the public would accept—albacore. It was called "tuna" or "tunny." It looked and tasted like chicken, and the phrase "Chicken of the Sea" first appeared on labels in 1912. Albacore, however, were available in California for only about 4 or 5 months per year, and almost not at all during some years.

The demand for canned tuna declined after World War I, but the fisheries for tunas continued. The fishery expanded to waters south of the US–Mexico border, where yellowfin and skipjack were available year-round. Larger, more efficient boats were built, which enabled the fishermen to travel further from their home ports in California and fish on previously unexploited populations of tunas. Most of the yellowfin and skipjack were caught by baitboats (Godsil, 1938) and purse seiners Shimada and Schaefer (1956: Table 2.3), and albacore were caught off California, Oregon, and Washington by both baitboats and trollers. The catches of yellowfin and skipjack far exceeded those of albacore and bluefin, and that is still the case in the early twenty-first century.

Most of the fishermen on US-flag vessels during this period were emigrants, or their sons or grandsons, from Japan, Italy, Portugal (including its offshore islands), and what is now Croatia (Gaffey, 2010).

2.3.2 World War II

During World War II, many tuna vessels were requisitioned by the US government for transporting food, other supplies, and troops to small islands in the Pacific Ocean. Many of these vessels were operated in combat zones by the fishermen who had operated them as fishing vessels before the war (Felando, 2008). During that period the US Navy constructed vessels similar to tuna boats, and some of these were later purchased by fishermen for conversion to fishing boats (Felando and Medina, 2011). Vessels that were owned by people of Japanese descent were confiscated at the beginning of World War II. Regrettably, when they were returned to their owners after the war, some were in very poor condition (Gaffey, 2010).

2.3.3 The Post-World War II Era

The fishery for tunas resumed after the end of World War II, but changes began to take place. Small vessels based in Latin America had been fishing for tunas in the EPO for many years, but interests in some of those nations wished to become more involved with the fishery. At about the same time, some of those nations announced that they were extending their territorial limits to 200 nautical miles (Joseph, 1989), and some US-flag vessels were seized for fishing within those limits. Also, the anchoveta, *Cetengraulis mysticetus*, an important bait species, nearly disappeared from the Gulf of Nicoya, in Costa Rica, which had a negative effect on the economy of the town of Puntarenas, which supplied the fishing boats with fuel and other supplies.

2.4 THE INTER-AMERICAN TROPICAL TUNA COMMISSION

The California Division of Fish and Game was the only organization performing research on the fishery for tunas in the EPO until 1950, but on May 31, 1949, a convention establishing the IATTC was signed. The IATTC consisted of representatives of two nations, Costa Rica and the United States, but research was performed by its staff, appointed by its director, Dr. Milner B. Schaefer, and working under his supervision. The IATTC's research program commenced in 1950 and is still in operation. The IATTC now has a membership of 21, including Belize, Canada, China, Colombia, Costa Rica, Ecuador, El Salvador, the European Union, France, Guatemala, Japan, Kiribati, Korea, Mexico, Nicaragua, Panama, Peru, Taiwan, the United States, Vanuatu, and Venezuela, plus four "cooperating nonmembers," Bolivia, Honduras, Indonesia, and Liberia. Its annual budget has exceeded 5 million US dollars in recent years.

As stated above, purse seining for tunas had been carried out since well before World War II. Also as stated above, two important developments, the invention of the power block and the availability of synthetic netting, which took place during the 1950s, made purse seining far more efficient than baitfishing. As a result, most of the baitboats with fish-carrying capacities greater than about 250 MT were converted to purse seiners during 1959–1961 and after that, during the 1960s, many larger purse seiners were built (Felando and Medina, 2011). This made an unforeseen expansion of the tuna fishery in the EPO, and the rest of the world, possible.

2.4.1 Regulations Adopted by the IATTC

Schaefer (1957), based on a study of catch, effort, and catch per unit of effort, had concluded that the area in which the fishery was conducted during the 1935–1954 period would support an annual catch of yellowfin of about 90 thousand metric tons. Beginning in 1961, the IATTC staff recommended to the members of the IATTC that they adopt quotas on the catches of yellowfin tuna, but the members were not able to agree to do that until 1966, when they agreed on a catch quota of 79,000 short tons (71,940 MT). The quota was less than the estimate of the maximum sustainable yield to permit the stock to recover from overfishing during the previous years. After that, the members were able to agree on a quota for each year of the 1967–1979 period. The quotas applied to an area called the Commission's Yellowfin Regulatory Area (CYRA), shown in Figure 2.1. The quotas increased over that period from 71,940 to 138,757 MT because the larger purse seiners were able to fish further offshore, exploiting populations of fish that were previously unexploited, or nearly so. An important feature of the regulations during that period was the "last open trip." A vessel that was at sea at the time that a period of regulated fishing was commenced was permitted to continue fishing without regulation until its wells were full and it had to return to port to unload. The vessel owners were able to anticipate when the closures would begin, so they instructed the captains of their vessels to return to port to unload so that they could return to sea shortly before the closure began, and then to commence their last open trips. After completing their last open trips, many vessels that would otherwise have been subject to regulation fished west of the CYRA, where they were not subject to regulation.

There were no regulations on tuna fishing in the EPO during the 1980–1997 period due to decreased fishing effort in that area. The decreased fishing effort was partly the result of the fleet emigration to other fishing areas due to the enforcement of the 200-mile limits by many coastal states and the 1982–1983 El Niño event, which caused the owners of many vessels to deactivate them or to transfer their operations to the western Pacific Ocean. The IATTC staff recommended quotas for yellowfin tuna during each year of that period except 1987, but in most cases the catches did not reach those levels. During this period the US participation in the fishery in the EPO decreased rapidly, from 126 out of 258 vessels in 1980 to 46 out of 172 vessels in 1990 to 13 out of 205 vessels in 2000, and the major tuna canneries in the United States all closed during that period (Felando and

Medina, 2011). (However, there were 36 to 39 US-flag vessels operating in the western Pacific Ocean between 2008 and 2012.) Now most of the canned tuna sold in the United States comes from fish caught in the western Pacific Ocean, but not all caught by US-flag vessels.

Regulation of the fishery in the EPO began again in 1998. The regulations for 1998–2001 applied only to the CYRA, but those after 2001 applied to the entire EPO. (The EPO was bounded at that time by the west coast of the Americas, 40°N, 150°W, and 40°S, but later, in 2010 with the entry in force of the Antigua Convention, its northern and southern boundaries were changed to 50°N and 50°S.) There were no more "last open trips," except for smaller vessels during 1998–2000, as with the adoption of the "1992 La Jolla Agreement" there was 100% observer coverage on Class-6 vessels, which by then made up the great majority of the fleet. Observers are biologists who accompany Class-6 purse seiners to collect scientific data and information on compliance with regulations. They will be discussed further in the next section. Also, there were some areas that were completely closed to fishing for tunas during specified periods after 1997. The resolutions of 1998–2000 did not mention exemptions from the regulations for any vessels, but after 2000 recreational fishing vessels were exempt in all years, and bait-boats, small purse seiners, trollers, and longliners were exempt in some of those years.

During the 1966–1979 period the objective of the regulations was to maintain the abundance of yellowfin tuna at a level that would support the maximum sustainable yield. Later, after 1997, there were limits on the catches of yellowfin (1998–2001) and bigeye (1998–2000) and closures that reduced the catches of both yellowfin and bigeye (and also skipjack).

It can be seen in Table 2.1 that the catches of yellowfin declined during the 2001–2012 period. Actually, the catches during 2001–2003 were the greatest on record; those of the years before 2001 are not much different from those of the years after 2003. Aires-da-Silva and Maunder (2012: 12 and 15–16) discuss the possibility that there are two, or possibly three, "productivity regimes" of yellowfin in the eastern Pacific Ocean, and stated on their page 4 that "the recent decrease may be a reversion to an intermediate productivity regime," but they did not express alarm about the situation.

2.4.2 Bycatches

During the late 1960s the public first became aware of the fact that purse seiners were making sets on tunas associated with dolphins (mostly spotted dolphins, *Stenella attenuata*, spinner dolphins, *S. longirostris*, and common dolphins, *Delphinus delphis* and/or *D. capensis*) and that dolphins were sometimes killed during the process. In response to public outcry, the US Marine Mammal Protection Act, which made it difficult for US-flag vessels to make sets on tunas associated with dolphins, was passed in 1972. This caused the owners of many vessels to sell them to non-US citizens, who transferred the registrations of those vessels to nations other than the United States, or to transfer their operations to the western Pacific Ocean, where tunas and dolphins do not associate with one another. In 1976 a resolution was adopted at the 33rd meeting of the IATTC authorizing the IATTC staff to become involved in dolphins research and management. Eventually, in 1998, a binding agreement, the Agreement on the International Dolphin Conservation Program (AIDCP), which established the International Dolphin Conservation Program (IDCP), was adopted. This agreement requires that every Class-6 vessel carry an observer, who collects scientific data on dolphins and, after the trip is completed, reports any violations that the captain or other crew members might have made to the secretariat of the IDCP (IATTC staff members) which, in turn, reports this information to the government of the nation in which that vessel is registered. In 1993 the duties of the observers were increased by requiring them to collect scientific data on discards of unmarketable fish and other animals. Since 1993 there has been virtually 100% observer coverage of Class-6 vessels, the only vessels permitted to make sets on tunas associated with dolphins. Colombia, Ecuador, Mexico, Nicaragua, Panama, and Venezuela have their own observer programs, and half of the observers on Colombian, Mexican, Nicaraguan, Panamanian, and Venezuelan vessels and one-third of those on Ecuadorian vessels are from the national programs. The vessels of other nations carry IATTC observers on all their trips. The national observers collect the same data as the IATTC observers and record the data on forms similar to those used by IATTC observers, and the data collected by the national observer programs, after processing, are made available to the IATTC staff. The IDCP has been outstandingly successful. Wade (1995) estimated that prior to 1973 the annual mortalities of dolphins ranged from roughly 226 to 559 thousand animals. It should be noted that the observer coverage during that period was minimal. During 2001–2012 the mortalities have ranged from 0.8 to 2.1 thousand animals, and never more than 0.08% of any stock.

Five species of sea turtles, olive Ridley (*Lepidochelys olivacea*), green (*Chelonia mydas*), loggerhead (*Caretta caretta*), hawksbill (*Eretmochelys imbricata*), and leatherback (*Dermochelys coriacea*) occur in the EPO, and these are caught incidentally by various types of fishing gear, including longlines and purse seines. At the request of several organizations in Ecuador, the IATTC staff undertook a 3-year program to search for ways to reduce the mortalities of sea turtles in the artisanal longline fishery of that nation. The program quickly spread to other nations adjacent to the EPO. It was found that the mortalities of sea turtles could be reduced by modifications of the types of hooks used and use of "de-hooking devices" to remove the hooks from the mouths of the turtles that were hooked with minimal injury to them (IATTC staff, 2013). Primary control of the sea turtle project was transferred from the IATTC to the World Wildlife Fund at the end of 2009.

Because of their low fecundity, sharks and rays are especially vulnerable to overfishing. Detailed data on the purse-seine catches of sharks in the EPO, by species, during 1993–2004 are given by Román Verdesota and Orozco-Zöller (2005). The IATTC staff has devoted most of its research effort on sharks toward two species, silky sharks, *Carcharhinus falciformis*, and oceanic whitetip sharks, *C. longimanus*. Blue sharks, *Prionace glauca*, are the "most abundant pelagic sharks species" (Nakano and Seki, 2003). The catch rates of blue sharks, exceed those of any other species of shark in the longline fishery of the western tropical Pacific Ocean (Williams, 1997: Figure 2), and the same appears to be the case for the EPO (Matsumoto and Bayliff, 2008: Figures 90 and 92). The IATTC staff has not performed any research on blue sharks, as research on that species is being conducted by other organizations (Walsh and Kleiber, 2001; Nakano and Seki, 2003).

Seabirds are rarely, if ever, caught by purse seines, but they are sometimes caught by longline gear when they attempt to eat the bait while the gear is being deployed or retrieved. Gilman et al. (2005), who reviewed the information available on the mortality of seabirds due to longline fishing, pointed out that many methods have been found that greatly reduce seabird mortality. They stated that "Given the existence of relevant legally binding accords, the level of understanding of the source and extent of longline mortality, and the availability of both effective and cost-saving seabird avoidance methods, the potential exists to minimize seabird mortality in longline fisheries to insignificant levels. ... To realize this potential, however, will require raising industry awareness of the operational benefits and economic incentives from using effective seabird avoidance methods, widely implementing relevant multilateral accords and initiatives, and establishing and enforcing effective seabird conservation measures." IATTC Resolutions C-11-02 and C-11-08, both adopted at the 82nd meeting of the IATTC in July 2011, called for observer coverage, in the EPO, of at least 5% of the trips of longline vessels with overall lengths greater than 20 m, beginning on January 1, 2013. The first of these resolutions deals almost entirely with configuration of the gear so as to minimize the mortalities of seabirds, but the second, wisely, states that "The main task of the scientific observers shall be to record any available biological information, the catches of targeted fish species, species composition and any available biological information as well as any interactions with non-target species such as sea turtles, seabirds and sharks."

2.4.3 The Current Situation in the EPO

2.4.3.1 Research

Many advances have been made in fisheries science since 1950, and the IATTC staff has kept abreast of all that are applicable to its work. Archival tags (Arnold and Dewar, 2001; Gunn and Block, 2001) are far superior to conventional tags, as they track the horizontal and vertical movements of the fish and record the ambient and internal temperatures of the fish at intervals of less than 1 min. Panella (1971) provided an important breakthrough in aging of fish by their hard parts when he discovered that increments are deposited on the otoliths of fish at daily intervals. Histological examination of the gonads was first used to study the maturation and spawning of tunas by Hunter et al. (1986). Stable isotope ratios (Peterson and Fry, 1987) provide useful information on the trophic levels of the various components of an ecosystem. Stock Synthesis, for example, Stock Synthesis II (Methot, 2005) provides a statistical framework for calibration of a population dynamics model using a diversity of fishery and survey data. It is designed to accommodate both age and size structure in the population and with multiple stock subareas. These are just a few of the advances in fisheries science that the IATTC staff has put to good use. IATTC staff members have made some important contributions to fisheries science, notably a book by Quinn and Deriso (1999) and papers by Schaefer (1954), Pella and Tomlinson (1969), Hall (1998), Olson and Watters (2003), and Maunder et al. (2006). Also a book by Joseph and Greenough (1979) was an important contribution to study of the law of the sea and exploitation of living marine resources. A person can simply use the table of contents of Joseph and Greenough's book as a directory to information on those subjects.

2.4.3.2 Management of the Fisheries

The situation regarding management of the fishery is mostly good. The members of the IATTC have agreed to regulations identical to or similar to those recommended by the IATTC staff since 1998. The concept of rights-based management on the high seas (Allen, 2010; Allen et al., 2010a,b) has been gaining in popularity in recent years. A few vessels that are registered in nations that are not members or cooperating nonmembers of the IATTC have been sighted fishing in the EPO, but these have nearly all been longliners, which catch far fewer fish than do purse seiners. It is now becoming increasingly difficult for vessels fishing illegally on the high seas to sell their catches.

In general, purse seine-caught fish are unloaded at a port to either a processing facility or a carrier vessel that will transport the fish to a processing facility, whereas longline-caught fish are unloaded to a carrier vessel at sea. IATTC Resolution C-98-10, adopted at the 62nd meeting of the IATTC in October 1998, called for prohibition of the "transshipment of tuna on the high seas by purse-seine vessels fishing for tunas in the EPO." A series of resolutions culminating in IATTC Resolution C-12-07, adopted at the 83rd meeting of the IATTC in June 2012, states that fish transferred from a large-scale longline fishing vessel to a carrier vessel in port must be accompanied by documentation that lets the buyer know whether the fish were caught legally. If fish are transferred to a carrier vessel at sea an observer must be on the carrier vessel to collect information on the fishing vessels from which the carrier vessel obtained the fish that it delivers to the buyer(s).

IATTC Resolution C-06-05, adopted at the 74th meeting of the IATTC in June 2006, called for "adoption of trade measures to promote compliance with the IATTC's efforts to prevent [Illegal, Unreported, and Unregulated] fishing in the EPO." It noted "that trade restrictive measures should be implemented only as a last resort, where other measures have proven unsuccessful to prevent, deter and eliminate any act or omission that diminishes the effectiveness of IATTC conservation and management measures."

The combination of regulation of transshipments and trade restrictions on illegally caught fish may eventually make it infeasible to fish for tunas and billfishes illegally in the open ocean, just as is now the case for fishing illegally for most or all species within the jurisdictions of many nations.

There are still some persistent problems. One is incidental catches of small bigeye and yellowfin tuna, and of other species, for example, sharks and dorado, *Coryphaena* spp., in sets directed at skipjack associated with floating objects. The bigeye and yellowfin can be sold to canneries unless they are extremely small, but it would be preferable if they were permitted to grow to greater sizes before being caught. Dagorn et al. (2012) examined "the relationships between the ratio of bycatch [including small tunas] to target catch across different set size classes in all oceans. Ratios were always highest when catches were small, with the smallest class of catches (23%–43%) while only contributing negligibly to the total target catch (3%–10%)." Fuller and Schaefer (2014) conducted experiments aboard a Class-6 purse seiner to search for methods that might lead to reduction of the bycatches. They found that "The captain's predictions [of the catch] were significantly related to the actual total catch and catch by species, but not to size categories by species." Both of these studies indicate that it may be possible to significantly reduce bycatches without significantly reducing the catches of the target species.

Another problem is excess fishing capacity (Joseph, 2003; Bayliff et al., 2005), which makes it necessary to enact closures of the fishery. This is economically wasteful, but not necessarily harmful to the utilization of the fish, although it could be if intensive fishing during some months and little or no fishing during other months causes some populations to be overfished and others to be underfished. IATTC Resolution C-98-06, adopted at the 61st meeting of the IATTC in June 1998, called for the establishment of what is now called the "Permanent Working Group on Fleet Capacity." This working group has held 16 meetings through the end of 2014. IATTC Resolution C-98-11, adopted at the 62nd meeting of the IATTC in October 1998, set limits for the fish-carrying capacities for the purse-seine vessels of 13 nations participating in the fishery, with a total limit of 158,837 MT. It is important to note (i) that these limits were to apply only to 1999 and (ii) that there are several caveats in the resolution regarding "the right of several states without vessels currently fishing in the EPO, but with a longstanding and significant interest in the EPO tuna fishery, to develop their own tuna fishing industries" and the rights of nations that "have expressed their interest in increasing the carrying capacity of their fleets licensed to fish for tuna in the EPO." Despite the fact that there is general agreement among the representatives of the members of the IATTC that the fish-carrying capacity of the fleet is too great, it continued to increase until 2007, reaching 266,008 MT in that year (but decreasing somewhat thereafter).

A third problem could be the bycatches of species for which there is little biological knowledge. Sharks and rays may be particularly vulnerable to the purse seine and longline fisheries. The IATTC staff has undertaken research on the two most important species in that fishery, silky sharks, *Carcharhinus falciformis*, and oceanic whitetip sharks, *C. longimanus*, but virtually nothing on any other species, and it would not be able to do so without additional funding. Shark-fin soup is considered to be a delicacy, particularly in some East Asian nations, so shark fins bring high prices to fishermen. Some fishermen, although not necessarily operating in the EPO, have removed the fins from recently caught sharks and discarded the rest of the sharks, a practice that is almost universally condemned. IATTC Resolution C-05-03, adopted at the 73rd meeting of the IATTC in June 2005, called for implementation of measures "to require that their fishers fully utilize any retained catches of sharks" and to "require their vessels to have onboard fins that total no more than 5% of the weight of sharks onboard."

2.5 OUTLOOK FOR THE FUTURE

The outlook for the future is uncertain. It is possible that fuel will become so expensive that it will become economically infeasible to fish far offshore. If that becomes the case, tuna might be caught mostly within the 200 miles of shore and the resource-adjacent nations would be able to charge high prices for licenses to fish within their Exclusive Economic Zones. If so, tuna might become a luxury food, instead of the economical food that it is today.

Skipjack is the principal species of tuna caught in association with floating objects. As mentioned above, Dagorn et al. (2012) showed that the percentages of small bigeye and yellowfin were lower in purse-seine sets on large schools of tunas than in sets on small schools, and Fuller and Schaefer (2014) showed that the captain of a purse seiner was able to predict the amounts of fish in a school that had been detected electronically before making a set on it. It seems possible that regulations will eventually be adopted that would penalize the owners or captains of vessels that catch relatively large amounts of small bigeye and/or yellowfin, which would be an incentive for vessel captains to make sets only on what they believe to be large schools of fish.

The crew of an average purse seiner probably spends more time searching for fish than deploying its net and bringing the catch aboard the vessel.

Some of the larger vessels carry helicopters that can search over a larger area in less time, and other vessels obtain information on the location of schools of fish from small land-based aircraft. Helicopters are expensive, and small, land-based aircraft cannot venture far from shore. Drones equipped with cameras and deployed from fishing vessels might be a good way to locate fish and increase the catches of the vessels that deploy them.

It is generally agreed that the world's tuna fishing capacity is greater than that necessary to harvest the maximum sustainable yields of tunas. This results in overfishing and/or inordinately long seasonal closures, both of which are economically wasteful. The fact that the fishing capacity of the vessels that fish for tunas in the EPO is greater than what is needed is recognized, but no agreement has been reached on actions that would reduce the size of the fleet.

Regardless of what happens, scientific investigations must continue, and representatives of the nations with stakes in the fishery must continue to cooperate with one another for the common good.

ACKNOWLEDGMENTS

Ms Joanne E. Boster furnished the data in the tables, Ms Jenny M. Suter provided information on US-flag vessels operating in the western Pacific Ocean, and Ms Christine A. Patnode prepared the figure.

REFERENCES

Aires-da-Silva, A., Maunder, M.N., 2012. Status of yellowfin tuna in the eastern Pacific Ocean in 2010 and outlook for the future. Inter-Amer. Trop. Tuna Comm. Stock Asses. Rep. 12, 3–111.

Allen, R., 2010. International management of tuna fisheries: arrangements, challenges and a way forward. FAO Fish. Tech. Pap. 536, vol. viii, 45 pp.

Allen, R., Joseph, J., Squires, D. (Eds.), 2010a. Conservation and Management of Transnational Tuna Fisheries. Wiley-Blackwell, Ames, Iowa, USA, xvi, 343 pp.

Allen, R., Joseph, J., Squires, D., 2010b. Managing world tuna fisheries with emphasis on rights-based management. In: Grafton, R.Q., Hilborn, R., Squires, D., Tait, M., Williams, M. (Eds.), Handbook of Marine Fisheries Conservation and Management. Oxford University Press, Oxford, UK, pp. 698–712.

Arnold, G., Dewar, H., 2001. Electronic tags in marine fisheries research: a 30-year perspective. In: Sibert, J.R., Nielson, J.L. (Eds.), Electronic Tagging and Tracking in Marine Fisheries. Kluwer Academic Publishers, Boston, pp. 7–64.

Bayliff, W.H., 1994. A review of the biology and fisheries for northern bluefin tuna, *Thunnus thynnus*, in the Pacific Ocean. FAO Fish. Tech. Pap. 336 2, 244–295.

Bayliff, W.H., de Leiva Moreno, J.I., Majkowski, J., (Eds.), 2005. Management of tuna fishing capacity: conservation and socio-economics. FAO Fish. Proc. 2, vol. xvi, 336 pp.

Ben-Yami, M., 1994. Purse Seining Manual. Fishing News Books, Oxford, UK, vol. x, 406 pp.

Bjordal, Å., Løkkeborg, S., 1996. Longlining. Fishing News Books, Oxford, UK, vol. ix, 156 pp.

Collette, B.B., Nauen, C.E., 1983. FAO species catalogue. Vol. 2. Scombrids of the world. FAO Fish. Synop. 125, vii, 137 pp.

Crockford, S.J., 1997. Archaeological evidence of large bluefin tuna, *Thunnus thynnus*, in coastal waters of British Columbia and northern Washington. U.S. Nat. Mar. Fish. Serv. Fish. Bull. 95 (1), 11–24.

Dagorn, L., Filmalter, J.D., Forget, F., Amandè, J.M., Hall, M.A., Williams, P., et al., 2012. Targeting bigger schools can reduce ecosystem impacts of fisheries. Canad. Jour. Fish. Aquatic. Sci. 69 (9), 1463–1467.

Estes, D., 1977. Kondo Masaharu and the best of all fisherman [sic.]. J. San Diego Hist. 23 (3), 1–13.

Felando, A.J., 2008. Tuna clippers & World War II. Mains'l Haul 44 (1-2), 28–45.

Felando, A., Medina, H., 2011. The Tuna/Porpoise Controversy: How Tuna Fishermen Were Caught in the Government's Net and Fought to Survive. Western Sky Press, vol. vii, 305 pp.

Felando, A., Medina, H., 2012. The origins of California's high-seas tuna fleet. J. San Diego Hist. 58 (1-2), 1–40.

Foreman, T.J., Ishizuka, Y., 1990. Giant bluefin tuna off Southern California, with a new California size record. Calif. Fish Game 76 (3), 181–186.

Fuller, D.W., Schaefer, K.M., 2014. Evaluation of a fishing captain's ability to predict species composition, sizes, and quantities of tunas associated with drifting fish-aggregating devices in the eastern Pacific Ocean. ICES Jour. Mar. Sci. 71 (7), 1774–1780.

Gaffey, J.T., II. 2010. Nets, Hooks and Boats: the San Pedro Bay Fishing Industry, 1850–2000. vol. iii. 321pp.

Gilman, E., Brothers, N., Kobayashi, D.R., 2005. Principles and approaches to abate seabird by-catch in longline fisheries. Fish and Fisheries 6 (1), 35–49.

Godsil, H.C., 1938. The high seas tuna fishery of California. Calif. Div. Fish Game, Fish Bull. 51, 41.

Gunn, J., Block, B., 2001. Advances in acoustic, archival, and satellite tagging of tunas. In: Block, B.A., Stevens, E.D. (Eds.), Tuna: Physiology, Ecology, and Evolution. Academic Press, San Diego, pp. 167–224.

Hall, M.A., 1998. An ecological view of the tuna-dolphin problem: impacts and trade-offs. Rev. Fish Biol. Fish. 8 (1), 1–34.

Hunter, J.R., Macewicz, B.J., Sibert, J.R., 1986. The spawning frequency of skipjack tuna, *Katsuwonus pelamis*, from the South Pacific. U.S. Nat. Mar. Fish. Serv., Fish. Bull. 84 (4), 895–903.

IATTC staff. 2013. IATTC Annual Report for 2009, 124 pp.

Joseph, J., 1989. Status of the tuna fishery in the eastern Pacific Ocean with regard to management. In: Miles, E.L. (Ed.), Management of World Tuna Fisheries: Implications of Extended Coastal Jurisdiction. University of Washington Press, Seattle and London, pp. 210–226.

Joseph, J., 2003. Managing fishing capacity of the world tuna fleet. FAO Fish. Circ. 982 xii, 67.

Joseph, J., Greenough, J.W., 1979. International Management of Tuna, Porpoise, and Billfish—Biological, Legal, and Political Aspects. University of Washington Press, Seattle and London, vol. xv, 253 pp.

Joseph, J., Klawe, W., Murphy, P., 1988. Tuna and Billfish: Fish without a Country. Inter-Amer. Trop. Tuna Comm., vol. ix, 69 pp.

Maunder, M.N., Sibert, J.R., Fonteneau, A., Hampton, J., Kleiber, P., Harley, S.J., 2006. Interpreting catch per unit of effort data to assess the status of individual stocks and communities. ICES J. Mar. Sci. 63 (8), 1373–1385.

Matsumoto, T., Bayliff, W.H., 2008. A review of the Japanese longline fishery for tunas and billfishes in the eastern Pacific Ocean, 1998–2003. Inter-Amer. Trop. Tuna Comm. Bull. 24 (1), 1–187.

Methot, R.D., 2005. Technical Description of the Stock Synthesis II Assessment Program: Version 1.17 – March 2005. NOAA Fisheries, 54 pp.

Miyake, M.P., 1968. Distribution of skipjack in the Pacific Ocean, based on records of incidental catches by the Japanese longline tuna fishery. Inter-Amer. Trop. Tuna Comm. Bull. 12 (7), 509–608.

Nakano, H., Seki, M.P., 2003. Synopsis of biological data on the blue shark, *Prionace glauca* Linnaeus. Bull. Fish. Res. Agency Japan 6, 18–55.

Neave, F., 1959. Records of fish from waters off the British Columbia coast. Fish. Res. Bd. Canada, J. 16 (3), 383–384.

Olson, R.J., Watters, G.M., 2003. A model of the pelagic ecosystem in the eastern tropical Pacific Ocean. Inter-Amer. Trop. Tuna Comm. Bull. 22 (3), 133–218.

Panella, G., 1971. Fish otoliths: daily growth layers and periodical patterns. Science 173 (4002), 1124–1127.

Pella, J.J., Tomlinson, P.K., 1969. A generalized stock production model. Inter-Amer. Trop. Tuna Comm., Bull. 13 (3), 419–496.

Peterson, B.J., Fry, B., 1987. Stable isotopes in ecosystem studies. Ann. Rev. Ecol. Syst. 18, 293–320.

Quinn II, T.J., Deriso, R.B., 1999. Quantitative Fish Dynamics. Oxford University Press, New York and London, vol. xv, 542 pp.

Radovich, J., 1961. Relationships of some marine organisms of the northeast Pacific to water temperatures particularly during 1957 through 1959. Calif. Dept. Fish Game Fish Bull. 112, 62.

Richardson, W.C., 1981. Fishermen of San Diego: the Italians. J. San Diego Hist. 27 (4), 1–7.

Román Verdesota, M., Orozco-Zöller, M., 2005. Bycatches of sharks in the tuna purse-seine fishery of the eastern Pacific Ocean reported by observers of the Inter-American Tropical Tuna Commission, 1993–2004. Inter-Amer. Trop. Tuna Comm. Spec. Rep. 11, ii, 67.

Sainsbury, J.C., 1996. Commercial Fishing Methods: an Introduction to Vessels and Gears, third ed. Fishing News Books, Oxford, UK, vol. viii, 359 pp.

Schaefer, M.B., 1954. Some aspects of the dynamics of populations important to the management of the commercial marine fisheries. Inter-Amer. Trop. Tuna Comm. Bull. 1 (2), 25–56.

Schaefer, M.B. 1957. A study of the dynamics of the fishery for yellowfin tuna in the eastern tropical Pacific Ocean. 2 (6), 245–285.

Scofield, W.L., 1951. Purse seines and other roundhaul nets in California. Calif. Dept. Fish. Game, Fish Bull. 81, 83.

Scofield, W.L., 1956. Trolling gear in California. Calif. Dept. Fish Game Fish Bull. 103, 45.

Shimada, B.M., Schaefer, M.B., 1956. A study of changes in fishing effort, abundance, and yield for yellowfin and skipjack tuna in the eastern tropical Pacific Ocean. Inter-Amer. Trop. Tuna Comm., Bull. 1 (7), 347–469.

Smith, A.F., 2012. American Tuna. University of California Press, Berkeley and Los Angeles, vol. xiv, 242 pp.

Wade, P.R., 1995. Revised estimates of incidental kill of dolphins (Delphinidae) by the purse-seine tuna fishery in the eastern tropical Pacific, 1959–1972. U.S. Nat. Mar. Fish. Serv., Fish. Bull. 93 (2), 345–354.

Walsh, W.A., Kleiber, P., 2001. Generalized additive model and regression tree analyses of blue shark (*Prionace glauca*) catch rates by the Hawaii-based commercial longline fishery. Fish. Res. 63 (2), 115–131.

Whitehead, S.S., 1931. Fishing methods for bluefin tuna (*Thunnus thynnus*) and an analysis of the catches. Calif. Dept. Fish Game, Bull. 33, 32.

Williams, P.G., 1997. Shark and related species catch in tuna fisheries of the tropical western and central Pacific Ocean. Case Study produced for the Food and Agriculture Organization of the United Nations in preparation for the FAO Technical Working Group meeting on the Conservation and Management of Sharks, Tokyo, Japan, April 23–27, 1998.

Zolezzi, J.H., Bradley Jr., L.D., 2008. The story of the San Diego tuna fleet. Mains'l Haul 44 (1-2), 8–27.

Chapter 3

Challenges Faced by Management of the Atlantic Bluefin Tuna Stock Related to the Development of Mediterranean Bluefin Tuna Farming

David J. Die
Department of Marine Ecosystems and Society, University of Miami, Coral Gables, FL, USA

3.1 INTRODUCTION

For the last 11,200 years, Atlantic bluefin tuna (ABFT) have been caught during their spawning migrations to the Mediterranean Sea (Di Natale, 2012). These fish enter this inland sea after spending years in the temperate and subtropical waters of the Atlantic Ocean, their primary feeding ground. During their migration, ABFT move close to shore in areas where the continental shelf is narrow, allowing fishermen to set traps or nets that have, for centuries, captured these fish (Di Natale, 2012; Marzano, 2013a). While it was the principal means to capture the schools of these fish, this fishing technique ensured an abundant supply of tuna, albeit subject to large fluctuations in catch (Ravier and Fromentin, 2001).

Today, tuna traps are mostly placed outside the Mediterranean Sea at the entrance of the Strait of Gibraltar. Of the 25 traps reported to ICCAT as currently being in use, only six are inside the Mediterranean, on the Italian coasts of Sardinia, Sicilia and Liguria, and in southeastern Spain (Figure 3.1). It is

D.D. Benetti, G.J. Partridge, A. Buentello (Eds): Advances in Tuna Aquaculture.
DOI: http://dx.doi.org/10.1016/B978-0-12-411459-3.00002-3

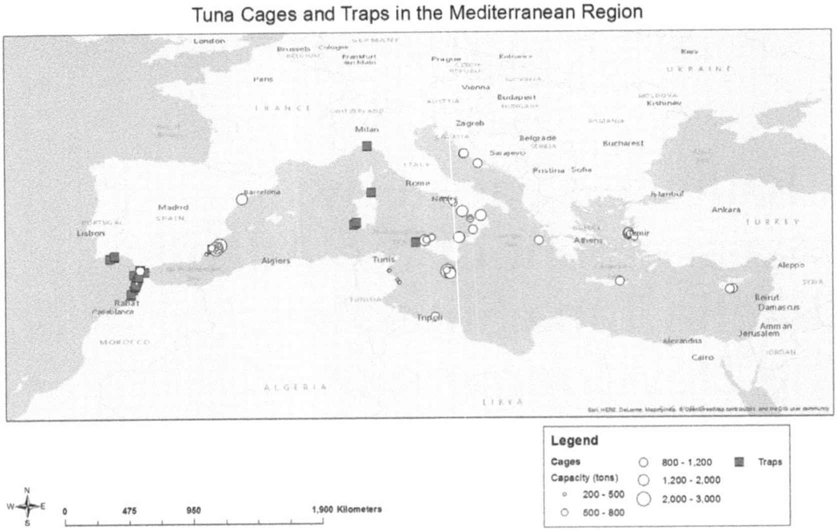

FIGURE 3.1 Location of working traps (squares) and ABFT farms (circles) in the Mediterranean Sea according to ICCAT records for 2015. Size of circle represents declared capacity of cages in farms (tons). It seems, however, that trap records held by ICCAT may not be completely up to date and that the only active traps inside the Mediterranean that hold quota for ABFT are those in southern Sardinia.

possible, however, that trap records held by ICCAT may not be completely up-to-date and that the only active traps inside the Mediterranean that hold quota for ABFT are those in Sardinia.[1]

The scenario changed after World War II when ABFT began to be targeted by a variety of fishing fleets, both inside and outside the Mediterranean (Mather et al., 1995). From 1950 to 1990, the ABFT catches in the Mediterranean gradually tripled from about 5000 to 15,000 tons. They doubled again in the next 5 years, reaching 35,000 tons in 1995, as the market for sashimi tuna expanded worldwide, fueled by increased demand from Japan (FAO, 2004) and in response to the higher prices obtained from fresh tuna in comparison to canned tuna (Miyake et al., 2010). The value of ABFT as a commodity has changed, from being a basic source of protein for soldiers of the Roman empire that conquered much of Europe and North Africa (Marzano, 2013a), to one of the most expensive seafood products, with individual fish being worth thousands of US dollars because of their high average price per kg (Collette et al., 2011) and large size. It must be noted that record auction prices paid in Tokyo at the beginning of the year are not indicative of real market value (Narula, 2014) but rather as indicators of social value. These values are probably driven by

1. A. Di Natale personal communication, ICCAT, Madrid.

perception of social status granted to those buying these fish, something that also probably occurred in Roman times, when high prices were paid for certain fish during special celebrations (Marzano, 2013b).

The first experimental farms for ABFT were started in the late 1960s in eastern Canada and in southern Spain, near Gibraltar, in the late 1970s (Miyake et al., 2003, 2010). Large-scale farming, however, did not start until the early 1990s in South Australia, where southern bluefin tuna were put in cages. Once in the cages, fish were fed for a few months primarily on a diet of oily fish, to increase the fat content of their meat, the size of fish and to control the time of harvest and delivery to market to maximize prices. This practice was then transferred to the Mediterranean in the mid-1990s, and the number of farms expanded rapidly in number and geographical extent within this region and by 1997 a new industry had become established (Miyake et al., 2010). In 2015, there are 53[2] different farm operations with cages in the shores of 10 Mediterranean countries and in two locations west of the Gibraltar straights in the Atlantic coasts of Morocco and Spain. In the last few years, the majority of the reported Mediterranean catch of ABFT, representing about two-thirds of the total catch of the eastern stock, has been transferred to farms and then exported overseas (ICCAT, 2015a,b). The declared farming capacity of the cages stands at more than 53,000 tons, greatly exceeding the current total allowable catch (TAC) for the eastern stock of ABFT which for 2015 was 16,142 tons. This capacity also exceeds most of the recent estimates of maximum sustainable yield that range from 23,000 to 74,000 tons (ICCAT, 2015a). The current recovery plan for this stock calls for TACs to be gradually increased to reach 23,155 tons by 2017 (ICCAT, 2015c), so it is unlikely that these farms can operate anywhere near their declared capacity for many years unless closed cycle hatchery production lessens their reliance on sourcing fish from the wild stock and therefore their dependence on the TAC.

Production capacity remains much greater in the western Mediterranean than in the eastern (Figure 3.1). Moreover, the ABFT cage culture business is not completely homogeneous, although most farms mainly increase fat content of muscle (fattening), those in Croatia also increase the size of the fish (farming[3]) so that they reach Japanese market sizes (Mylonas et al., 2010). These differences result in a strongly bimodal size distribution for fish put in cages, with modes at 130 and 230 cm (Ortiz, 2015). Average weight increases for fish are difficult to measure but they range from 20% to 40% for fattening operations (Mylonas et al., 2010) to the doubling in weight that is achieved in Croatian farms. Croatian fish are put in cages at 8–20 kg and harvested at between 30 and 50 kg (Ticina et al., 2007).

2. According to the records on the ICCAT database that can be downloaded from http://iccat.int.
3. Elsewhere in this chapter we will refer to farms and farming as to include both types of practices fattening and farming, unless otherwise noted.

3.2 MANAGEMENT CHALLENGES WITH THE DAWN OF A NEW INDUSTRY

The successful development of ABFT farming resulted in a number of new challenges for ICCAT, an organization that was designed to manage wild fishery resources. The challenges are of two kinds: those related to sustainability of ABFT stocks and those related to ecosystem effects related to tuna aquaculture. Challenges related to the management of wild stocks of ABFT have attracted the most attention from ICCAT because ecosystem-based management within the ICCAT area is still in its infancy and has yet to be enshrined in the ICCAT convention texts, even though it is being considered as a desirable management paradigm (ICCAT, 2015c).

The most obvious effect of the development of ABFT farms are firstly that it challenged ICCAT's capacity of monitoring harvest levels in the Mediterranean, and secondly that the high demand for wild fish to be put in farms may have partially fueled increases in catch that ultimately led to excessive harvesting levels during the late 1990s and early 2000s.

3.3 THE NEW MONITORING NEEDS

For decades, ABFT management had worked on the premise you could define a catch quota (TAC), and such a quota would determine the maximum catch to be removed by fishermen. Along with the acceptance of such a quota, fishermen were obliged to provide information on the fish they caught. With the development of cage culture, many Mediterranean fishermen shifted from landing their tuna catch to capturing them live and towing it to the cages where they would sell it to the tuna farmers. Because fish were not landed nor killed, initially commercial fishermen did not report such sales in the expectations that it was for farmers to declare it at the time of harvest. For a few years, ICCAT had difficulties determining how to track these sales, however, since 2000 a series of resolutions started to facilitate ICCAT's job of ensuring that accurate records of removals from the wild population were maintained (ICCAT, 2015d). Currently, there is an obligation to use video to record transfers of fish from the towing cage to the farm cage as well as a requirement to estimate the number and weight of fish caught by fishermen and where they were captured, including estimates of losses of fish occurring during the transport or transfer to the cage. Farmers are also obliged to report quantities of fish harvested from the cage. Since 2006, ICCAT has required that farms authorized to keep ABFT must maintain a record of their operations, including location and holding capacity (in tons). To determine the average size of fish at harvest, farms are also required to collect and measure, during harvesting operations, either 100 fish for every 100 tons of live fish, or a 10% sample of the total number of caged fish (ICCAT, 2015d).

The current monitoring system for ABFT is so demanding that it requires individual records of harvest and sale for each fish that enters the market (ICCAT, 2015e,f). This ensures that for each fish sold its producer has the rights to extract it from the wild stock. Additionally, the records are used to identify whether the fish was produced in a farm. In spite of this monitoring, there are reports that suggest that farm production of ABFT is not reported consistently (Metian et al., 2014). This is partially expected due to the time lags between removal from the ocean, harvest from the farm and sale in the market, which can generate discrepancies when such quantities are expressed on an annual basis.

3.4 POPULATION DECLINES

Cage culture has revolutionized the fishery for ABFT largely because it has led to overall increases in the average price obtained for individual fish (Metian et al., 2014). This increase in price has created not only incentives for larger catches and for investment in cage infrastructure but it has also spurred Illegal, Unreported and Unregulated (IUU) fishing. Between 1999 and 2006, ICCAT set a TAC of 32,000 tons and 29,500 tons in 2007 and reported catches were around these TACs for many years. The scientific committee of ICCAT, however, estimated that real annual catches of ABFT for that period were between 50,000 and 60,000 tons, much greater than those reported (Figure 3.2). Such unreported catches would represent large overruns of the TAC and one of the reasons why the stock of ABFT continued to decline during this period (ICCAT, 2015g).

The high price of ABFT (Collette et al., 2011), continued increase in demand of this product (Ottolenghi, 2008), and the success provided by the development of cage culture are contributors to the challenges that ICCAT had to properly manage this resource. At the same time, the economic value

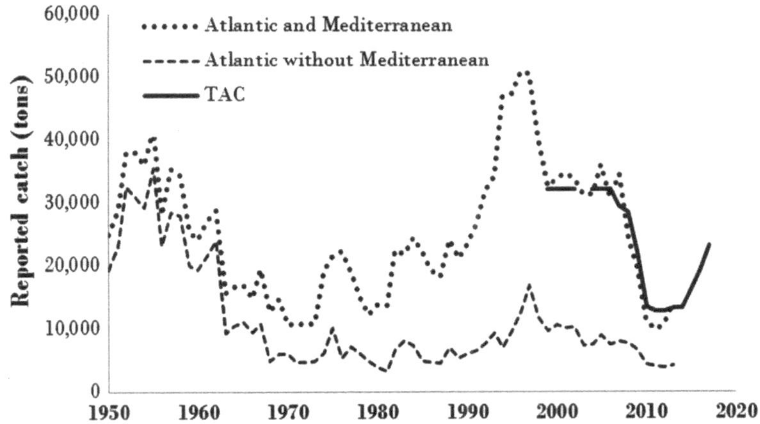

FIGURE 3.2 Reported eastern ABFT catches by region and TACs set by ICCAT.

of the fishery and aquaculture operations as well as the high visibility of this species in the eyes of Nongovernmental Organizations (NGOs) also contributed to the high degree of attention paid to the management of ABFT. In 2009, Monaco proposed to list ABFT under appendix two of the Convention on International Trade of Endangered Species (CITES). A listing under CITES would mean severe restrictions in the trade of ABFT, affect the farming operations and be an acknowledgment in the eyes of many that ICCAT was not able to manage the stock properly. After evidence was presented to the CITES Committee on the case for and against listing, the proposal was voted down by the CITES members attending the 2010 convention (CITES, 2010a,b). That same year, ICCAT made a number of important decisions regarding the stock, including an additional large reduction in the TAC and enhanced monitoring of catches. This enhanced monitoring and ICCAT's continued strong support for dedicated ABFT research (ICCAT, 2010, 2015g) aim to significantly reduce the uncertainty associated with the estimates of ABFT stock status.

3.5 A REBUILDING POPULATION

Two years after ICCAT reduced the TAC to its lowest historical value, the spawning stock started showing signs of rebuilding, riding a wave of high recruitment that probably started from strong year classes produced in 2004–2007 (ICCAT, 2015a). There is more recent evidence of highly successful larval production and high larval survival in 2012 (Ingram et al., 2015) which could have resulted in strong year classes, but recent estimates of recruitment remain uncertain because of their reliance on fishery dependent indices representing older fish (ICCAT, 2015a). Although there is still considerable disagreement over whether the increase in recruitment is solely a result of reductions in fishery quotas, it seems likely that the reductions in quota have allowed these year classes to flourish from the lowered fishing mortality (ICCAT, 2015a), something that may not have happened in the absence of such drastic management measures as those imposed by ICCAT since 2009 (ICCAT, 2015g).

Last year ICCAT agreed to gradually increase catch quotas in the next 5 years while monitoring the stock (ICCAT, 2015c). During 2017, ICCAT will conduct another assessment of the status of the stock, at which point there is hope that there will be more certainty about the reasons behind the rebuilding, its magnitude and the appropriateness of agreed TACs.

3.6 ECOSYSTEM IMPACTS OF FISHING AND FARMING

The changes introduced by the development of the tuna farming industry may have had a number of consequences on the Mediterranean ecosystem. Most of these changes are difficult to measure and few have been studied, but it is possible to describe what those changes may have been and the information available to assess them:

1. Changes in ecosystem components caused by the demands for seafood required to sustain farmed ABFT
3. Changes in the gear used to harvest ABFT to be put in cages
3. Changes in the ecological communities surrounding ABFT cages

3.7 DEMAND FOR SEAFOOD REQUIRED TO SUSTAIN ABFT IN FARMS

The development of farming and fattening has meant that ABFT are no longer exclusively supported in their growth by predation on wild prey items. The fish that are transferred to cages are fed for months on a diet based on marine fish harvested by humans. This represents a shift in the diet of tuna during those months in captivity. Metian et al. (2014) estimated the global demand for forage fish to supply ABFT farms to be between 168 and 362 thousand metric tons. Forrestal et al. (2012) investigated the possible ecosystem effects of cage culture by using an ecosystem model to estimate the changes created by the development of ABFT cage culture in the western Mediterranean. They concluded that effects may arise from the increased catches incentivized by the transformation to a more valuable product and by the shift in tuna diet from the natural diet of wild tuna to the human-provided diet used in cage culture. The majority of the tuna diet used in farms is made of forage fish and its species composition is much less diverse than that encountered by wild tuna. The most common species used as feed are small forage fish of the genus *Sardinella*, *Sardina*, *Clupea*, *Scomber*, *Trachurus*, the sparid *Boops boops*, and some cephalopods (Vita et al., 2004; Aguado et al., 2004).

The relative changes in harvesting of these species resulting from the demand for tuna feed may affect the balance which the Mediterranean ecosystem would reach under different scenarios. Although Forrestal et al. (2012) considered many different plausible scenarios for different levels of forage fish fishing and tuna farming, none included scenarios where the ABFT stock was strongly increasing as seems to be occurring in recent years. It is unclear, therefore, whether the main conclusions of the Forrestal et al. (2012) study would hold in the presence of such stock rebuilding. On the other hand, this study only looked at the type of farming conducted in the western Mediterranean, where fish are fed for only few months. In other farms, fish are kept for more than a year and therefore the shift in diet happens over longer periods, possibly increasing the type of ecosystem effects identified by Forrestal et al. (2012).

3.8 CHANGES IN THE GEAR USED TO HARVEST ABFT

Different gears have different impacts on ecosystems, with some gear producing high rates of bycatch, such as longlines and gillnets, while others much less, like purse seines. Some gears also have significant effects on marine habitats, such as trawling, while others much less such as hand lines. Bjordal (2009) proposed a general framework for rating different types of gears based on their overall impact on the ecosystem. Bycatch rates are also dependent on the specific operations of each fleet, not just on gear type. So changes in the proportion of ABFT caught by different types of gears and fleets could increase or lessen the impact of tuna fishing on other components of the ecosystem.

Three gears have dominated the catch of ABFT in the Mediterranean Sea: purse seine, trap, and longline. In the 1950s and 1960s, less than 25% of ABFT were caught by purse seines, but during the 1970s, the amount grew to 70%. In the mid-1990s, the purse seine catch grew again and now represents more than 80% of the ABFT catch. Conversely, traps were used to catch between 50% and 60% of ABFT in the 1950s and 1960s, but between the 1970s and 1980s the trap catches declined to 8% of the total. Since the 1990s, it has been 3% or less. Longline catches began in the 1960s when they represented 5% of the total, grew to 20% by 1996, and declined by half in the late 1990s. Since the 2000s, longline catches have represented about 10% of the total Mediterranean ABFT catch (Figure 3.3).

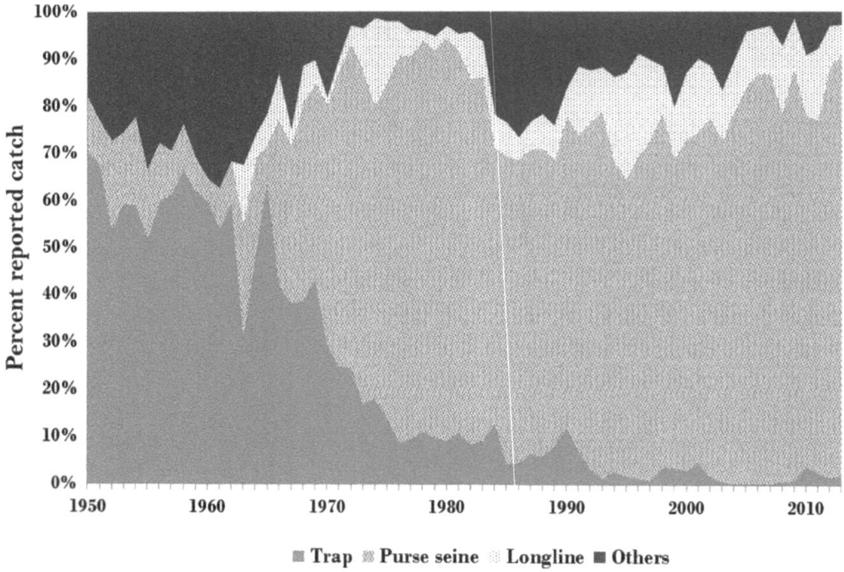

FIGURE 3.3 Proportion by fishing gear of ABFT catches reported to ICCAT from the Mediterranean.

The majority of the reduction in longline catches and more recent increases in purse seine catches coincided with the time tuna farming developed in the Mediterranean. Changes in overall longline catches, however, mask large changes in the composition of the longline fleets. Until 1985, only Spain and Japan reported significant catches of ABFT on longlines. Japan has stopped reporting longline catches in the Mediterranean since 2010. From 1986 to 1992, it is estimated that between one quarter and two-thirds of longline catches came from IUU vessels. From 1993 to 2008, the number of countries reporting longline catches grew considerably. Since 2009, the majority of longline catch is reported by a few European countries and Morocco.

From the 1950s to the 1970s, France, Italy, and Yugoslavia reported most of the catch of ABFT made from purse seine. In the late 1980s, Spain and Tunisia started reporting purse seine catches and in the 1990s a number of other countries began also reporting purse seine catches: Turkey, Libya, Croatia, and Greece. In the 2000s, the list has continued to grow to include Korea, Algeria, Egypt, Morocco, and Malta. With the success of farming ABFT, more of the catch quota is being shifted to purse seine and away from longlines and bait boats. For example, in 2012 and 2013 the Spanish bait boat fleet operating in the North Atlantic transferred its quota to other Spanish fleets in the Mediterranean that supply farms.

Historical changes in tuna catches from traps have been tied to many factors including coastal development, pollution, management actions, and competition from other gears (Di Natale and Idrissi, 2012). Since the 1960s, however, these catches have been reduced by a factor of ten.

Tuna traps retain relatively little bycatch in comparison to other gear types. For instance, Storai et al. (2011) reported only 42 elasmobranchs caught in the course of 20 years in six Italian traps. In comparison, western Mediterranean longlines that target tuna and swordfish have bycatch rates of seabirds that may affect the populations of some species such as Cory's shearwater, *Calonectris diomedea* (Garcia Barcelona et al., 2010; Baez et al., 2014a). Longlines also catch sea turtles in many areas of the Mediterranean (Tudela, 2004). In the western Mediterranean, longlines catch large number of loggerhead sea turtles (Alvarez et al., 2010), however, their rates of capture are a function of many variables including bait type, location of deployment, night or day fishing, and large-scale oceanographic conditions (Baez et al., 2014b). Sharks are a significant proportion of the longline catch, but are rarely discarded, because of their market value and, are almost always landed (Megalofonou et al., 2005).

In the past, driftnets targeting swordfish and tunas also caught dolphins (Aguilar, 2000; Bearzi et al., 2003; Bearzi et al., 2008) and sharks (Tudela, 2004; Cortes et al., 2010) but, since 2004, these gears have been banned from the Mediterranean (ICCAT, 2004). Purse seines that target ABFT for the farms catch marine mammals sporadically but it is reported that most marine mammals are released alive from the net (Tudela, 2004). The purse seine fishery that targets small pelagic fish, some of which are directed to tuna farms, may have greater impacts, especially on dolphin populations (Tudela, 2004).

It can, therefore, be concluded that there is evidence that farming may have shifted fishing effort directed to ABFT away from longlines and bait boats to purse seines. To date there has been no evaluation on what the net effect of such shift of gear types has had on total bycatch associated with ABFT fisheries. It is clear, however, that many of the strongest shifts in gear composition and area fished occurred before the development of tuna farming.

3.9 CHANGES IN THE ECOLOGICAL COMMUNITIES SURROUNDING ABFT CAGES

Tuna farming can have a number of potential effects on the marine communities that are located close to the cages. Among the most commonly cited are effects derived from the presence of the structures themselves, those related to uneaten feed and those related to the metabolic waste produced by tuna.

Cages used in marine aquaculture farms act as fish aggregating devices (Sanchez-Jerez et al., 2011) and ABFT farms are no different. There are recent reports of a greater number of ABFT sightings around these farms than there were prior to the farms' existence (Arechavala-Lopez et al., 2015). These aggregations attract sport fishermen to areas in the vicinity of the cage to target large pelagic fish. Some of the ABFT found around the cages may have been released by the farmers to comply with ICCAT quotas, and it is possible that they remain around the cage because their behavior has been modified by their capture and the time spent in the cage (Arechavala-Lopez et al., 2015). These comparisons, however, do not take into account temporal changes in the Atlantic-wide population abundance, as estimated by ICCAT assessments.

Both excess tuna feed and tuna waste can enrich benthic communities around the cages. The intensity of organic enrichment of benthos around marine farms is generally dependent on hydrographic conditions, the size of the farm and to a lesser extent husbandry methods (Borja et al., 2009). Studies that have looked at impacts on benthic communities near tuna cages have not detected significant changes close to cages that are located in deep and offshore waters (Vezzulli et al., 2008; Borja et al., 2009; Moraitis et al., 2013), but have detected impacts from cages placed in shallower and more coastal areas (Vita and Marin, 2007). Such differences may reflect differences in water mixing rates but also the length of time cages lay fallow, so that cages that are void of tuna for longer periods probably have lesser impacts on benthos (Moraitis et al., 2013).

3.10 DISCUSSION

It has been extensively reported that part of the large increase in catches from the eastern ABFT stock was due in part to the development of fattening and farming operations (ICCAT, 2009; Miyake et al., 2010). The fact remains that spawning stock biomass of the eastern stock declined gradually by two-thirds from 1975 to 2005 (ICCAT, 2015g). Farming and fattening, however, only began in earnest in the mid-1990s, so only part of this biomass decline can be associated with the development of such operations and more specifically only those declines that occurred in the last 20 years. Conversely, if farming is only one of the many contributors to changes in population size of ABFT, basin-wide ecosystem effects of changes in ABFT abundance can only partially be tied to tuna farming.

ICCAT has strived to evaluate the changes that have affected the ABFT stock, and those are reported in the results of the periodic assessments of the eastern ABFT stock (ICCAT, 2015g). Among the most obvious changes on the stock are the increased pressures on the spawning stock (the main source of tuna for the farms) and the reduction in the harvest of small ABFT, as quota holders that traditionally fished small fish have sold their share to those that can capture larger ones to be put in farms. Another change has been the shift in fishing mortality from the Atlantic Ocean to the Mediterranean. This shift, however, precedes the development of farms. It began in the mid-1960s, when less than 20% of the total reported catch for the eastern ABFT stock was caught in the Mediterranean. By the late 1980s, already 70% of the catch of the eastern stock came from the Mediterranean. Since the mid-1990s when farming fully developed, up to 85% of the reported catch came from the Mediterranean, but this proportion has again stabilized at about 70% (ICCAT, 2015g). It has been widely reported that catches of the eastern Atlantic stock were significantly underreported in the period 1997–2007 (ICCAT, 2015a). It seems that the underreporting in 2007 was equally significant in the Atlantic as in the Mediterranean (ICCAT, 2009), possibly implying that the percentage of the ABFT catch from the Mediterranean for the period 1998–2007 was not affected by underreporting. If that is true, the development of tuna farming can only have created a relatively small incentive to shift catches from the Atlantic into the Mediterranean, in comparison to other factors such as those related to quota allocations to different countries. It is likely, however, that the present dependence of farms on ABFT caught in the Mediterranean is unlikely to change; thus, much of the TAC for the eastern stock will continue to be directed to supply tuna farming.

The current management paradigm for ABFT assumes that sustainability can be ensured if harvest levels are controlled so that there are enough spawning fish to provide for future recruitment and enough surviving adults to be the next generation of spawners. Given that current farming operations depend exclusively on the collection of wild fish, sustainability of aquaculture operations is tied to the success of this management paradigm. Scientists have shown, however, that recruitment of large pelagic fish, including ABFT, is only weakly predicted by the abundance of the spawning stock, and that environmental conditions have a strong effect on recruitment (Cury et al., 2014). This leads to large inter-annual variations in recruitment, which are often translated into populations of adults that are dominated by a few year classes. Ravier and Fromentin (2001) have also hypothesized that multi decadal fluctuations in catches may be the result of resonant effects created by short-term inter-annual variation in recruitment. Although current management, based on establishment of TACs, seeks to dampen these fluctuations to ensure a more reliable supply of fish for capture, recruitment variability will also determine the sizes of fish that are available for capture in any particular year and area. Additionally, the current objective of ICCAT to achieve maximum yield is also translated into TAC management that seeks to reduce harvests when the stock is not at the level that should produce maximum yield. Strong reductions in TAC imposed by ICCAT since 2008 to rebuild the stock reduced the amount of fish available for farms (ICCAT, 2015a). On the other hand, the current TAC schedule that calls for gradual increases in TAC over the next 5 years (ICCAT, 2015e) should have the opposite effect and somewhat increase the amount of fish available for farms.

The fact remains that under the current management paradigm, farms have to adapt to both fluctuations related to TAC management and environment-related changes in the availability of fish to be captured. Only the development of industrial-scale production of juvenile tuna in hatcheries could provide farms the ability to be independent from such fluctuations. This independence may also allow harvest of wild tuna to be reduced, allow the ABFT stock to grow larger, and reduce impacts on bycatch associated with Mediterranean tuna fisheries. Current developments in ABFT hatchery technologies are described in Chapter 1 of this book.

Even if hatcheries were able to help supply enough fish for the cages to have ABFT farming mostly independent of the wild stock, industry should continue to pursue the development of diets that are less dependent on seafood to reduce ecosystem effects of fishing forage fish. If lessons from farming other carnivorous marine species are any indication, however, it is unlikely that forage fish will ever be completely replaced from tuna farm diets. This topic is thoroughly discussed in Chapter 12 of this book.

Some ecosystem threats of tuna farming are unlikely to disappear and will have to be managed accordingly. That will be the case with localized impacts of tuna farms, such as the fact that they act as fish aggregating devices and as point source for nutrients from uneaten tuna feed and tuna waste. These effects, however, seem to be manageable by placing farms in high energy locations with strong currents and away from coastal areas, and also by limiting the size and possibly the number of farms so as to not overwhelm ecosystem balance.

ACKNOWLEDGMENTS

This research was carried out under the auspices of the Cooperative Institute for Marine and Atmospheric Studies (CIMAS), a Cooperative Institute of the University of Miami and the National Oceanic and Atmospheric Administration, cooperative agreement NA10OAR4320143. A. Di Natale provided useful comments to an earlier version of the manuscript. E. Starnes prepared Figure 3.1.

REFERENCES

Aguado, F., Martinez, F.J., Garcia-Garcia, B., 2004. In vivo total nitrogen and total phosphorous digestibility in Atlantic bluefin tuna (*Thunnus thynnus thynnus* Linnaeus, 1758) under industrially intensive fattening conditions in Southeast Spain Mediterranean coastal waters. Aquacult. Nutr. 10, 413–419.

Aguilar, A., 2000. Population biology, conservation threats and status of Mediterranean striped dolphins (*Stenella coeruleoalba*). J. Cetacean Res. Manage. 2 (1), 17–26.

Alvarez, D.Q., Cardona, L., De Haro, A., Pubill, E., Aguilar, A., 2010. Sources of bycatch of loggerhead sea turtles in the western Mediterranean other than drifting longlines. ICES J. Mar. Sci. 67 (4), 677–685.

Arechavala-Lopez, P., Borg, J.A., Segvic-Bubic, T., Tomassetti, P., Oezguel, A., Sanchez-Jerez, P., 2015. Aggregations of wild Atlantic bluefin tuna (*Thunnus thynnus* L.) at Mediterranean offshore fish farm sites: environmental and management considerations. Fish. Res. 164, 178–184.

Bearzi, G., Reeves, R.R., Notarbartolo di Sciara, G., Politi, E., Canadas, A., Frantzis, A., et al., 2003. Ecology, status and conservation of short-beaked common dolphins (*Delphinus delphis*) in the Mediterranean Sea. Mammal Rev. 33 (3), 224–252.

Bearzi, G., Fortuna, C.M., Reeves, R.R., 2008. Ecology and conservation of common bottlenose dolphins *Tursiops truncatus* in the Mediterranean Sea. Mammal Rev. 39 (2), 92–123.

Baez, J.C., Garcia-Barcelona, S., Mendoza, M., Ortiz de Urbina, J.M., Real, R., Macias, D., 2014a. Cory's shearwater by-catch in the Mediterranean Spanish commercial longline fishery: implications for management. Biodivers. Conserv. 23 (3), 661–681.

Baez, J.C., Macias, D., Garcia-Barcelona, S., Real, R., 2014b. Interannual differences for sea turtles bycatch in Spanish longliners from western Mediterranean Sea. Sci. World J. 2014, 861396.

Bjordal, A., 2009. Regulation of fishing gears and methods. In: Cochrane, K.L., Garcia, S.M. (Eds.), A Fishery Manager's Guidebook. Food and Agriculture Organization of the United Nations and Blackwell Publishing, pp. 167–195.

Borja, Á., Rodríguez, J.G., Black, K.D., Bodoy, A., Emblow, C., Fernandes, T.F., et al., 2009. Assessing the suitability of a range of benthic indices in the evaluation of environmental impact of fin and shellfish aquaculture located in sites across Europe. Aquaculture 293, 231–240.

CITES, 2010a. Proposal to include Atlantic Bluefin Tuna (*Thunnus thynnus* (Linnaeus, 1758)) on Appendix I of CITES in accordance with Article II 1 of the Convention. Convention on International Trade in Endangered Species of Wild Fauna and Flora Fifteenth meeting of the Conference of the Parties, Doha, Qatar, March 13–25, 2010, consideration of proposals for amendment of appendices I and II. 34p.

CITES, 2010b. Convention on International Trade in Endangered Species of Wild Fauna and Flora Fifteenth meeting of the Conference of the Parties, Doha, Qatar, March 13–25, 2010, consideration of proposals for amendment of appendices I and II. 34p.

Collette, B.B., Carpenter K.E., Polidoro B.A., Juan-Jordá M.J., Boustany, A., Die D.J., et al. 2011. High value and long life—double jeopardy for tunas and billfishes. *Science*, 333, 292–293.

Cortes, E., Arocha, F., Beerkircher, L., Carvalho, F., Domingo, A., Heupel, M., et al., 2010. Ecological risk assessment of pelagic sharks caught in Atlantic pelagic longline fisheries. Aquat. Living Resour. 23 (1), 25–34.

Cury, P.M., Fromentin, J.-M., Figuet, S., Bonhommeau, S., 2014. Resolving Hjort's dilemma: how is recruitment related to spawning stock biomass in marine fish? Oceanography 27 (4), 42–47.

Di Natale, A., 2012. An iconography of tuna traps, essential information for the understanding of the technological evolution of this ancient fishery. Col. Vol. Sci. Pap. ICCAT 67, 33–74.

Di Natale, A., Idrissi, M., 2012. Factors to be taken into account for a correct reading of tuna trap catch series. Col. Vol. Sci. Pap. ICCAT 67, 242–261.

FAO, 2004. World Tuna Markets. Globefish Research Program, 74. FAO, Rome, 107 p.

Forrestal, F., Coll, M., Die, D.J., Christensen, V., 2012. Ecosystem effects of bluefin tuna (*Thunnus thynnus thynnus*) aquaculture in the North-Western Mediterranean Sea. Mar. Ecol. Prog. Ser. 456, 215–231.

Garcia Barcelona, S., Ortiz de Urbina, J.M., de la Serna, J.M., Alot, E., Macias, D., 2010. Seabird bycatch in Spanish Mediterranean large pelagic longline fisheries, 2000–2008. Aquatic Living Resour. 23 (4), 363–371.

ICCAT, 2004. Recommendation by ICCAT relating to Mediterranean swordfish. ICCAT recommendation [03–04], 1p.

ICCAT, 2009. Report of the 2008 Atlantic bluefin tuna stock assessment session. Col. Vol. Sci. Pap. ICCAT 64, 1–352.

ICCAT, 2010. ICCAT Atlantic wide research programme for Bluefin Tuna. In: Report for biennial period, 2010–11 PART I 2010 – Vol. 2. pp. 238–242.

ICCAT 2013. Recommendation amending the recommendation by ICCAT to establish a multi-annual recovery plan for bluefin tuna in the eastern Atlantic and Mediterranean. In: 2013 Compendium management recommendations and resolutions adopted by ICCAT for the conservation of Atlantic tunas and tuna-like species. Rec [12-03] pp. 48–78.

ICCAT, 2015a. Executive Summary Bluefin tuna east. In: Report for biennial period, 2014–15 PART I 2014 – Vol. 2, pp. 83–101.

ICCAT 2015b. Recommendation by ICCAT amending the recommendation 13-07 by ICCAT to establish a multi-annual recovery plan for bluefin tuna in the eastern Atlantic and Mediterranean. In: 2015 Compendium management recommendations and resolutions adopted by ICCAT for the conservation of Atlantic tunas and tuna-like species. Rec [14-04] pp. 47–82.

ICCAT, 2015c. Report of the second meeting of the working group on convention amendment. In: Report for biennial period, 2014–15 PART I 2014 – Vol. 1, pp. 147–168.

ICCAT 2015d. Recommendation by ICCAT on Bluefin Tuna Farming. In: 2015 Compendium management recommendations and resolutions adopted by ICCAT for the conservation of Atlantic tunas and tuna-like species. Rec [06-07] pp. 41–42.

ICCAT 2015e. Recommendation by ICCAT amending recommendation 09-11 on an ICCAT bluefin tuna catch documentation program. In: 2015 Compendium management recommendations and resolutions adopted by ICCAT for the conservation of Atlantic tunas and tuna-like species. Rec [11-20] pp. 225–242.

ICCAT, 2015f. 2015–2020 SCRS science strategic plan. In: Report for biennial period, 2014–15 PART I 2014 – Vol. 2, pp. 323–343.

ICCAT, 2015g. Report of the 2014 Atlantic bluefin tuna stock assessment session. Col. Vol. Sci. Pap. ICCAT 71, 604–691.

Ingram Jr., G.W., Alvarez-Berastegui, D., García, A., Pollack, A.G., López-Jurado, J.L., Alemany, F., 2015. Development of indices of larval bluefin tuna (*Thunnus thynnus*) in the western Mediterranean Sea. Collect. Vol. Sci. Pap. ICCAT 71, 1279–1296.

Marzano, A., 2013a. Large-scale fishing. Harvesting the Sea: the Exploitation of Marine Resources in the Roman Mediterranean. Oxford University Press.

Marzano, A., 2013b. Demand, process and distribution. Harvesting the Sea: The Exploitation of Marine Resources in the Roman Mediterranean. Oxford University Press.

Mather, F.J., Mason, J.M. Jr, Jones A.C. 1995. Life History and Fisheries of Atlantic Bluefin Tuna. NOAA Technical Memorandum NMFS-SEFSC-370. p. 165.

Megalofonou, P., Yannopoulos, C., Damalas, D., De Metrio, G., Deflorio, M., de la Serna, J.M., et al., 2005. Incidental catch and estimated discards of pelagic sharks from the swordfish and tuna fisheries in the Mediterranean Sea. Fish. Bull. 103 (4), 620–634.

Metian, M., Pouil, S., Boustany, A., Troell, M., 2014. Farming of bluefin tuna–reconsidering global estimates and sustainability concerns. Rev. Fish. Sci. Aquacult. 22, 184–192.

Miyake, P.M., De La Serna, J.M., Di Natale, A., Farrugia, A., Katavic, I., Miyabe, N., et al., 2003. General review of bluefin tuna farming in the Mediterranean area. Col. Vol. Sci. Pap., ICCAT 55, 115–124.

Miyake, M., Guillotreau, P., Sun, C.-H., Ishimura, G., 2010. Recent Developments in the Tuna Industry: Stocks, Fisheries, Management, Processing, Trade and Markets. FAO Fisheries and Aquaculture Technical Paper. No. 543. FAO, Rome.

Moraitis, M., Papageorgiou, N., Dimitriou, P.D., Petrou, A., Karakassis, I., 2013. Effects of offshore tuna farming on benthic assemblages in the Eastern Mediterranean. Aquacult. Environ. Interact. 4, 41–51.

Mylonas, C., De la Gandara, F., Corriero, A., Belmonte Rios, A., 2010. Atlantic bluefin tuna (*Thunnus thynnus*) farming and fattening in the Mediterranean Sea. Rev. Fish. Sci. 18 (3), 266–280.

Narula, S.K., 2014. Sushinomics: How Bluefin Tuna Became a Million-Dollar Fish. The Atlantic. Jan 5, 2014.

Ortiz, M., 2015. Preliminary review of bluefin tuna (*Thunnus thynnus*) size and weight measures taken with stereo video cameras at caging operations in the Mediterranean Sea 2014. Col. Vol. Sci. Pap. ICCAT 71, 1417–1425.

Ottolenghi, F., 2008. Capture-based aquaculture of bluefin tuna. In: Lovatelli, A., Holthus, P.F. (Eds.), Capture-based Aquaculture. Global Overview. FAO Fisheries Technical Paper. No. 508. FAO, Rome, pp. 169–182.

Ravier, C., Fromentin, J.M., 2001. Long-term fluctuations in the eastern Atlantic and Mediterranean bluefin tuna population. ICES J. Mar. Sci. 58, 1299–1317.

Sanchez-Jerez, P., Fernández-Jover, D., Uglem, I., Arechavala-López, P., Dempster, T., Bayle-Sempere, J.T., et al., 2011. Coastal fish farms as fish aggregation devices (FADs). In: Bortone, S.A., PereiraBrandini, F., Fabi, G., Otake, S. (Eds.), Artificial Reefs in Fishery Management. CRC Press, Taylor & Francis Group, Boca Raton, FL, pp. 187–208.

Storai, T., Zinzula, L., Repetto, S., Zuffa, M., Morgan, A., Mandelman, J., 2011. Bycatch of large elasmobranchs in the traditional tuna traps (tonnare) of Sardinia from 1990 to 2009. Fish. Res. 109 (1), 74–79.

Ticina, V., Katavic, I., Grubisic, L., 2007. Growth indices of small northern bluefin tuna (*Thunnus thynnus*, L.) in growth-out rearing cages. Aquaculture 269, 538–5432007.

Tudela, S., 2004. Ecosystem effects of fishing in the Mediterranean: an analysis of the major threats of fishing gear and practices to biodiversity and marine habitats. Studies and Reviews. General Fisheries Commission for the Mediterranean. No. 74. FAO, Rome.

Vezzulli, L., Moreno, M., Marin, V., Pezzati, E., Bartoli, M., Fabiano, M., 2008. Organic waste impact of capture-based Atlantic bluefin tuna aquaculture at an exposed site in the Mediterranean Sea. Estuar Coast Shelf Sci. 78, 369–384.

Vita, R., Marin, A., 2007. Environmental impact of capture based bluefin tuna aquaculture on benthic communities in the western Mediterranean. Aquacult. Res. 38, 331–339.

Vita, R., Marin, A., Jimenez-Brinquis, B., Cesar, A., Marın-Guirao, L., Borredat, M., 2004. Aquaculture of bluefin tuna in the Mediterranean: evaluation of organic particulate wastes. Aquacult. Res. 35, 1384–1387.

Chapter 4

From Ocean to Farm: Capture-Based Aquaculture of Bluefin Tuna in the Eastern Mediterranean Sea

F. Saadet Karakulak, Bayram Öztürk and Taner Yıldız
Faculty of Fisheries, Istanbul University, Laleli, Istanbul, Turkey

4.1 INTRODUCTION

Atlantic bluefin tuna (*Thunnus thynnus*—ABFT) is one of the most important species in the Mediterranean, attracting the highest level of commercial interest from the fisheries industry. It is regarded as an economically appealing species with respect to foreign currency inflow, providing added-value, generating employment, and supporting tuna farms within eastern Mediterranean countries. Farming has become increasingly attractive to investors due largely to the potential for high returns. This expansion is directly related to the Japanese market. The ABFT culture activities in the region are specifically aimed at producing tuna with the optimal fat content demanded by the sushi and sashimi markets, and both fresh and frozen farmed products are exported to Japan.

Archaeological excavations have shown that fishing for ABFT has occurred in the Mediterranean since the seventh millennium BC (Desse and Desse-Berset, 1994). Most ancient authors mentioned ABFT in both the Istanbul Strait and the Dardanelles. Bonito and tuna constituted the most important natural resources and source of income to the Byzantine people, with the exception of taxes from vessels passing Bosporus. Under the Roman Empire during the first three centuries AD, these fish were inscribed on bronze coins (Schünert-Geiss, 1972). Many ancient authors, such as Homer (eighth century BC), Plinius (first century AD), and Athenaios (second to third centuries AD), wrote about bonito and ABFT captured both in the Dardanelles and Istanbul Straits (Tekin, 2000).

D.D. Benetti, G.J. Partridge, A. Buentello (Eds): Advances in Tuna Aquaculture.
DOI: http://dx.doi.org/10.1016/B978-0-12-411459-3.00020-5

Aristotle (384–322 BC) considered the Black Sea to be the most important spawning area in the Mediterranean Sea for tuna (Smith and Ross, 1910) and he emphasized the importance of ABFT migration into and out of the Black Sea, and its use as a spawning ground at the time of the summer solstice. In the 1930s to 1950s, tuna eggs and larvae were encountered in ichthyoplankton samplings in the northern Black Sea (Vodyanitsky, 1936; Vodyanitsky and Kazanova, 1954; Oven, 1959; Duka, 1959; Zaitsev, 1959). As the salinity of the Black Sea is approximately half that of the Mediterranean Sea (MacKenzie and Mariani 2012), tuna reproduction in the Black Sea must have required specific adaptations of egg buoyancy, fertilization, and development for reproductive success. However, ABFT larvae were not encountered in ichthyoplankton samplings in the southern Black Sea in 1990s (Piccinetti-Manfrin et al., 1995). Nowadays, ABFT appears to be absent from the Black Sea. Reasons for their disappearance are unclear, and might have resulted from various factors including overfishing, environmental changes, eutrophication, disappearance of mackerel (which represent a critical component of tuna diet), intense ship traffic in the narrow Istanbul strait, and noise from ships.

Three defined ABFT spawning areas are known in the west and middle Mediterranean (Piccinetti et al., 1997; Nishida et al., 1998; Garcia et al., 2003): the Balearic Sea, around Malta Island, and defined marine areas to the east coast of Sicily and South Tyrrhenian Sea (Figure 4.1). Some researchers suggest that a spawning area exists in the eastern Mediterranean; however, no evidence has been provided to support this theory (Piccinetti and Piccinetti-Manfrin, 1994; Piccinetti et al., 1997; Tsuji et al., 1997).

Karakulak et al. (2004) submitted the first evidence that there might be a spawning area in the eastern Mediterranean in the Levant Sea based on studies about the spawning biology of wild-caught ABFT. It was found that spawning starts in middle to late May, and occurs 1 month earlier in the north Levant Sea when compared with other areas (Figure 4.1) (Oray and Karakulak, 2005; Oray et al., 2005; Heinisch et al., 2008). Tagging studies

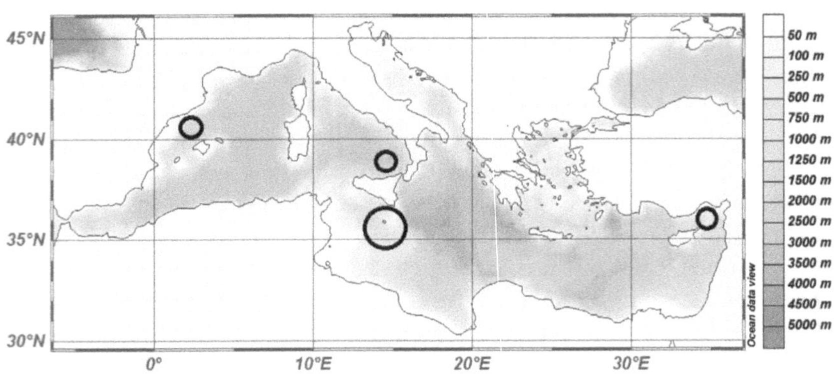

FIGURE 4.1 Known spawning areas in the Mediterranean Sea.

in the eastern Mediterranean showed that some ABFT do not migrate toward the Gibraltar Strait; instead, they stay in this area (De Metrio et al., 2004, 2005). According to these studies, hypotheses have been developed concerning an ABFT spawning stock that remains permanently in the eastern Mediterranean. More tagging and genetic studies are needed to fully understand migration routes and behavior of ABFT.

4.2 ABFT FISHERIES

The fisheries for ABFT are currently distributed from the Aegean Sea to the Levant Sea in the eastern Mediterranean Sea. ABFT fishing has had no significant activity in the Black Sea since the late 1980s (Zaitsev and Mamaev, 1997; Karakulak, 2004).

The eastern Mediterranean catch of about 1000 to 3000 tons accounted for the bulk of captures during the 1980s and 1990s. From the mid-1990s, it increased to 6000 tons, but decreased thereafter, mostly due to catch regulations (Figure 4.2).

Purse seiners from the Turkish fleet land the most ABFT in the eastern Mediterranean Sea, with Cyprus and Greece utilizing mainly longliners (ICCAT, 2014). In 2008, the number of fishing vessels in the ABFT fishery in the eastern Mediterranean consisted of 96 purse-seine vessels in Turkey (> 20 m), 22 purse seine, and 104 longline/other line vessels in Greece (< 24 m), and 9 longline and 1 purse-seine vessel in Cyprus (Rademaekers et al., 2010; Karakulak, 2012; Karakulak et al., 2012).

The International Commission for the Conservation of Atlantic Tunas (ICCAT) recommends that fishing capacity of member countries should be in proportion with allocated quotas. According to this recommendation, guidelines for reducing ABFT fishing capacity were established at a

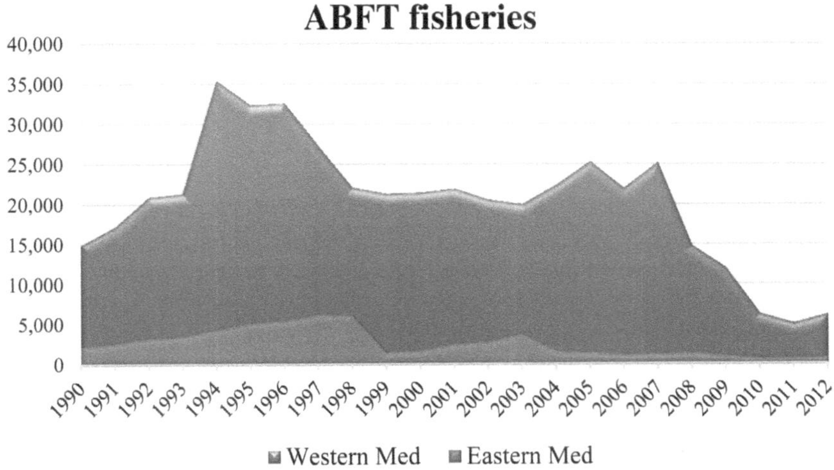

FIGURE 4.2 Catches of ABFT in the western and eastern Mediterranean Sea (ICCAT, 2014).

minimum of 25% in 2010, 50% in 2011, 20% in 2012, and 5% in 2013 (ICCAT, 2012). In that context, fishing capacity in Turkey was 683 tons and there were 57 purse-seine vessels fishing for ABFT in 2009. As a result of ICCAT recommendations, fishing capacity was gradually reduced to a fishing quota of 535 tons and 11 vessels in 2012 (Karakulak et al., 2012). As per ICCAT recommendations, purse-seine fishing may take place from May 26 to June 24, large-scale pelagic longline vessels over 24 m may fish from January 1 to May 31, and recreational and sport fishing may take place from June 16 to October 14 in the Mediterranean.

Trap fisheries already existed in ancient Greece and Rome. One of the major ABFT fisheries was the trap fishery in the Sea of Marmara (The Straits of Istanbul and Canakkale). However, these catches peaked in 1967 (1488 tons) and then declined in the late 1980s (Karakulak, 2003; Karakulak and Oray, 2009). Turkey began utilizing purse seiners in the 1950s (Iyigüngör, 1957) and further expanded its fleet in the early 1980s. Since then, the ABFT catch has kept increasing with a slowdown in 1999. Purse-seine fishing has been carried out intensively in the Levant Sea. The catch increased rapidly to 5899 tons in 1998. It then decreased steadily to around 500 tons by 2010 due to the quota agreed by ICCAT (Table 4.1).

Conversely, the large pelagic fleet from Greece is composed mostly of small vessels (average length of 12 m) with which they exploit mainly the Aegean, Ionian, and Cretan Seas, but occasionally extend their activities to the eastern Levantine basin. The main target species include ABFT, swordfish, and albacore. Most of the boats targeting exclusively ABFT are scattered in the North Aegean Sea, and use mainly hand lines. ABFT catches during 1990s fluctuated from 200 to 1200 tons, and continued to decline to around 176 tons by 2012, due to the established quotas (Table 4.1).

The ABFT was not a particularly abundant fish species in the Black Sea, although schools of 30–40 specimens could be found in the north-western part. Bulgarian coastal waters averaged annual catches of ABFT between 1941 and 1960 of 1.5 to 0.4 tons (Golenchenko, 1952). Subsequently, the catch of ABFT ceased (Zaitsev and Mamaev, 1997).

Another important coastal fishery method for ABFT in the eastern Mediterranean Sea was gillnet fishing (Miyake et al., 2004; Karakulak et al., 2007; Akyol et al., 2008; Ceyhan et al., 2011); however, it has been curtailed significantly after the United Nations General Fisheries Commission for the Mediterranean (GFCM) and ICCAT Recommendations and Resolutions forced a ban in drift gillnet fisheries. Even though artisanal harpoon fisheries still exist on the North Aegean coast of Turkey and Greece, tuna catches are minor and likely underreported.

French and Spanish fishing vessels also fish in the eastern Mediterranean. Reliable data on catch quantities from these vessels are unavailable, although the World Wildlife Fund (WWF, 2006) reported that Capture Per Unit Effort (CPUE) in the Levant Sea were higher than those in the western Mediterranean (Figure 4.3).

TABLE 4.1 Total Catch (Tons) of ABFT (*Thunnus thynnus*) by Flag and Year in the Eastern Mediterranean Sea (ICCAT, 2014)

Year	Egypt	Cyprus	Greece	Israel	Syria	Turkey
1990	–	10	201	–	–	2059
1991	–	10	175	–	–	2459
1992	–	10	447	–	–	2817
1993	–	14	439	–	–	3084
1994	–	10	886	–	–	3466
1994	–	10	1004	–	–	4220
1996	–	10	874	14	–	4616
1997	–	10	1217	–	–	5093
1998	–	21	286	–	–	5899
1999	–	31	248	–	–	1200
2000	–	61	622	–	–	1070
2001	–	85	361	–	–	2100
2002	–	91	438	–	–	2300
2003	–	79	422	–	–	3300
2004	–	105	389	–	–	1075
2005	–	149	318	–	–	990
2006	–	110	255	–	–	806
2007	–	1	285	–	50	918
2008	–	132	350	–	41	879
2009	–	2	373	–	–	665
2010	–	3	224	–	34	409
2011	–	10	172	–	–	528
2012	64	18	176	–	–	536

4.3 CAPTURE-BASED AQUACULTURE (FARMING AND FATTENING) OF ABFT

In the eastern Mediterranean, ABFT farming, or fattening, started in Turkey in 2002 with four companies and a total capacity of 2097 tons/year (Colak et al., 2003). The number of farms increased to 9 in 2005 and their capacity was reported at 9440 tons/year (Altinbulak, 2005). In 2015, the six

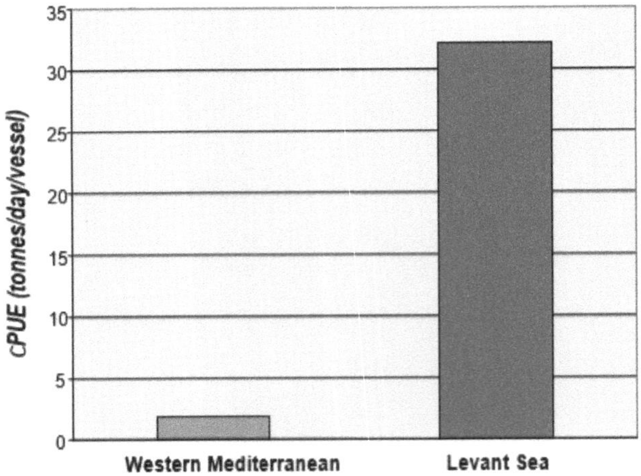

FIGURE 4.3 CPUE for French and Spanish Purse Seiners in Western Mediterranean and Levant Sea (WWF, 2006).

production centers on the ICCAT list for Turkey are concentrated in Izmir (Aegean Sea) (ICCAT, 2015). ABFT were caught by Turkish purse-seine fleets operating in the Levant Sea fishing grounds. Tuna companies decreased their capacity each year depending on the established quotas. In the current situation, to increase capacity utilization, companies buy live tuna from other Mediterranean countries and transfer by towing them to their grow-out sites in Turkey.

In Greece, capture-based aquaculture of ABFT began in 2003. There are two centers for ABFT on the ICCAT list. One of them is in the Echinades Islands (Prefecture Kefallonian-Ithaki Islands, Cephalonia), and the other is in the Messaras Gulf, Prefecture of Heraklion (Crete) (Figure 4.4). Greece has an installed capacity of 2100 tons/year (Martín, 2007; ICCAT, 2015). ABFT are still caught from purse-seine fleets operating in the Central Mediterranean fishing grounds between Malta and Libya and towed to farms located around 500 miles north east of the fishing grounds in the Ionian Sea (Tzoumas et al., 2010).

In Cyprus, there are three ABFT grow-out farms, each with a capacity of 1000 tons/year. Two of the farms are located in Vasiliko and one in Limassol. The ABFT farms work together with Spanish and French companies using five vessels for purse seining (Martín, 2007; ICCAT, 2015). A Turkish company began fattening operations of tuna in the north part of Cyprus in 2002. Later, the on-growing center was moved to Famagusta (Gazimagosa) with a capacity of 3000 tons. This company was closed down in 2004 and its fattening operation does not appear on the ICCAT list of tuna farms (Martín, 2007; Table 4.2).

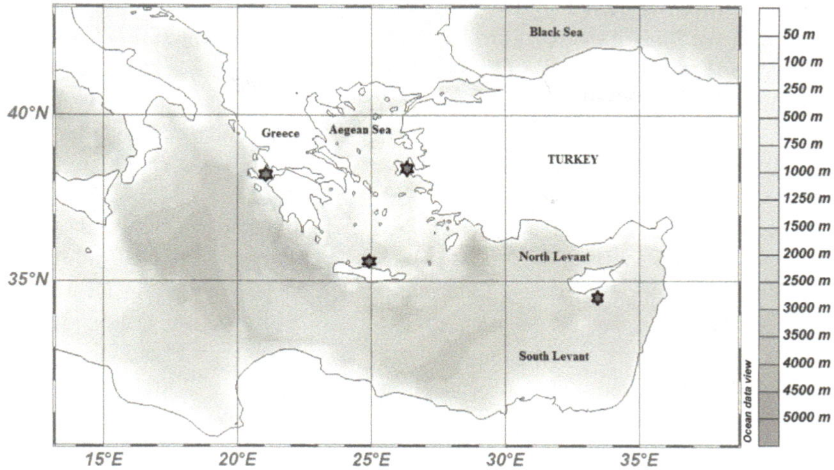

FIGURE 4.4 ABFT farms in the eastern Mediterranean Sea.

TABLE 4.2 Number of Registered and Active ABFT Farming in the Eastern Mediterranean in 2015

	Cyprus	Greece	Turkey
Numbered registered	3	2	6
Cumulated capacity (tons)	3000	2100	6140
Number active	2	1	6

Most ABFT farming companies from countries in the European Union have relied on support from the EU by way of the European Fishery Fund. The European Fishery Fund played an important role in the establishment and modernization of the purse-seine fleet, as well as supporting the development of tuna farming. In addition to European Fishery Fund support, there are both national and regional funds depending upon the politics of individual Mediterranean countries (Sumalia and Huang, 2010). Tudela and García (2004) reported that at least 19 million Euro from EU funds were given to the development of tuna farming. Three quarters (75%) of farming enterprises and tuna fishing fleets benefitted from this assistance. As EU member countries, Cyprus and Greece will utilize these funds as of 2015. While Turkish tuna companies compete with other companies and other countries without government support, they have exemption from tax for imported feeds and purse-seine vessels (Bilgin Topcu, 2012).

ABFT farming is generally practiced by transferring captured fish to cages and feeding them for various periods of time. Production conditions are highly dependent upon both the amount of wild tuna caught and the amount of fish obtained to feed them. Both are limiting factors. Catching of small pelagic fish used as food in tuna farming is performed by purse seiners. ABFT held in the fattening cages encompass a wide range of sizes, from relatively small juvenile fish (30–50 kg) to large fish (240–250 kg).

The tuna are caught in May–June and are harvested 4–8 months later. During this time, the tuna are fed well such that they command the highest possible market price. Japanese tuna market demand is highest in the winter. ABFT catches, transfer into towing cages, transfer into permanent cages and feeding is conducted as previously described in this book and elsewhere (Karakulak, 2007).

In Turkey, the ABFT fattening cages are made of high-density polyethylene pipes (500 mm Ø) with a diameter of 50 m. Nets with a depth of 20–30 m and a mesh size of 80–110 mm are bound to these round cages. These cages are installed in areas with high currents and depths of at least 50 m and 1000–3000 m from shore. The cages are fixed to the sea floor with an anchoring system that is dependent on the regional conditions. Generally, plow-type anchors with a weight of 750–1000 kg are used. The volumes of the surface buoys used in towing cages are between 260 and 1250 lt. In Turkey, a stocking density of 2–3 kg/m^3, corresponding to up to 130 tons of fish, are stocked in each cage (Karakulak, 2007). In Greece, the stocking densities range from 1.2 to 2.5 kg/m^3 at the beginning of the farming period and 2.5–4.0 kg/m^3 at the end (Tzoumas et al., 2010).

After being transferred to the fattening cages, ABFT will start to be fed within 20–40 days once accustomed to captivity. Daily diving monitors the status of fish. Dead fish are immediately removed from the cage. During feeding, divers check whether the fish consume the feed. In Turkey, ABFT are fed imported mackerel (*Scomber scomber*) for the most part. In addition, herring (*Clupea harengus*), capelin (*Mallotus villosus*), shad (*Alosa alosa*), sardine (*Sardina pilchardus*, *Sardinella aurita*), horse mackerel (*Trachurus* sp.), anchovy (*Engraulis encrasicolus*), and imported squid are used to a lesser extent (Yıldırım, 2004; Karakulak, 2007). Tuna feed up to 5–8% of their body weight daily. Feeding occurs once a day (in the morning), 6 days a week in all Turkish companies (Yildirim, 2004). However, the food is distributed twice a day (in the morning and early afternoon), 6 days a week in Greece (Tzoumas et al., 2010).

Harvesting is performed in December and January. Daily maximum harvesting capacity of companies varies between 33 and 42 tons (Yildirim, 2004). ABFT are sold as fresh or frozen, and most of the fresh tuna are sent to the Japanese market by air. Vessels from east Asian countries and those in the Mediterranean Sea buy tuna from the ABFT farms and sell it as frozen tuna in the Asian market.

4.4 PROBLEMS

The most important problem for tuna fattening activities in Turkey and Greece is obtaining wild tuna to stock the cages, as a result of reduced fishing and quota distribution. Efforts to obtain juveniles from eggs obtained from the cages have only recently started in Turkey, and it is widely recognized that these efforts should be pursued further. It is anticipated that, in the future, tuna juvenile supply from hatcheries will partially replace wild caught tuna for grow-out operations.

There are also problems in estimating the quantity and weight of captured ABFT. This weight calculation is the most important criteria in feeding the fish. Accurate calculation of this weight will minimize feed waste. Petrell et al. (1997), Shieh and Petrell (1998), and Martinez de Dios et al. (2003) reported that stereo-video camera systems were a viable method in fish farming for estimation of biomass. These systems have been used in ABFT farming (Harvey et al., 2003; Costa et al., 2009; Ramfos et al., 2012). Proliferation of stereoscopic cameras used in tuna farming might provide opportunity to increase knowledge and data directed to global tuna population (Aguado-Giménez and García-García, 2005).

When making decisions concerning farming areas, fluctuations in sea water temperature, currents, waves, and meteorological parameters are particularly relevant. Waves are the most important external vector that causes deformation of net cage systems. Deformations were observed in Turkey in Antalya (Antalya Bay, North Levant Sea) due to waves in the 2000s. (Yildirim, 2004). In addition, mortality of 700 tons of ABFT was reported in 2003 as a result of a sudden change in water quality (e.g., salinity and turbidity) caused by floodwaters in Antalya (Karakulak, 2007). After this event, Turkish companies transferred their Antalya farms to İzmir and moved further offshore.

Concentrating fattening activities in areas suitable for farming in the Aegean Sea has occasionally led to competing user conflicts with the tourism sector. Also, the use of small pelagic fish as feed sources for tuna farming has posed problems for worldwide tuna farms from both a sustainability perspective and the spread of diseases. In this context, it is recognized that the development of alternative artificial feeds to replace small pelagics is required to maintain and support the sustainable expansion of the tuna aquaculture industry. Also importantly, a standard control system is needed to prevent the occurrence of potential pathogens and to warrant the quality of feed fish (Ottolenghi, 2008).

Although Tanrıkul and Perçin (2005) reported that there is limited availability of knowledge on tuna diseases in Turkish tuna farming, diseases have had only a very limited impact in ABFT farming. This is due to the fact that mature tuna are resistant to bacterial diseases and that farming has been generally carried out in high energy sites with adequate currents and high dissolved oxygen levels and using stocking densities lower than those used for other fish species. Besides, large individuals with developed immune systems are stocked in cages, and the farming cycle is restricted in 4–8 months (Munday et al., 2003; Di Maio and Mladineo, 2008; Hayward et al., 2008; Johnston et al., 2008). An update on tuna diseases is provided in Chapter 11 of this book.

It has been reported that fish mortalities occur in Turkey primarily due to net entanglement and stress but these mortalities have been as low as 1% (Bilgin and Topçu, 2012). Similar cumulative mortalities have been reported to occur during the towing and transfer of fish and acclimation to farm conditions in other Mediterranean studies (Katavic et al., 2003). Fish are not fed during transfer to cages from the fishing grounds in order to minimize effects from stress, reducing the mortality rate to a minimum. In addition, Perçin et al. (2009) reported that some companies attached the transportation cages directly to the mooring systems, thus decreasing fish stress as well as time and labor. Mortalities during transfer were significantly reduced from 10% in Turkey and 21% in Spain in the early years when operations began (Oray and Karakulak, 2003; Norita, 2003) to 1–4% today.

4.5 MANAGEMENT AND CONSERVATION

The ABFT fishery is regulated by ICCAT. Since 1982, the Commission has managed ABFT as two separate stocks: a western Atlantic stock and an eastern Atlantic stock that includes the Mediterranean. ICCAT established a total allowable catch (TAC) for both stocks.

Tuna farming in the Mediterranean gained momentum in the second half of the 1990s, with increasing number of farms in a short period of time that essentially led to overfishing. The Commission set a quota at 32,000 tons for the east Atlantic and Mediterranean stock for the years 2003, 2004, 2005, and 2006 (ICCAT, 2007). In 2006, The Standing Committee on Research and Statistics (SCRS) of ICCAT indicated that the spawning stock biomass had been declining rapidly in recent years while fishing mortality had been increasing rapidly, especially for large ABFT (i.e., ages 8 +). The increase in mortality for large ABFT is consistent with a shift toward targeting larger individuals destined for fattening/farming. SCRS recommended decreasing the quota with no fishing in the spawning season in order to assist in recovering the stock (ICCAT, 2007). However, ICCAT was not able to control illegal fishing in the Mediterranean and was criticized for not taking into consideration scientific evidence and failing to curb Illegal, Unreported and Unregulated (IUU) fishing (Greenpeace, 2007). Indeed, ICCAT's resolutions and recommendations are often controversial and frequently challenged by one or more of the various stakeholders directly or indirectly involved in the tuna fishing and aquaculture industries.

Following suit, the Government of Monaco put forward a proposal to list ABFT in Appendix 1 of the Convention on International Trade in Endangered Species (CITES) in 2009 so as to forbid international trade. Countries such as the United States, Norway, Switzerland, and England supported the initiative (Greenpeace 2010). The matter of whether tuna would be included in the endangered list was voted on during The Parties Conference (COP15) of CITES 15 in Doha/Qatar, and the proposal was rejected by a majority vote. Had it been accepted, ICCAT would have gotten permission to harvest the ABFT stock only under an approved management plan, and CITES would have forbidden the trade of ABFT, which is essentially the equivalent of a fishing ban. As a result, in 2008, ICCAT established a multiannual recovery plan for ABFT in the eastern Atlantic and Mediterranean. The fishing quota was set at 22,000 tons in 2009, 19,950 tons in 2010 and 13,500 tons in 2011. In addition to this, recommendations for reduction of fishing capacity, and monitoring of both vessels and farms have been proposed.

According to regulations set forth by ICCAT, since 2008 all member countries must comply with the Bluefin Tuna Catch Documentation Program (BCD form), disclosing all data about exported fishing and farmed tuna to Japan. Outside of these documents, it is difficult to define the quantity of tuna being supplied to the Japanese market from nonmember states to ICCAT, as statistics are irregular and there are contradictions between references (Ottolenghi, 2008).

Fattening ratios of mature individuals (over 60 kg) in most ABFT farms located in the Mediterranean (excluding Croatia) were estimated to vary between 25% and 38% (Gordoa, 2010; Galaz, 2012). When investigating the fattening ratio of ABFT from BCD forms submitted to ICCAT of Mediterranean countries, the fattening values reported vary between 6% and 289%, mostly between 40% and 90% (Trápaga et al., 2014). These authors noted that a fattening ratio above 40% for adult tuna would be difficult to explain biologically and may alert to potential measuring or reporting errors (Trápaga et al., 2014).

Length and weight of fish cannot be determined accurately due to the fact that most captured tunas are transferred to cages. This situation affects the validity of available data for stock assessment. Video cameras are used for counting the fish but a more effective method is needed to accurately determine length and weight. The relevant Regional Fisheries Management Organizations (RFMOs) as well as the supervising authorities have been promoting during the last years the development and application of new methodologies for the estimation of the live ABFT biomass during transfers that would work independently and with improved accuracy.

IUU fishing is widely recognized as one of the biggest concerns on ABFT management in the Mediterranean Sea and other Atlantic Ocean areas. The catch reported in 2004 (34,500 ton) was slightly over the TAC of 32,000 ton, but SCRS estimated it to be about 50,000 ton. The World Wildlife Fund (WWF, 2006) reported that EU (mostly French) and Libyan fleets are largely responsible for most of the IUU catches. In Egypt, fisheries authorities are well-informed on

IUU fishing activities and observers are placed at fishing ports and on board of ABFT fishing vessels (Oztürk, 2015). On the other hand, fishing of ABFT by non-ICCAT members forms another big barrier to the successful management of their stocks by ICCAT. Indeed, ICCAT has declared IUU fishing vessels every year on its website. One of the reasons for ICCAT's failure to prevent IUU fishing is lack of a powerful reinforcement and punishment system. It seems obvious that ICCAT should be more persuasive and that serious penalties and fines should be applied to responsible countries in order to reduce IUU fishing.

Oztürk (2009) stated that, for various reasons, there is no high sea legal regulatory regime enforced in the international waters of the Aegean and Mediterranean Seas. This has recently caused several management problems in the region, such as IUU fisheries of the ABFT (*Thunnus thynnus*), albacore (*Thunnus alalunga*), bullet tuna (*Auxis rochei*), and swordfish (*Xphias gladius*). The stocks of these species are depleted and cetaceans are entangled in offshore driftnets even though this kind of fisheries is forbidden in the Mediterranean Sea. There is no appropriate management tool for sustainable fisheries of highly migratory and large pelagic species in the area. The General Fisheries Commission for the Mediterranean (GFCM) has addressed issues related to IUU fishing on a number of occasions over the past decade. Fishing activities of GFCM nonmember countries were reported (Aranda et al. 2010; Jolly, 2012). With regard to fishing management, being a member or collaborating with GFCM is very important for countries that have fisheries in the Mediterranean.

For the eastern Atlantic and Mediterranean, the results of the stock assessment in 2012 indicate that the spawning stock of ABFT showed clear signs of increased biomass, probably due to decrease in catch effort and fishing capacity established by ICCAT (ICCAT, 2012). Although this evaluation is very positive, verification is needed with fishery-independent surveys (large-scale tagging program, aerial survey, larval survey) to corroborate the results. In addition, information on catch composition, effort and spatial distribution of the purse-seine fisheries (which represent more than 60% of the total recent reported catch) is needed.

4.6 CONCLUSIONS

The management of the Mediterranean stock is essential to the future of ABFT, as it represents the majority of this species' global population. High priority also needs to be given to protecting spawning stocks because this region represents a critical refuge to this species.

Habitat protection is a key issue for vulnerable species such as ABFT. The eastern Mediterranean part of Turkey is one of the spawning area of this species (Oray and Karakulak, 2005; Oray et al., 2005). The largest concentration of the ABFT larvae is typically found in the Taseli Strait area (Mersin Bay) (Orek et al., 2007). The eggs and larvae of ABFT are very small compared to the adult size of the species, and development of the larvae is very rapid. Larvae usually hatch within 24 h of spawning, so that the occurrence of the larvae in any area indicates that the adults spawned in the close vicinity. Gücü and Oztürk (2010) suggested the designation of the Taseli Straits as a Marine Protected Area (MPA) in the eastern Mediterranean Sea. One of the reasons marine protected areas would be effective is because migration patterns of the ABFT are well understood and, consequently, it is well known when and where they aggregate to spawn. ABFT are most vulnerable when they are spawning. If MPAs are created at their spawning grounds, the fishing pressures would be greatly decreased and would allow populations to recover (Sumalia and Huang, 2010). In MPAs, human activity is monitored and restricted, this way animals passing through them have abundant food supplies and are safe from human activities.

ABFT was listed as an endangered species in the World Conservation Union (IUCN) Red List of Threatened Species (criterion A2bd) in 2011 (Collette et al., 2011). Because ABFT stocks are depleted in the Mediterranean Sea, there is a need to elaborate and implement robust and efficient recovery plans to combat IUU fishing. To this end, cooperation between GFCM and ICCAT should be strengthened. This should include sustainable aquaculture from eggs of ABFT, which should be further promoted by both organizations in light of their expertise and considering that aquaculture could be one of the best solutions for stock recovery and mitigation of IUU fishing activities in the Mediterranean Sea. Encouraging precedents exist as in Japan they have succeeded to breed Pacific bluefin tuna and have recently begun harvesting tuna grown from hatchery-reared juveniles. This achievement is opening new doors for the aquaculture industry and conservationists. It is anticipated that as these technologies spread to other regions around the world, the existing pressure on wild ABFT populations can be reduced, allowing for their recovery in the near future. It has also been reported that some European companies already have started breeding ABFT. For example, Kiliç A.S. has recently succeeded in producing hatchery raised ABFT in Turkey, where the company continues advancing the artificial propagation of this species in the eastern Mediterranean. These achievements may help the expansion of sustainable fisheries and aquaculture practices of the ABFT in the eastern Mediterranean region.

REFERENCES

Aguado-Giménez, F., García-García, B., 2005. Changes in some morphometric relations in Atlantic bluefin tuna (*Thunnus thynnus thynnus* Linnaeus, 1758) as a result of fattening process. Aquaculture 249, 303–309.

Akyol, O., Karakulak, F.S., Ceyhan, T., Dede, A., 2008. Driftnets, used in Turkish Seas and the regulations. Ege J. Fish Aqua. Sci. 25 (2), 153–157.

Altinbulak, B., 2005. Investigation of bluefin tuna (*Thunnus thynnus* L., 1758) ranching in Turkey sea as micro and macro scales. Master Thesis, Izmir, Aegean University, p. 84.

Aranda, M., Bruyn de. P., Murua, H. 2010. A report review of the tuna RFMO's CCSBT, IATTC, IOTC, ICCAT and WCPFC. EU FP7 Project. No: 212188 TXOTX, Deliverable 2(2).

Bilgin Topcu, B., 2012. Socio-economic analysis of the bluefin tuna (*Thunnus thynnus* L.1758) farming sector in Turkey during the harmonization process with European Union. Ankara University, Doctoral Thesis, Ankara, p. 170.

Ceyhan, T., Akyol, O., Karakulak, F.S., 2011. The albacore fishery in Turkey. Col. Vol. Sci. Pap. ICCAT 66 (5), 1867–1871.

Colak, A., Colak, S., Guven, E., 2003. Primary Applications on Bluefin Tuna (*Thunnus thynnus*). "Workshop on Farming, Management and Conservation of Bluefin Tuna" Turkish Marine Research Foundation, Publication No. 13, Istanbul, pp. 40–50.

Collette, B., Amorim, A.F., Boustany, A., Carpenter, K.E., de Oliveira Leite Jr., N., Di Natale, A., et al., 2011. *Thunnus thynnus*. In: IUCN 2012. IUCN Red List of Threatened Species. Version 2012.2. www.iucnredlist.org. Accessed in April 2013.

Costa, C., Scardi, M., Vitalini, V., Cataudella, S., 2009. A dual camera system for counting and sizing Northern Bluefin Tuna (*Thunnus thynnus*; Linnaeus, 1758) stock, during transfer to aquaculture cages, with a semi-automatic Artificial Neural Network tool. Aquaculture 291, 161–167.

De Metrio, G., Arnold, G.P., De La Serna, J.M., Cort, J.L., Block, B.A., Megalofonou, P., et al., 2005. Movements of bluefin tuna (*Thunnus thynnus* L.) tagged in the Mediterranean Sea with pop-up satellite tags. Col. Vol. Sci. Pap. ICCAT 58 (4), 1337–1340.

De Metrio, G., Oray, I., Arnold, G.P., Lutcavage, M., Deflorio, M., Cort, J.L., et al., 2004. Joint Turkish-Italian research in the eastern Mediterranean: bluefin tuna tagging with pop-up satellite tags. Col. Vol. Sci. Pap. ICCAT 56 (3), 1163–1167.

Desse, J., Desse-Berset, N., 1994. Stratégies de pêche au 8ème millénaire: les poissons de Cap Andreas Kastros (Chypre). In: Le Brun, A. (Ed.), Fouilles récentes à Khirokitia. Editions Recherche sur Civilisations, Paris, pp. 335–360.

Di Maio, A., Mladineo, I., 2008. Ultrastructure of Didtmocystis semiglobularis (Didymozoidae, Digenea) cysts in the gills of Pacific Bluefin Tuna (*Thunnus orientalis*). Parasitol. Res. 103, 641–657.

Duka, K.F., 1959. About the spawning of fishes in the Sebastopol Bay. Tr. Sevastop. Biol. St. 11, 297–317 (in Russian).

Galaz, T., 2012. Eleven years –1995–2005- of experience on growth of bluefin tuna (*Thunnus thynnus*) in farms. Col. Vol. Sci. Pap. ICCAT 68 (1), 163–175.

Garcia, A., Alemany, F., Velez-Belchi, P., Rodriguez, J.M., Lopez Jurado, J.L., Gonzalez Pola, C., et al., 2003. Bluefin and frigate tuna spawning off the Balearic archipelago in the environmental conditions observed during the 2002 spawning season. Col. Vol. Sci. Pap. ICCAT 55 (3), 1261–1270.

Golenchenko, A.P., 1952. To organize the tuna fishing in the Black Sea. Rybnoe khozyaistvo 10, 18–19 (in Russian).

Gordoa, A., 2010. Estimating the fattening factor of Atlantic Bluefin Tuna (*Thunnus thynnus*) on tuna farms: The Ametlla de Mar facility as a case Study. Col. Vol. Sci. Pap. ICCAT 65 (3), 848–857.

Greenpeace, 2007. Pirate booty: how ICCAT is failing to curb IUU Fishing, Greenpeace Spain, Spain, p. 50.

Greenpeace, 2010. Time and tuna running out. http://www.greenpeace.org/international/PageFiles/209195/Time%20and%20Tuna%20Timeline%20v3.pdf, 08.10.2011.

Gücü, A.C., Oztürk, B., 2010. Scientific rationale for the proposed CIESM Levant Marine Park. Marine Peace Parks in the Mediterranean – a CIESM proposal. CIESM workshop monographs 41, 61–68.

Harvey, E., Cappo, M., Shortis, M., Robson, S., Buchanan, J., Speare, P., 2003. The accuracy and precision of underwater measurement of length and maximum body depth of southern bluefin tuna (*Thunnus maccoyii*) with a stereo-video camera system. Fish. Res. 63, 315–326.

Hayward, C.J., Aiken, H.M., Nowak, B.F., 2008. An epizootic of *Caligus chiastos* on farmed southern bluefin tuna *Thunnus maccoyii* off South Australia. Dis. Aquat. Anim. 79, 57–63.

Heinisch, G., Corriero, A., Medina, A., Abascal, F.J., de la Serna, J.M., Vassallo-Agius, R., et al., 2008. Spatio-temporal pattern of bluefin tuna (*Thunnus thynnus* L. 1758) gonad maturation across the Mediterranean Sea. Mar. Biol. 154, 623–630.

ICCAT, 2007. Bluefin tuna executive summary in report for biennial period 2006–2007, Part 1, vol. 2, Madrid, p. 240.

ICCAT, 2012. Report of the 2012 Atlantic Bluefin Tuna Stock Assessment Session. In: Meeting Report of the Standing Committee on Research and Statistics (SCRS), International Commission for the Conservation of Atlantic Tunas (ICCAT), October 2012, Madrid, Spain (Doc. Num. SCI-033/2012, p. 124).

ICCAT, 2014. Stat. Bull. 42 (1), 1960–2012, ICCAT, June 2014, Madrid, p. 159.

ICCAT, 2015. ICCAT Record of ABFT Farming Facilities. http://iccat.int/en/ffb.asp, May 2015.

Iyigüngör, D., 1957. Méthods et Moyens de Pêche au Thon Actuellement en Usage en Turquie. Gen. Fish. Coun. Medit. 33, 251–255.

Johnston, C.J., Deveney, M.R., Bayly, T., Nowak, B.F., 2008. Gross and histopathological characteristics of two lipomas and nuerofibrosarcoma detected in aquacultured southern bluefin tuna, *Thunnus maccoyii* (Castelnau), in South Australia. J. Fish Dis. 31, 241–247.

Jolly, D., 2012. As regulators meet, fishing boats thumb their noses. NY Times1 June 2012.

Karakulak, F.S., 2004. Catch and effort of the bluefin tuna purse-seine fishery in Turkish waters. Fish. Res. 68, 361–366.

Karakulak, F.S., 2007. Fishing and farming of the bluefin tuna (*Thunnus thynnus* L.). In: Candan, A., Karatas, S., Kucuktas, H., Okumus, I. (Eds.), Marine Aquaculture in Turkey, Turkish Marine Research Foundation (TUDAV) Publication Number 27, Istanbul, ISBN-978-975-8825-18-9, pp. 61–70.

Karakulak, F.S., 2012. Atlantic bluefin tuna (*Thunnus thynnus thynnus* Linnaeus, 1758) fishery in Turkish waters. In: Tokac, A., Gucu, A.C., Ozturk, B. (Eds.), The state of the Turkish Fisheries, Turkish Marine Research Foundation (TUDAV), Istanbul, Publication Number 34, pp. 382–390, ISBN-978-975-8825-26-4.

Karakulak, F.S., Bilgin, B., Gökoğlu, M., 2007. Albacore (*Thunnus alalunga* Bonnaterre 1788) fishery in the Antalya Bay (Levantine Basin). Rapp. Comm. Int. Mer Medit. 38, 512, April 09–13, 2007, Istanbul.

Karakulak, F.S., Oray, I.K., 2009. Remarks on the fluctuations of bluefin tuna catches in Turkish waters. In: World Symposium for the study into the stock fluctuation of northern bluefin tunas (*Thunnus thynnus and Thunnus orientalis*), Including the historic periods, April 22–24, 2008, Santander. Col. Vol. Sci. Pap. ICCAT 63 (1):153–160.

Karakulak, F.S., Yıldız, T., Bilgin Topçu, B., 2012. Turkish bluefin tuna purse seine fleet from 1987 to 2012 and fisheries management. Ege J. Fish Aqua. Sci. 29 (4), 187–192.

Karakulak, S., 2003. Bluefin tuna fishery in Turkey. In: Oray, I.K. and Karakulak, F.S. (Eds.), Workshop on Farming, Management and Conservation of Bluefin Tuna. April 5–7, 2003, Turkish Marine Research Foundation (TUDAV). Publication Number 13, Istanbul, p. 120–133.

Karakulak, S., Oray, I., Corriero, A., Deflorio, M., Santamaria, N., Desantis, S., et al., 2004. Evidence of Spawning Area for the Bluefin Tuna (*Thunnus thynnus* L.) in the Eastern Mediterranean. J. Appl. Ichtyol. 20, 318–320.

Katavic, I., Ticina,V., Grubišic,L., Franicevic, V. 2003. Tuna farming as a new achievement in mariculture of Croatia. Workshop on Farming, Management and Conservation of Bluefin Tuna. Turkish Marine Research Foundation, Istanbul, Publication No: 13; pp. 10–20.

MacKenzie, B.R., Mariani, P., 2012. Spawning of bluefin tuna in the Black Sea: historical evidence, environmental constraints and population plasticity. PLoS One 7 (7), e39998.

Martín, M.J.I., 2007. Aquaculture in the eastern Mediterranean: Greece, Turkey and Cyprus. European Parliament Directorate General for Internal Policies of the Union, Policy Department Structural and Cohesion Policies, March 2007, Brussels, p. 28.

Martinez de Rios, J.R., Serna, C., Ollero, A., 2003. Computer vision and robotic techniques in fish farms. Robotica 21, 233–243.

Miyake, M.P., Miyabe, N., Nakano, H., 2004. Historical trends of tuna catches in the world. FAO Fisheries Technical Paper. No. 467. Rome, FAO, p. 74.

Munday, B.L., Sawada, Y., Cribb, T., Hayward, C.J., 2003. Diseases of Tunas, *Thunnus* spp. J. Fish Dis. 26, 187–206.

Nishida, T., Tsuji, S., Segawa, K., 1998. Spatial data analyses of Atlantic bluefin tuna larval surveys in the 1994 ICCAT BYP. Col. Vol. Sci. Pap. ICCAT 48, 107–110.

Norita, T., 2003. Feeding of bluefin tuna: experiences in Japan and Spain. In: Bridges, C. Gordin, H., Garcia, A. (Eds.), Cahiers Options Méditerranéennes. Vol. 60, Domestication of the Bluefin Tuna *Thunnus thynnus thynnus*, CIHEAM, Instituto Agronomico de Zaragoza, Zaragoza, Spain, pp. 153156.

Oray, I., Karakulak, S., Alıçlı, Z., Ateş, C., Kahraman, A., 2005. First evidence of spawning in the eastern Mediterranean Sea: preliminary results of TUNALEV larval survey in 2004. Coll. Vol. Sci. Pap. ICCAT 58 (4), 1341–1347, Madrid.

Oray, I.K., Karakulak, F.S., 2003. General review of bluefin tuna farming in Turkey. In: Oray, I. K., Karakulak, F.S. (Eds.), Workshop on Farming, Management and Conservation of Bluefin Tuna, April 5–7, 2003, Turkish Marine Research Foundation (TUDAV), Publication Number 13, Istanbul, p. 32–35.

Oray, I.K., Karakulak, F.S., 2005. Further evidence of spawning of bluefin tuna (*Thunnus thynnus thynnus* L., 1758) and the tuna species (*Auxis rochei* Ris., 1810, *Euthynnus alletteratus* Raf., 1810) in the eastern Mediterranean Sea: preliminary results of TUNALEV larval survey in 2004. J. Appl. Ichthyol. 21, 236–240.

Orek, A.Y., Uysal, Z., Hossucu, B., 2007. Abundance of Scombrid larvae in Mersin Bay (NE Mediterranean). Rapp. Comm. Int. Mer. Medit. 38, 411.

Ottolenghi, R., 2008. Capture-based aquaculture of bluefin tuna. In: Lovatelli, A., Holthus, P.F. (Eds), Capture-based aquaculture. Global overview. FAO Fisheries technical Paper. No. 508. Rome, FAO, pp. 169–182.

Oven, L.S., 1959. Pelagic eggs of fishes in the Black Sea near Karadag. Tr. Karadag. Biol. Stn. 15, 17–21 (in Russian).

Öztürk, B., 2009. Marine protected areas in the high seas of the Aegean and Eastern Mediterranean Seas, some proposals. Black Sea/Mediterr. Environ. 15, 69–82.

Oztürk, B., 2015. Nature and extend of the illegal, unreported and unregulated (IUU) fishing in the Mediterranean Sea. J. Black Sea/Mediterr. Environ. 21 (1), 67–91.

Perçin, F., Fırat, K., Saka, Ş., 2009. Bluefin tuna (*Thunnus thynnus* L., 1758): fishing, towing and the problems during transportation to the fattening. Ege J. Fish Aqua. Sci. 26 (3), 35–241.

Petrell, R.J., Shi, X., Ward, R.K., Naiberg, A., Savage, C.R., 1997. Determining fish size and swimming speed in cages and tanks using simple video techniques. Aquaculture Eng. 16, 63–84.

Piccinetti, C., Piccinetti-Manfrin, G., 1994. Distribution des larves de Thonidés en Méditerranée. FAO Fish. Rep. 494, 186–206.

Piccinetti, C., Piccinetti-Manfrin, G., Soro, S., 1997. Résultats d'une campagne de recherche sur les larves de thonidés en Mediterranée. Col. Vol. Sci. Pap. ICCAT 46, 207–214.

Piccinetti-Manfrin, G., Marano, G., De Metrio, G., Piccinetti, G., 1995. An attempt to find eggs and larvae of bluefin tuna (*Thunnus thynnus*) in the Black Sea. Coll. Vol. Sci. Pap. ICCAT 44, 316–317.

Ramfos, A., Tzoumas, A., Alaveras, P., Katselis, G., 2012. Pilot study of a dual camera system for the determination of size of Atlantic bluefin tuna (*Thunnus thynnus* L. 1758) during caging. Col. Vol. Sci. Pap. ICCAT 68 (1), 230–235.

Schünert-Geiss, E., 1972. Die Münzpragung von Byzantion. Teill II: Kaiserzeit, Adolf M. Hakkert-Amsterdam.

Shieh, A.C.R., Petrell, R.J., 1998. Measurement of size in Atlantic Salmon (*Salmo salar* L.) cages using stereographic video techniques. Aquaculture Eng. 17, 29–43.

Smith, J.A., Ross, W.D., 1910. The works of Aristotle Translated into English: the Revised Oxford Translation. Historia Animalium, by D'Arcy Wentworth Thompson. Bekker (543b35). Oxford at the Clarendon Press, London, England.

Sumalia, R., Huang, L., 2010. Managing Bluefin Tuna in the Mediterranean Sea. Marine Policy. 36: 502–511.

Tanrıkul, T.T., Perçin, F., 2005. Infectious diseases of tunas. Ege J. Fish Aqua. Sci. 22 (3–4), 463–468.

Tekin, O., 2000. Golden Horn and the Tunas of Istanbul. Istanbul J. 32, 92–94, Golden Horn Istanbul. (in Turkish).

Trápaga, S.S., Tudela, S., Quílez-Badia, G., 2014. Bluefin tuna farming growth rates in the Mediterranean. Col. Vol. Sci. Pap. ICCATSCRS/2013/208.

Tsuji, S., Nishikawa, Y., Segawa, K., Hiroe, Y., 1997. Distribution and abundance of *Thunnus* larvae and their relation to the oceanographic condition in the Gulf of Mexico and the Mediterranean Sea during May through August of 1994. Col. Vol. Sci. Pap. ICCAT 46 (2), 161–176.

Tudela, S., García, R., 2004. Tuna Farming in the Mediterranean: The Bluefin Tuna Stock at Stake. WWF Mediterranean Program Office, Rome, Italy, p. 32.

Tzoumas, A., Ramfos, A., De Metrio, G., Corriero, A., Spinos, E., Vavassis, C., et al., 2010. Weight growth of Atlantic Bluefin tuna (*Thunnus thynnus* L.1758) as a results of a 6–7 months fattenings process in the central Mediterranean. Col. Vol. Sci. Pap. ICCAT 65 (3), 787–800.

Vodyanitsky, V.A., 1936. Observations on pelagic eggs of Black Sea fishes. Travaux de la Station Biologique de Sebastopol, Tome V, pp. 3–40 (in Russian).

Vodyanitsky, V.A., Kazanova, I.I., 1954. Key of pelagic eggs and larvae of the Black Sea fishes. Tr. VNIRO 28, 240–323 (in Russian).

WWF, 2006. The plunder of bluefin tuna in the Mediterranean and East Atlantic in 2004 and 2005, uncovering the real story, the collapse of fisheries management. May 2006, p. 91.

Yildirim, S., 2004. A study on some area, system and activity features of bluefin tuna (*Thunnus thynnus* Linnaeus, 1758) farms in Turkey. EU J. Fish. Aquat. Sci. 21 (3–4), 301–305.

Zaitsev, Y.P., 1959. New data about the ichthyoplankton of the north-western part of the Black Sea. Naukovy Zapiski Odesskoi Biol. Stantsii. 1, 77–87 (in Ukrainian).

Zaitsev, Y.P., Mamaev, V., 1997. Marine Biological Diversity in the Black Sea. A Study of Change and Decline. GEF Black Sea Environmental Programme. United Nations Publications, New York.

Chapter 5

Research on the Reproductive Biology and Early Life History of Yellowfin Tuna *Thunnus albacares* in Panama

Daniel Margulies[1], Vernon P. Scholey[2], Jeanne B. Wexler[1] and Maria S. Stein[1]

[1]Inter-American Tropical Tuna Commission, La Jolla, CA, USA, [2]Laboratorio Achotines, Comision Interamericana del Atun Tropical, Correo General, Las Tablas, Provincia de Los Santos, Republica de Panama

5.1 INTRODUCTION

There is increasing interest worldwide in the development of full-life-cycle aquaculture of tuna (Sawada et al., 2005; Reglero et al., 2013). Due to their high value in global fresh-fish markets and their many favorable biological traits, tuna are under intensive study for their aquaculture potential (Masuma et al., 2008). Yellowfin tuna (*Thunnus albacares*) (YFT), although less studied for cultivation than its congeners Pacific bluefin tuna (*Thunnus orientalis*) (PBFT), southern bluefin tuna (*Thunnus maccoyii*) (SBFT), and Atlantic bluefin tuna (*Thunnus thynnus*) (ABFT), are an attractive candidate species for full-life-cycle aquaculture. YFT exhibit high fecundity (Schaefer, 2001), reach sexual maturity in as little as 2 years in tropical waters (Schaefer, 1998; Margulies et al., 2007a), and grow rapidly as larvae, juveniles, and adults (Wild, 1986; Wexler et al., 2003; Margulies et al., 2007b).

Although decades of research have provided considerable information on the populations of adult tuna, relatively little was known about the vital rates (growth, survival) of the early life stages of tuna through the mid-1980s. These considerations motivated the Inter-American Tropical Tuna Commission (IATTC) to establish a research facility in 1985 at Achotines Bay in the Republic of Panama for the purpose of studying the early life histories and reproductive biology of tropical tuna (Scholey, 1997).

D.D. Benetti, G.J. Partridge, A. Buentello (Eds): Advances in Tuna Aquaculture.
DOI: http://dx.doi.org/10.1016/B978-0-12-411459-3.00004-7

The Achotines laboratory was the idea of the late Dr. James Joseph, Director of the IATTC from 1969 to 1999.

From 1996 to present, the focus of research at the Achotines laboratory has centered on the reproductive biology and early life history of YFT, utilizing fertilized eggs spawned by captive broodstock. The YFT research has been focused on ecological questions aimed at understanding the recruitment process and the factors that affect it. Much of the IATTC's ecological research conducted on YFT also has strong application to tuna aquaculture, providing important insights into reproductive biology in captivity and patterns of growth and survival during early life stages.

In this chapter, we summarize the IATTC's research program on YFT conducted at the Achotines laboratory. We provide a short history of the development of the laboratory infrastructure designed to support YFT studies in captivity, and we discuss important aspects of the research that have application to the development of full-life-cycle aquaculture programs for YFT and other tuna species.

5.1.1 The IATTC's Achotines Laboratory

The Achotines laboratory is located on the southern coast of the Azuero Peninsula in the Los Santos province of the Republic of Panama (Figure 5.1). This region is in the northwestern portion of the Panama Bight. The continental shelf is quite narrow at this location; the 200-m depth contour occurs only 6 to 10 km from shore. The annual range of sea surface temperature in the region is approximately 21 to 29 °C. From 1984 to 1995, both field-based and laboratory studies were conducted at the Achotines laboratory on coastal species of scombrids, mainly black skipjack *Euthynnus lineatus*, bullet and/or frigate tuna *Auxis* spp., sierra *Scomberomorus sierra*, and striped bonito *Sarda orientalis*. These studies included field surveys and collection at sea of scombrid larvae and juveniles, plus laboratory experiments that investigated the growth, nutrition, and physiology of coastal scombrids (Lauth and Olson, 1996; Olson and Scholey, 1990; Margulies, 1993; Wexler, 1993). Methods were developed for the collection and husbandry of late-larval and juvenile scombrids. Early- stage scombrids were collected at sea by night lighting and transported alive to the wet laboratory for rearing and experimentation. The life cycle for black skipjack tuna was completed in captivity by rearing field-caught larvae to reproductive maturity over a period of 1 year. This was the first time that a species of tuna was reared in captivity from the larval stage to reproductive size. During 1993–1994, a black skipjack broodstock population was developed in an above-ground culture tank (38 m^3 volume) and the broodstock spawned almost daily for 2 years, providing eggs and larvae for experimental studies (Margulies et al., 2007b).

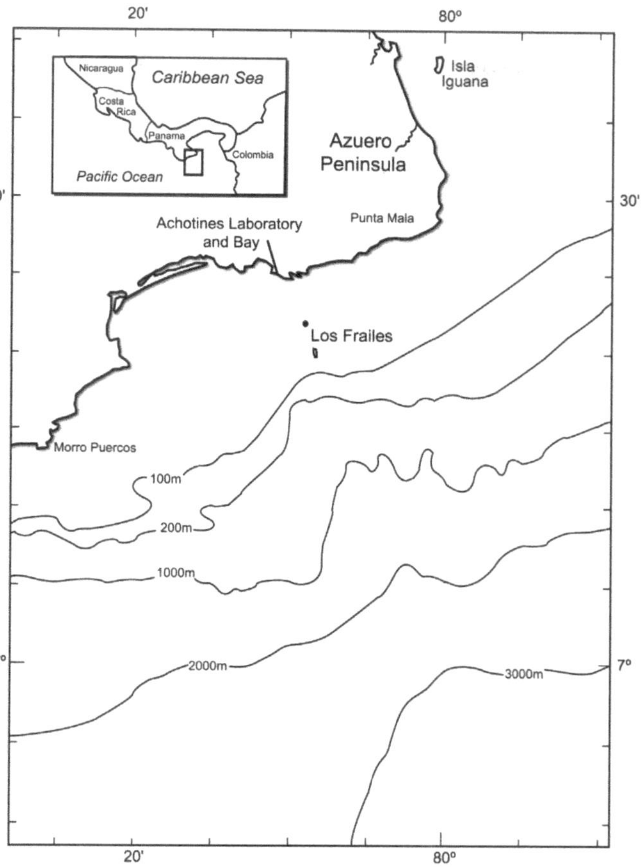

FIGURE 5.1 Location of the Inter-American Tropical Tuna Commission's Achotines laboratory, Republic of Panama.

5.1.2 YFT Research at the Achotines Laboratory

In 1990, discussions were initiated between the IATTC and Japanese agencies to conduct collaborative studies of the spawning and early life history of YFT, in order to combine the research expertise of the various organizations. Based on the success of joint studies of YFT conducted in Japan in 1992 and 1993 (Margulies et al., 2001; Wexler et al., 2001) and the development of successful methods for rearing young tuna, including YFT, at the Achotines laboratory, in December 1993 the IATTC, the Overseas Fishery Cooperation Agency (OFCF) of Japan, and the government of the Republic of Panama agreed to undertake a joint study of YFT at the Achotines laboratory. The 8-year project, funded mostly by the OFCF and the IATTC, extended from 1993 through 2001. The project was designed to study the reproductive biology and early life history of YFT, by developing a spawning population of adults and studying the egg, larval, and juvenile stages in the laboratory (Wexler et al., 2003; Margulies et al., 2007a,b).

Expansion of and improvements to the infrastructure of the Achotines laboratory were necessary to carry out the objectives of the joint YFT project. A major expansion of the laboratory was initiated in 1994 with funding provided primarily by the OFCF. Construction of new broodstock and rearing tanks and a large expansion of the seawater intake system was carried out from 1994 through 1996, and the broodstock and rearing tanks were operational by early 1996 (Wexler et al., 2003; Margulies et al., 2007b). A concrete pier and boat ramp were constructed in 1999 to facilitate boat operations.

From 1996 to present, the IATTC has conducted research on the reproductive biology in captivity and early life history of YFT. The majority of the YFT research has been conducted by the Early Life History Research Group of the IATTC, but important investigations have also been conducted in collaboration with Japanese scientists, the University of Miami's Aquaculture Program, and other academic and nongovernmental researchers. The results of this research are summarized in a series of publications listed on the IATTC website (http://www.iattc.org/AchotinesLab/AchotinesPublicationsENG.htm).

In 2011, the IATTC, Kinki University (KU) of Japan, and the Autoridad de los Recursos Acuáticos de Panama (ARAP) began a 5-year comparative study of the reproductive biology and early life history of YFT and PBFT (Science and Technology Research Partnership for Sustainable Development, SATREPS). The joint research project is funded by the Japan International Cooperation Agency (JICA) and Japan Science and Technology Agency (JST), and is being conducted mostly at the Achotines laboratory and the Fisheries Laboratories of Kinki University in Wakayama Prefecture, Japan. The studies are the first in the world to investigate important comparative aspects of the reproductive biology, genetics, and early life histories of PBFT and YFT. Although Pacific bluefin are temperate to subtropical and YFT are tropical to subtropical in their adult life histories, the early life stages of both species require warm water ecosystems as nursery grounds, thus providing a common background for comparative studies. An additional objective of the project is to develop technologies for the successful aquaculture of juvenile YFT, including sea-cage culture.

The experimental program at the Achotines laboratory has led to advances in husbandry of adult YFT and has provided a long-term opportunity to monitor and study captive spawning patterns of YFT. Research on early life history stages has led to advances in rearing methods for larvae and early juveniles, and has provided insights into the optimal physical and biological conditions for early growth and survival of YFT. These experimental results have strong application to the future development of full-life-cycle aquaculture of YFT. The following is a summary of aquaculture-related aspects of the IATTC's research program with YFT.

5.2 BROODSTOCK DEVELOPMENT AND REPRODUCTIVE BIOLOGY IN CAPTIVITY

5.2.1 Infrastructure Development

Extensive infrastructure is required to maintain broodstock YFT in captivity. During the aforementioned IATTC-OFCF-Panamanian Government project, the construction of in-ground concrete YFT broodstock tanks was completed in early 1996. Housed in a 1300-m^2 roofed, open-walled building, the main broodstock tank (Tank 1) measures 17 m in diameter and 6 m deep and was designed to be large enough to minimize the stress of captivity and enhance the chances of spawning. Tuna purse-seine netting was hung around the perimeter of the tank to prevent fish from jumping out of the tank and translucent panels were placed in the roof above the tank to allow exposure to the natural photoperiod. An aeration tower to aerate and degas both the makeup and recirculated water entering Tank 1 uses two fiberglass pumps to pump water from the surface of the tank sump to the top of the tower (Figure 5.2). That water cascades through the tower's packed column and returns by gravity to the tank via four outflow pipes. About 10% of the water pumped to the aeration tower is diverted through a foam fractionator to remove small particles and dissolved organic material, and then returns to the aeration tower. Bio-filtration for Tank 1 is provided by drawing water from the bottom of the sump with a fiberglass pump that delivers to a slow sand bio-filter. The filtered water returns by gravity to the sump with solids retained by the bio-filter and flushed away daily by backwashing. Filtered and UV-sterilized makeup water, drawn through a seawater intake outside of Achotines Bay, is continuously added to Tank 1 to replace 3–10% of the water daily.

Housed in the same building, the design of a 170 m^3 reserve broodstock tank, Tank 2, is similar to that of Tank 1 but on a reduced scale. The principal difference is the lack of an aeration tower and foam fractionator with air stones providing aeration. Tank 2 holds smaller tuna as reserve broodstock and for brief periods has held newly mature spawning YFT involved in genetic studies.

As part of the ongoing SATREPS project, several upgrades to the Achotines laboratory have been completed during 2011–2014. The upgrades

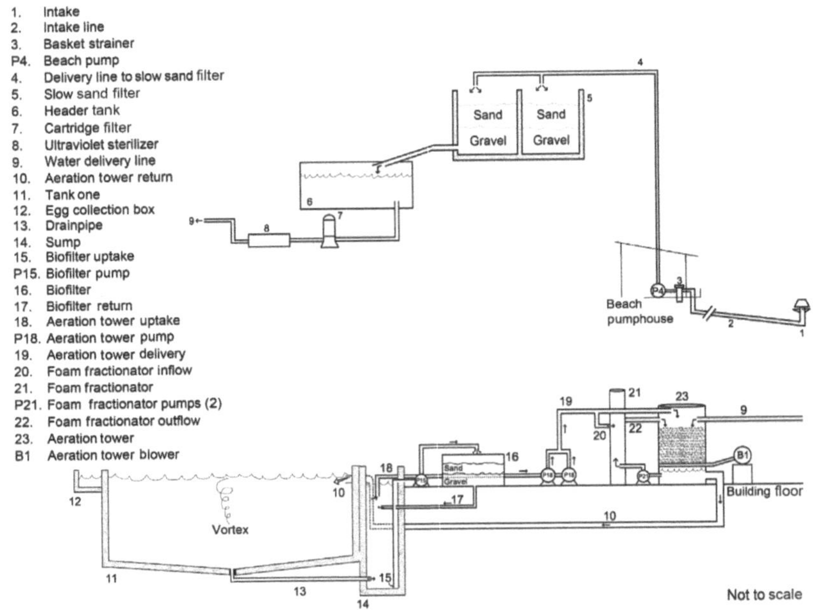

FIGURE 5.2 Seawater delivery and filtration systems (upper diagram) and broodstock tank (Tank 1) systems (lower diagram) at the Achotines laboratory. The reserve holding tank (Tank 2) also has biofiltration (lower diagram) but water is not circulated through an aeration tower or foam fractionator. *(From Wexler et al., 2003)*

include a new cold storage facility, construction of additional office and library and conference space, an enclosed algae and rotifer culture annex, and the delivery of a new research vessel, which was custom-built to transport live tuna.

5.2.2 Development and Growth of Adult YFT

The main (Tank 1) and reserve (Tank 2) broodstock tanks were initially stocked with YFT captured near the Frailes Islands (Figure 5.1) between April and June of 1996 (Wexler et al., 2003; Margulies et al., 2007a). A 7.3 m panga-style boat with an integral live tank was the primary capture vessel. On average, 50% of the YFT caught survived capture and handling, and approximately 30% became broodstock in Tank 1 or Tank 2. Husbandry protocols have remained consistent to present day. Each fish is tagged with a microchip implant tag, and then weighed, measured, and injected with oxytetracycline (OTC). Injection of OTC into the musculature establishes a temporal mark in the otoliths and vertebrae to aid in aging studies. Growth, survival, and feeding and energy requirements of the broodstock from 1996 through 1999 are described by Wexler et al. (2003). These results provided the first long-term broodstock information for tropical tuna in a land-based system. A diet of 50% squid and 50% fish, such as thread herring (*Opisthonema* spp.) or anchoveta (*Cetengraulis mysticetus*), provides adequate nutrition for broodstock YFT and supports almost continuous spawning. The average daily ration fed to the broodstock ranges from 2.5% to 4.8% body weight/day. The inclusion of squid as 50% of the daily ration provides essential fatty acids (HUFAs) important for reproductive performance and egg quality (Izquierdo et al., 2001; Watanabe and Vassalb-Agius, 2003) in addition to a lower-fat diet that aids in reducing fatty tissue deposits in the broodstock fish. Vitamins, at 0.5% to 1.5% of the weight of the food, are mixed in a gelatin base and placed in body cavities of the squid and finfish fed to the broodstock. Since 1999, the broodstock have been periodically replenished with younger, smaller fish to ensure continuous spawning.

Growth of the broodstock has been estimated during two overlapping time periods, 1996–2001 and 1999–2014 (the time periods overlap due to occurrence of individual fish in the broodstock during both time periods) (Figure 5.3). For captive YFT between 1996 and 2001, growth equations and model parameters are described in Wexler et al. (2003) and in Table 5.1. For captive YFT between 1999 and 2014, model parameters that provide the best fit to the length and weight data are also listed in Table 5.1. Parameters of the weight data were first estimated in 2006 and continue to provide the best fit to the current weight data. The captive growth rates were compared with those of wild-caught YFT used in an aging and growth study (Wild, 1986; Table 5.1, Figures 5.3 and 5.4). Estimates of the growth rates in length of the captive fish decreased with increasing lengths of the fish. The mean growth rates in length of fish held in captivity more than 1 year ranged from 18 to 37 cm/yr during 1996–2001 and from 11 to 62 cm/yr during 1999–2014. The mean growth rates in weight ranged from 11 to 26 kg/yr during 1996–2001 and from 4 to 36 kg/yr during 1999–2014. Survival of the YFT originally stocked in Tank 1 was high during the first year of captivity and decreased with increasing biomass; most mortalities were associated with striking the tank wall (Wexler et al., 2003). The broodstock YFT have survived for up to approximately 6 years in captivity with an estimated age of about 7 years (Figures 5.3 and 5.4). Based on the survival and growth of our initial broodstock, we continue to maintain our current and future long-term broodstock tuna in Tank 1 with initial stocking densities of no more than 0.50 to 0.75 kg/m^3 after the first year in captivity.

While Tank 2 temporarily holds reserve broodstock fish and has been emptied for cleaning or maintenance periodically, Tank 1 has been populated with YFT and in continuous operation since June 1996. The initial estimated growth rates both in length and weight are similar between wild and captive

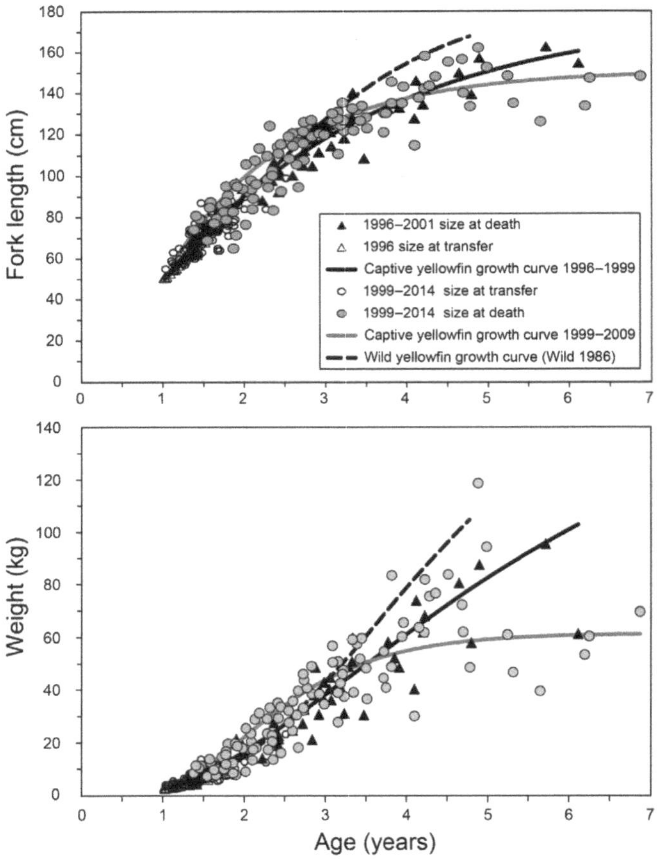

FIGURE 5.3 Relationships between length (top panel) and weight (bottom panel) and estimated ages of captive YFT in Tank 1 and wild YFT in the eastern Pacific Ocean (Wild, 1986). The von Bertalanffy model for the captive YFT and Richards model for wild YFT (see Table 5.1 for model parameters) were applied to the length data (top panel). The Richards model for captive fish and the Gompertz model for wild YFT (Table 5.1) were applied to the weight data (bottom panel). Ages were assigned to each of the captive fish by applying Eq. (6) of Wild (1986) (Table 5.1) to their lengths and the corresponding weights at capture.

YFT until they reach about 3 years of age, when the captive fish begin to grow more slowly than wild fish (Figure 5.3). Although the size of the tank probably contributes to limiting growth in captivity, the energy allocation for continuous spawning may also take priority over that for growth in YFT that have reached reproductive size. The captive broodstock fish have been fed enough to maintain good health and support continuous spawning, but they probably consume far less food than wild YFT. In an aquaculture operation where diet composition and quantity is maximized for production, captive

TABLE 5.1 Parameter Estimates for: a) Growth Equations for Size Measurements (Y_t), and b) Predicting Age (t) from Y_t

	Equation	Parameters			
		Y_∞	K	t_*	m
a. Dependent variable Y_t					
FL (cm) Wild YFT	(1) in Wild (1986)	188.2	0.724	1.825	1.434
WT (kg) Wild YFT	(2) in Wild (1986)	178.4	0.555	3.638	
FL (cm) Captive YFT 1996–1999	(1) in Wexler et al. (2003)	179.2	0.3761		
WT (kg) Captive YFT 1996–1999	(1) in Wexler et al. (2003)	169.3	0.6671		0.5682
FL (cm) Captive YFT 1999–2009	(1) in Wexler et al. (2003)	150.8	0.6744		
WT (kg) Captive YFT 1999–2006	(1) in Wexler et al. (2003)	61.6	532.756		0.9979
b. Independent variable and applicable size range					
FL (30.2–167.9 cm)	(6) in Wild (1986)	211.7	0.521	1.726	

The applicable *t* range for wild YFT is 0.455–4.762 years (Wild 1986). Legend: FL = Fork length, WT = weight, YFT = Yellowfin Tuna.

growth rates could match or possibly exceed growth rates of wild tuna, provided growth is not limited by enclosure size.

In the study of wild tuna referenced in Figures 5.3 and 5.4 (Wild, 1986), 196 YFT from the tropical eastern Pacific were measured, aged, and analyzed. In that study, sexually dimorphic growth was noted, with females attaining maximum sizes of 143 cm and 62 kg, and living to 3.5 years of age, while the largest males were 168 cm and 105 kg, and reached 5 years of age. In our captive broodstock fish, a similar trend in maximum growth was seen with the largest male reaching 162 cm and 119 kg and the largest female 156 cm and 80 kg. In captivity, female YFT are presumably investing much greater levels of energy into ovary development and egg production than the males are in development of testes and milt production. There may be a physiological or genetic limit to growth in weight in reproductively active females. In our captive broodstock population, there has been little evidence for greater longevity in males, as exhibited in wild YFT. Sixteen of the 27 oldest broodstock fish (4–7 yrs of age) have been females, including the two oldest fish.

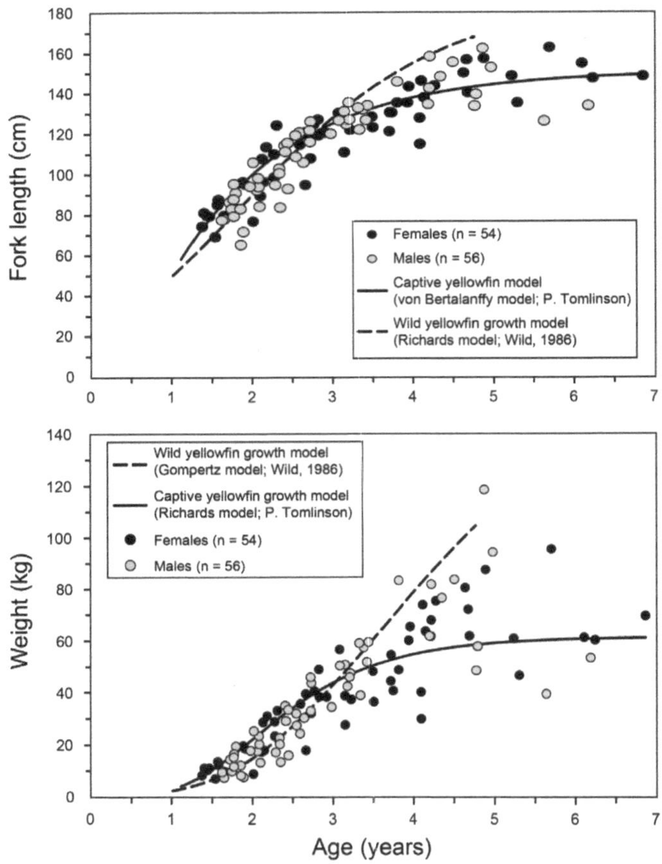

FIGURE 5.4 Relationships between length (top panel) and weight (bottom panel) and estimated ages of captive YFT in Tank 1 and wild YFT in the eastern Pacific Ocean (Wild, 1986), plotted by sex. Size data for captive YFT are reported for the time period 1999–2014 and the growth model parameters are listed in Table 5.1.

5.2.3 Spawning Dynamics and Egg Studies

The development of the spawning population of YFT at the Achotines laboratory has provided an opportunity to study the daily spawning dynamics of YFT over multiple years. Since 1996, the courtship and spawning behaviors of the captive YFT, their spawning periodicity, the influence of physical and biological factors on spawning and hatching, and the egg and early larval development have been observed and described. These aspects of YFT spawning dynamics for the 1996–2000 period are summarized in Margulies et al. (2007a).

The spawning by the YFT broodstock at the Achotines laboratory, beginning in 1996, represents the first occurrence worldwide of sustained

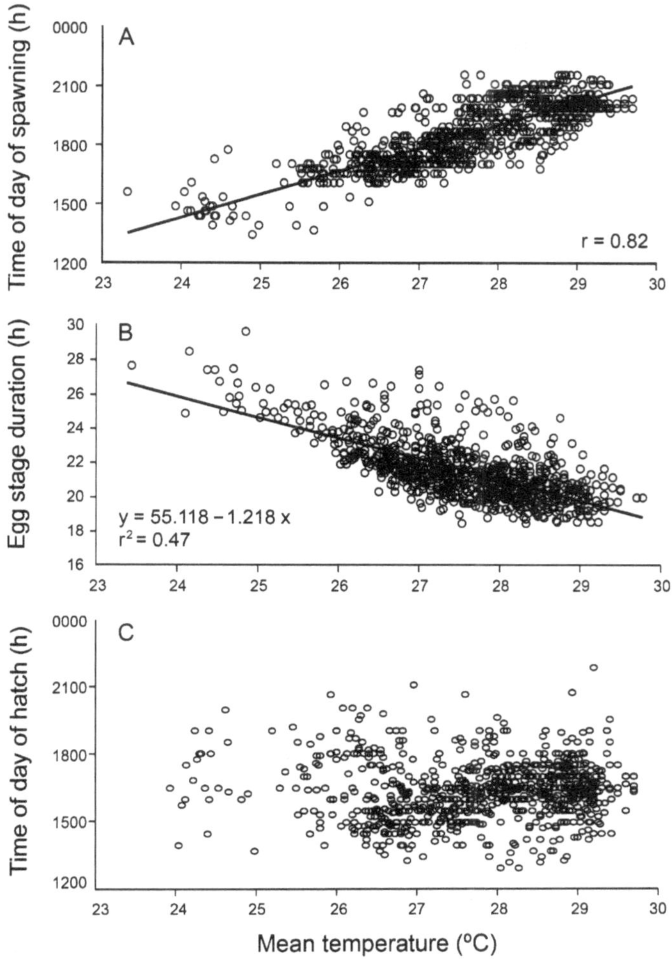

FIGURE 5.5 The relationship between mean daily water temperature and (A) time of day of spawning, (B) egg-stage duration, and (C) time of day of hatching for YFT in captivity, during the period 1996–2000. Plotted values are daily estimates. *(From Margulies et al., 2007a)*

spawning by YFT in land-based facilities (Margulies et al., 2007a). During the first 4 years, YFT spawning was near-daily over extended periods, and occurred at a water temperature range of 23.3 °C to 29.7 °C (Figure 5.5), while spawning occurred over a similar water temperature range of 22.8–29.7 °C from 2001 through 2014 (there were a few anomalous spawning events at 22.8 °C during 2012). Spawning events are preceded by courtship behaviors for 1 to 4 h prior to spawning. Spawning groups consist of one female and one to five males, and typically two to eight groups spawn simultaneously in the tank, depending on total numbers of fish. During the

1996–2000 period, the time of day of spawning was strongly and directly correlated, and egg stage duration inversely correlated, with mean daily water temperature, resulting in a narrow range for the time of day at hatching, 95% of which occurred between 1500 and 1900 h (Figure 5.5). During subsequent years (2001–2014), the broodstock generally exhibited the same pattern for the diel timing of spawning and hatching, except for two anomalous years, when repeated, rapid changes in water temperature altered the normal patterns of diel timing of spawning.

From late 1996 through 2012, the captive YFT broodstock spawned on a near-daily basis but there were periods of no spawning activity. Daily spawning occurred about 90% of the time during this 15-year period when water temperatures were conducive to spawning. There were cessations in spawning lasting more than a week during approximately 10% of that period. Data collected on tank water quality, broodstock size distribution, and sex ratios in the broodstock tank since 1996 were analyzed to look for possible factors to explain periods of no spawning. Water temperature appears to be the main exogenous factor controlling the occurrence and timing of spawning, with spawning occurring as long as broodstock receive adequate daily food rations and water temperature is above 23.3 °C (Margulies et al., 2007a). During several periods, spawning cessations were also associated with the occurrence of ≤ 6 mature females in the broodstock group. Preliminary analyses do not show any other factors exhibiting strong associations with spawning cessations.

The age of captive YFT broodstock at first spawning for the 1996–2000 period was estimated to be 1.3–2.8 years (12–28 kg in weight), averaging slightly less than 2 years (Margulies et al., 2007a). Over short time periods (<1 month), spawning females can increase their egg production by 30% to 234% in response to short-term increases in daily ration of 9% to 33%. Egg production peaks from 4 to 21 days (average of 12 days) after increasing rations. Hatch rates average 83%. Fertilized eggs average 0.97 mm in diameter (range of 0.85 to 1.13 mm) and contain a single oil globule averaging 0.22 mm in diameter (range of 0.15 to 0.28 mm). Water temperature is significantly, inversely related to egg size, egg stage duration, larval size at hatch, and yolksac larval duration. The growth potential from early-stage larva to size at recruitment (30 cm, 6 months of age) is estimated to be very high, approaching 10^6 to 10^7 times.

5.2.4 Shipments of YFT Eggs and Larvae

There is worldwide interest in studying the biology and ecology of tuna eggs, larvae, and juveniles in both the commercial sector and for educational and institutional research. Wild capture of some life stages of tuna is possible but expensive and impractical. As interest in tuna aquaculture increases, the development of successful shipping techniques of tuna during the early life stages is of great interest in the aquaculture community. The initial investment and maintenance of a broodstock population can be cost prohibitive and logistically difficult due to environmental factors and/or highly restrictive regulatory guidelines. Several different projects at the Achotines laboratory have investigated the feasibility of international shipments of eggs and larvae of YFT.

Simulated air shipment trials of live YFT eggs and larvae were first carried out as part of a joint IATTC–University of Miami workshop in June 2005 (Zink et al., 2011). The results of these trials were promising enough to consider actual shipments. In January of 2010, as part of an IATTC–Hubbs Sea World Research Institute (HSWRI) joint project, three insulated boxes of live YFT larvae in plastic bags filled with seawater were shipped from the Achotines laboratory to HSWRI in San Diego, California, USA. The total transit time was about 24 h and survival of larvae during shipment ranged from 24% to 80%. From 2010 to 2014, multiple shipments were made from Achotines laboratory to HSWRI. Eggs were shipped at densities of 1300–2000/L, and yolksac larvae were shipped at densities of 200–350/L. Shipping water was supersaturated with pure oxygen to approximately 200% saturation, and Trizma® and Amquel+ ® were added to the water to buffer pH and ammonia, respectively. Frozen gel packs were packed inside each insulated shipping box to stabilize temperature during shipment. These efforts demonstrated both the feasibility and challenges involved in such shipments. While survival of YFT larvae on arrival to HSWRI was as high as 80%, with subsequent survival in culture tanks to 9 d post-hatch (IATTC-HSWRI, unpublished data), there were also shipments with 100% mortality due to airline handling errors or paperwork issues. During 2014, air shipments of YFT larvae to Baltimore, Maryland, USA, were also conducted as part of a joint project between the IATTC and the University of Maryland Baltimore County (UMBC), Institute of Marine and Environmental Technology. Survival results were similar to the IATTC-HSWRI shipments. While the feasibility of air shipment of YFT eggs and larvae from Panama to the United States has been demonstrated, additional research will be required before international air shipment can be considered a routine procedure.

5.2.5 Genetic Analysis of Adult YFT

Genetic monitoring of captive YFT broodstock for spawning ecology, using mitochondrial DNA, was reported by Niwa et al. (2003). In that study, 27 genotypes were identified in a population of 28 broodstock individuals and 18 genotypes that established a single individual's identity. It was recommended that future research utilize short DNA fragments from the D-Loop with single-strand conformation polymorphism (SSCP) analysis for improved real-time monitoring of genetic variation of offspring and determination of individual spawning profiles of YFT. As part of the ongoing SATREPS study, DNA samples of captive broodstock YFT, plus DNA extracted from eggs spawned by these broodstock, were collected from 2012 to 2014, and will be compared using the aforementioned SSCP analysis. The genetic variation of the mtDNA *D-loop* region of YFT appears to be so high that it is probably useful for both identification of individual fish and investigations of population structures in the wild.

It may be desirable to maintain a certain sex ratio of captive broodstock tuna, but sex determination via external characteristics of late juvenile and early adult YFT captured from the wild and stocked into broodstock tanks is not possible. As part of the SATREPS project, two possible methods of non-invasive sex determination of live YFT broodstock are being investigated. Genetic techniques developed by KU to determine the sex of offspring of captive Pacific bluefin (Agawa et al., 2015) are being applied to determine the sex of captive YFT broodstock and their offspring at the Achotines laboratory. Sex determination via immunochemical analyses (Susca et al., 2001) is also being investigated.

Additional genetic analysis of YFT broodstock has been conducted. Utilizing DNA obtained from a single captive spawning pair of YFT and the resulting offspring, Mendelian inheritance of the ribosomal protein gene intron variations was confirmed (Chow et al., 2001) and contributed to a study confirming Mendelian inheritance of microsatellites (Takagi et al., 2003).

5.2.6 Visual Development of Juvenile and Adult YFT and Considerations for Aquaculture

Scombrid fishes are highly active predators and rely mostly on vision to detect, track, and capture prey (Magnuson, 1963). The retina in tuna is highly developed having rods, single and double cones (Margulies, 1997; Loew et al., 2002). The major visual axis is directed upward or upper-fore, and visual resolution along this line of sight is the highest known in teleost fishes (Tamura and Wisby, 1963; Nakamura, 1969). Yet, holding juvenile and adult tuna in confined spaces of land-based tanks and sea cages does create special challenges for visually based feeding and swimming. Tank walls and sea cages must be visually identified by the captive fishes (visual acuity) and successful visual adaptations must be maintained in response to changes in light and photoperiod (mediated by retinomotor responses in the fish retina).

Important aspects of the IATTC's husbandry program with YFT have been developed, taking into consideration the visual development of juvenile and adult YFT. The land- based broodstock tanks at the Achotines laboratory are circular, in-ground concrete tanks, with the internal tank walls painted grey. To ensure maximum visual detection of the tank walls by the captive YFT, vertical black stripes spaced 46 cm apart were painted on the tank walls. The spacing of the vertical stripes was estimated to ensure minimum visual acuity perception (Margulies, 1997; Loew et al., 2002) of adjoining stripes by broodstock fish to provide a sense of the physical barrier. This system has worked well for over 18 years, although tank wall strikes by the broodstock fish, as mentioned previously, do occur periodically. There is a lag time in the retinomotor response in tuna retinas (Margulies, 1997; Fukuda et al., 2010) which allows the fish to adjust sensitivity from scotopic (dark) to photopic (light) vision. Since the majority of the YFT tank wall strikes at the Achotines laboratory occur around dawn, it is assumed that the wall strikes result from insufficient time to adapt from scotopic to photopic vision.

Microspectrophotometric (MSP) studies of YFT adults at the Achotines laboratory have revealed that their retinas contain visual pigments with strong sensitivity in the blue- to blue-green range of the color spectrum (peaks at 483–485 nm), but that they also contain a violet-sensitive pigment at 426 nm. This suggests that adult YFT may have the ability for limited hue discrimination in the blue-green to violet regions of the spectrum (i.e., they are not isochromats and may have the capacity for color vision) (Loew et al., 2002). This spectral sensitivity of adult YFT may be important in the choice of colors and contrasting hues for broodstock tanks and sea cages.

Another important aspect of YFT husbandry has been to minimize the occurrence of sudden changes in light intensity for all life stages reared at the Achotines laboratory. Introduction of bright white light during the nighttime can cause catastrophic panic and burst swimming leading to wall strikes, especially in juvenile and adult YFT. Therefore, ambient light intensity is allowed to change naturally with daily photoperiod, and at night all staff activities and egg collections are conducted with only red lights, which are outside the spectral sensitivity of YFT (Loew et al., 2002).

5.3 LABORATORY STUDIES OF YFT LARVAE

5.3.1 Development, Survival and Growth Dynamics of Larvae

Since 1997, the IATTC has studied the development, survival, and growth of YFT early life stages reared from eggs spawned by our YFT broodstock. Laboratory investigations have been conducted on the effects of biological and physical factors on larval development and vital rates (survival, growth). Although the laboratory studies have been focused on ecology and pre-recruit survival patterns, most of the experimental results have direct applicability to aquaculture of YFT through the development of optimal rearing protocols for larvae and early juveniles. The following sections present important aquaculture-related findings from the IATTC's investigations of YFT larvae and early juveniles.

5.3.1.1 Yolksac Stage

At hatching, yolksac YFT larvae average 2.5 mm (range 2.0 to 2.9 mm) in notochord length (NL) and 30 μg (range 25 to 41 μg) in dry weight. Larvae at hatching have unpigmented eyes, no alimentary tract nor mouth, and exhibit a large, elliptical yolk mass containing a single posterior oil globule. Mean larval length at hatching is positively correlated with mean egg diameter, and negatively correlated with mean incubation temperature (Margulies et al., 2007a).

The duration of the yolksac larval stage is inversely (but weakly) related to water temperature and ranges from 56 h at 29.0 °C to 65 h at 23.5 °C. Eye pigmentation and mouth formation in larvae occur almost simultaneously, and at the stage of mouth formation there is usually only a trace of yolk remaining. First-feeding larvae average 3.3 mm NL (range 2.7 to 3.9 mm) and weigh on average 22 μg dry weight (range 14 to 30 μg). The mouth width of first-feeding larvae ranges from 225 to 350 μm, averaging 262 μm (Margulies et al., 2007a).

5.3.1.2 Survival Patterns

Rearing of YFT larvae is conducted in circular, fiberglass tanks ranging in water volume from 700 to 1000 L. A flow-through water system is maintained with daily tank turnover rates of 2.5 to 3.5 exchanges/day. Natural mortality rates of YFT larvae during early feeding stages are high (Lang et al., 1994), and larval mortality in IATTC laboratory-rearing trials is substantial. Both the yolksac and first-feeding stages are periods of highest absolute mortality. Laboratory survival rates during the yolksac stage (from hatching until 2.3 to 3.0 days after hatching (dah)) generally range from 50% to 80%, averaging 65–70%. Since the focus of IATTC early life history studies is to simulate natural conditions, YFT yolksac larvae have been reared without any artificial manipulation of tank conditions. Typically in tuna aquaculture systems, an oil film layer is introduced during the yolksac stage to minimize contact with the surface layer to reduce surface adhesion mortality of larvae. YFT larvae can also exhibit mortality due to surface adhesion (Honryo et al., 2014). These types of tank manipulations are not necessary, given the IATTC approach for maintaining natural conditions for rearing, but they could be used in an aquaculture or commercial-scale system for YFT.

Mortality during the early feeding stages of YFT in the laboratory is very high. In IATTC laboratory experiments and rearing trials (conducted mostly at water temperatures of 26–30 °C), absolute mortality is highest during the first 2 days of feeding. Typically, IATTC experiments and rearing trials are conducted using relatively low to moderate concentrations of prey (generally 1–3 rotifers/mL during the first week of feeding) and lighting that simulates natural photoperiod (12 h light, 12 h dark). Use of extended photoperiod for rearing of YFT larvae up to 9 dah has demonstrated both significantly improved survival and growth (at higher light intensities of 30 μmoles/m^2/s during the night phase) and no significant differences in survival or growth (at lower light intensities of 10 μmoles/m^2/s during the night phase) (Partridge et al., 2011). Comparisons of extended photoperiod with natural photoperiod in IATTC trials have produced mixed results for growth and survival (IATTC, unpublished data). The food levels used in IATTC rearing trials are designed to simulate natural feeding conditions and are 3–10 × lower than those routinely used in tuna aquaculture systems (Tanaka et al., 2014). High mortality of YFT larvae occurs during the first 2 days of feeding despite the extremely high incidence of first-feeding success (typically > 90% occurrence). Presumably the majority of early-feeding mortality is the result of malnutrition and insufficient daily rations while subject to very high metabolic demands at high developmental temperatures (Margulies et al., 2007b; Wexler et al., 2007). Daily mortality continues at a steadily decreasing rate during the first week of feeding, with 80–95% of feeding-stage mortality occurring by 10 dah. Another period of relatively high mortality begins during the early notochord-post flexion stage (generally 14–16 dah), when larvae become cannibalistic in rearing systems. This period of cannibalism-related mortality continues through juvenile metamorphosis (generally 24–30 dah).

Survival to the completion of the larval stage in the laboratory is variable, but generally low. From hatching stage (1 dah) to juvenile metamorphosis, overall survival rate is generally 1–3%. In approximately 25% of rearing trials, survival is higher (3–5%), and in 10% of rearing trials, survival is exceptionally high (5–10%). Subtle differences in physical conditions (light intensity, current patterns, micro-turbulence) and feeding levels among rearing tanks often produce surprisingly large differences in survival, particularly during the first week of feeding.

5.3.1.3 Effects of Food Type and Abundance on Larval Growth

Since 1997, IATTC scientists have studied growth in the laboratory of YFT larvae and juveniles reared from eggs from our YFT broodstock. The effects of food type and availability, water temperature, and other physical factors on the growth of YFT larvae and juveniles have been investigated up to 100 dah (Margulies et al., 2007b; Wexler et al., 2007).

Early-larval growth in the laboratory (the first 7–11 days of feeding) is exponential in length and weight (<0.35 mm/day in length and 20–35% dry weight/d), but growth increases significantly during the late-larval and

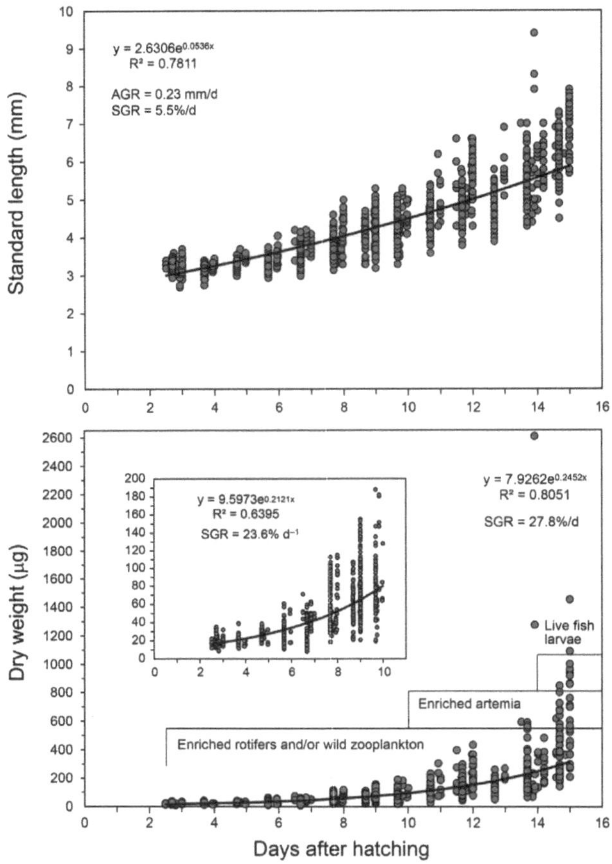

FIGURE 5.6 Growth in length (top panel) and dry weight (bottom panel) of YFT larvae from 3 to 15 days after hatching (dah) in the laboratory. AGR = average growth rate estimated from differentiation of the exponential model parameters. SGR = specific growth rate (%/day). The types of food provided sequentially during development are indicated in the bottom panel.

early-juvenile stages (>0.6 mm/day in length and 30–50% dry weight/d) (Figures 5.6 and 5.7). These growth patterns are based on a sequential diet of enriched rotifers fed at 1000–3000/L (3–14 dah), enriched *Artemia* fed at 100–250/L (10–18 dah), newly hatched YFT larvae fed at 1–10/L (14–40 dah), and minced fish fed at a daily ration of 50–125% of body weight (25–100 dah) (Figures 5.6 and 5.7). For feeding of YFT larvae, enrichment of cultured rotifers and *Artemia* has been successful using either AlgaMac 2000 or AlgaMac Enrich (Aquafauna BioMarine, Inc., Hawthorne, CA, USA). These growth patterns are also based on rearing trials conducted at water temperatures of 26–30 °C. YFT larvae become piscivorous at around 6.5 mm standard length (SL) (14–16 dah), and the timing of the

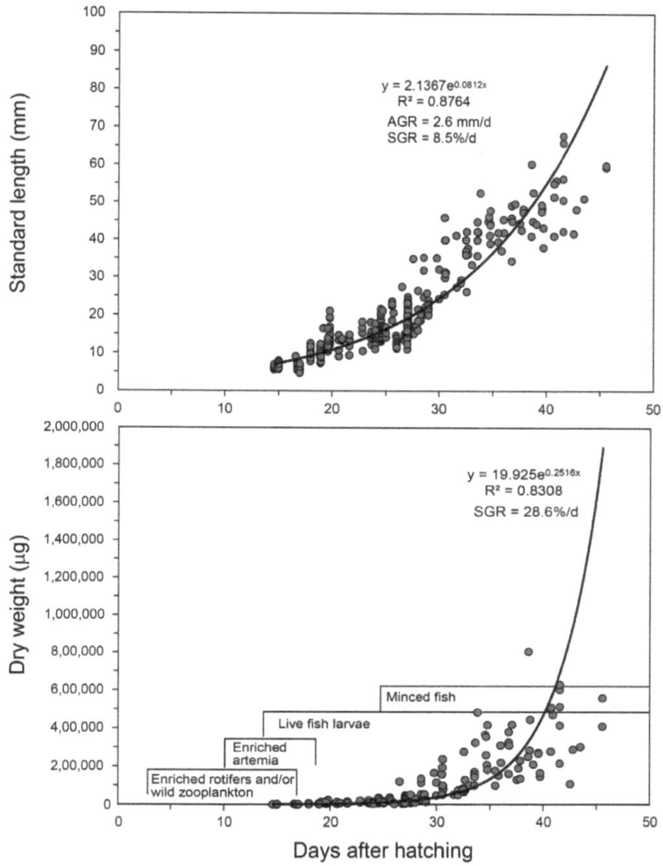

FIGURE 5.7 Growth in length (top panel) and dry weight (bottom panel) of YFT larvae and early-juveniles from 15 to 45 days after hatching (dah) in the laboratory. AGR = average growth rate estimated from differentiation of the exponential model parameters. SGR = specific growth rate (%/day). The types of food provided sequentially during development are indicated in the bottom panel.

onset of piscivory seems to determine, in part, an individual's growth potential. Laboratory cohorts that are early piscivores (~6.2–7.0 mm SL) grow more rapidly, and the onset of a piscivorous diet corresponds with a substantial increase in growth rate (Figure 5.7). The nutritional advantages of a fish diet are numerous, providing higher quantities of calories/prey item (Margulies et al., 2001), essential fatty acids (Tocher et al., 2008), and free amino acids for promotion of protein synthesis (Buentello et al., 2011). Larvae that remain zooplanktivorous or are slow to transition to a piscivorous diet lag behind in growth and are usually cannibalized in rearing tanks.

Copepods, the natural prey of first-feeding YFT larvae (Uotani et al., 1981), are rich in highly unsaturated fatty acids (HUFAs), particularly docosahexaenoic acid (DHA), a critical requirement of marine fish larvae for development and growth (Tocher et al., 2008). YFT larvae exhibit strong positive selection for copepods when fed in prey assemblages with rotifers (Margulies et al., 2001). During the larval stage, even small amounts of wild microzooplankton in the diet may promote more rapid growth rates when background food levels of cultured prey are low. Although natural zooplankton (particularly copepods) would be the preferred food type for rearing of YFT larvae, the routine use of copepods or other natural zooplankton prey for sustained rearing of YFT is difficult. Our research group has conducted pilot studies for culture of copepods as larval food, but copepod cultures often produce variable results. Our group has also collected wild plankton as a food source for experimental-scale studies of YFT larvae, but field collections are weather-dependent and labor-intensive. Our approach for most experimental investigations and rearing of YFT larvae has been to use nutritionally enriched cultured prey (rotifers and *Artemia*) with a subsequent transition to newly hatched YFT larvae as a standard diet for YFT. This diet sequence produces acceptable growth patterns in YFT larvae. In comparison with growth rates of field-collected YFT of a similar age range, our most rapid laboratory growth rates to date have been 20% less than those of the slowest-growing group in the field; however, the sizes-at-age of the laboratory fish approach those of the field-collected fish shortly after a piscivorous diet is introduced at 18 dah (Margulies et al., 2007b; Wexler et al., 2007). Laboratory growth rates of pre-flexion larvae are probably lower than comparable growth rates *in situ* due to a higher occurrence of slow-growing larvae in the laboratory; most malnourished or slow-growing larvae appear to be removed rapidly via starvation and predation in the wild (Margulies, 1993; Wexler et al., 2007).

Rapid growth of YFT larvae can be promoted in the laboratory under intermediate (1000–3000 rotifers/L) to higher (>3000 rotifers/L) cultured-prey levels (the cultured prey must be nutritionally enriched). Preliminary evidence suggests that a copepod diet (either primary or supplemental) may promote fast growth as well. The optimal water temperature range to promote fast growth in YFT larvae through juvenile transformation is approximately 26–30 °C.

5.3.1.4 Density-Dependent Growth

Physiological and morphological changes during early development of YFT influence their feeding behavior and their abilities to efficiently feed and assimilate food in a dynamic ocean environment. Growth of YFT is variable during the first three weeks of feeding, and is strongly influenced by food availability and composition, rearing water temperatures, and stocking densities. The magnitude of density effects on growth during the early life stages of YFT appears to be stage-specific. In laboratory rearing trials, there is a strong early onset of density-dependent growth during the pre-flexion stage (during the first week of feeding), regardless of background food levels, but density effects may be more pronounced at lower food levels (Jenkins et al., 1991). Results of experiments conducted at the Achotines laboratory to date indicate significantly slower growth rates (ANCOVA, $P < 0.01$) in both weight and length for larvae during the first week of feeding stocked at two- to four-fold higher densities and fed at relatively high mean daily background food levels of 1600 to 3600 rotifers/L. Two- to four-fold increases in larval density during the first 6 days of feeding (3−9 days post hatching) result in growth deficits as high as 56% in weight and 47% in length, but growth rates are more variable for larvae stocked at lower densities. This magnitude of difference in growth deficits at higher densities can persist in a subtle pattern into the early juvenile stage and affect mean biomass of a juvenile cohort.

Density effects on growth are most pronounced during the early-feeding stages, and persist into the early-juvenile stage, but the relative magnitude of density effects diminishes with larval size and is more strongly influenced by background food levels during the older stages of larval development (i.e., flexion and post flexion larval stages and early-stage juveniles) (IATTC unpublished data). A two-fold increase in density results in mean growth deficits of 19−28% in length and 18−30% in weight for 9- to 15-day-old larvae when fed 84−87% lower food levels (i.e., daily mean food levels of 375 zooplankters/L). Growth rates and the adjusted mean sizes at age are significantly higher (ANCOVA, $P < 0.001$) for larvae stocked at lower densities. When 9- to 15-day-old larvae are fed higher food concentrations (i.e., daily mean food levels of 2400−2700 zooplankters/L), growth rates are not significantly different between higher and lower stocking densities. A three- to four-fold increase in densities for post-flexion larvae and early-stage juveniles (14−23 days post hatching) resulted in minimal growth deficits when fed a diet of fish larvae, but there is some evidence that density effects can persist in older transitioning larvae (i.e., $\geq$18 days post hatching) when daily mean food levels become limiting. Even subtle density effects on growth during the relatively long juvenile phase of development in YFT could have a "fine-tuning" effect on recruitment and the standardized mean biomass of a cohort (estimated from the survival and mean final weight). The mean cohort biomass is 29−37% greater for early-stage juveniles raised at lower densities (*i.e.*, at two- to four-fold differences in stocking densities).

The survival of slower-growing individuals between 3 and 23 dah does not appear to be impacted by larval density in a laboratory environment and in the absence of predation during the first 2 weeks after hatching. A limited amount of mortality does occur with the onset of cannibalism at around 14 to 16 dah but has little impact on the final survival estimates regardless of stocking density. Percentage survival estimates for larvae and juveniles in density trials are not significantly different at the end of each experimental period (ANOVA, $P > 0.10$) (IATTC, unpublished data).

Supporting high survival and relatively rapid growth of larvae hatched from fertilized eggs poses major challenges for mass production of a tuna species in aquaculture. Depending on the rearing objectives of a cohort, it may be preferable to stock first-feeding larvae at low to intermediate density levels to provide a more balanced approach for optimal survival and rapid growth. In order to avoid significant growth deficits at higher stocking densities, relatively high background food levels of stage-specific prey types (Figures 5.6 and 5.7) should be provided during all early life stages of YFT.

5.3.2 Effects of Physical Factors on Survival and Growth

5.3.2.1 Effects of Water Temperature and Dissolved Oxygen

The optimal water temperature and dissolved oxygen ranges for survival, development, and growth of YFT fertilized eggs and yolksac and first-feeding larvae have been determined from a series of experiments conducted at the Achotines laboratory (Wexler et al., 2011, 2012). Based on the results of these experiments, yolksac and first-feeding YFT larvae exhibit lethal limits for their survival at temperatures less than 21 °C and greater than 33 °C (Figure 5.8). Embryos hatched alive at all temperatures tested except 36 °C; however, larvae were malformed after hatching at temperatures <20 °C and ≥ 34 °C. The optimal range of temperatures for rapid growth and moderate to high survival in first-feeding larvae is from about 26–31 °C; the mean specific growth rates in weight increase from 21% to 45%/d within this range.

YFT larvae appear to be more sensitive to hypoxic conditions than are most other species of marine fish larvae (Miller et al., 2002; Ishibashi et al., 2005). Lethal conditions (100% mortality) for post-hatching and first-feeding YFT larvae occur at dissolved oxygen concentrations of <2.2 mg O_2/L (~<34% oxygen saturation) at temperatures between 26 °C and 29 °C. Significantly lower survival of first-feeding larvae first occurred when they were exposed to dissolved oxygen concentrations of 2.7 mg O_2/L (~40% oxygen saturation). Tuna larvae during the yolksac stage and first feeding may be more sensitive to low levels of dissolved oxygen at high water temperatures due to their higher metabolic rates (Miyashita et al., 1999), the inability to regulate respiration and make physiological adjustments during low dissolved oxygen conditions (Pelster, 1999; Ishibashi et al., 2007), and because they are adapted to an oxygen-rich environment in the upper mixed layer where they naturally occur (Davis and Clementson, 1989; Lauth and Olson, 1996; Owen, 1997).

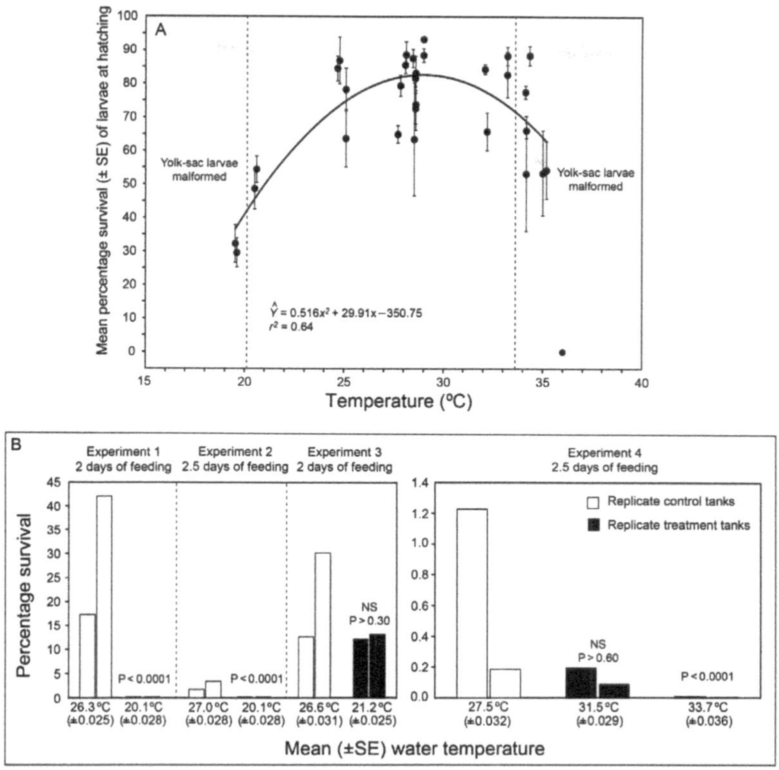

FIGURE 5.8 The effect of water temperature on survival of YFT yolksac and first-feeding larvae. (A) Percentage survival of YFT larvae at hatching incubated at different mean temperatures. The quadratic function fitted to the data describes the percentage survival over the mean water temperature range at which hatching occurred (19.5–35.2 °C). Plotted values are means ± SE (B) Percentage survival of YFT larvae after 2.0–2.5 days of feeding in four different temperature experiments. Target water temperature treatments were 20, 21, 32, and 34 °C. Significant and nonsignificant (NS) differences between mean percentage survivals of a treatment group and control group are indicated.

Minimizing mortality during the early stages of development in YFT larvae is essential for successful aquaculture production. Based on the physical requirements for optimal survival and growth of YFT larvae, yolksac and first-feeding larvae should be maintained at rearing water temperatures between about 26 °C and 31 °C and at dissolved oxygen concentrations >2.7 mg O_2/L.

5.3.2.2 *Effects of Micro-Turbulence on Survival and Growth*

Feeding success of marine fish larvae can be influenced by the levels of micro-scale turbulence in the feeding environment (Rothschild and Osborn, 1988; Dower et al., 1997). The probability of prey encounters and feeding success of larvae may increase directly with micro-scale turbulence up to an asymptotic turbulence level, and then decrease at higher levels of turbulence (MacKenzie et al., 1994). Early studies by our IATTC group and the Japan Sea Farming Association (JASFA) in Japan in 1992 indicated a strong potential for the influence of micro-scale turbulence on the feeding success of YFT larvae (Margulies et al., 2001). We expanded these investigations during 1997–2000 in a series of laboratory experiments at the Achotines laboratory which examined the survival of YFT larvae during the first week of feeding under conditions of variable micro-turbulence. Turbulence in the experimental tanks was measured as the mean horizontal velocity of a neutrally buoyant surface drogue; in 1999 and 2000, these velocities were calibrated against velocities measured at depth with a micro-acoustic Doppler current meter (ADV). Turbulent dissipation rates encountered by larvae were estimated with a three-dimensional model of turbulent energy dissipation/kinematic viscosity based on current velocities measured by the ADV (Kimura et al., 2004). Prey levels maintained during the experiments were low to intermediate (250 to 3000 rotifers/L) by tuna aquaculture standards.

Survival of YFT larvae during the first week of feeding shows a dome-shaped pattern in relation to micro-turbulence (Figure 5.9). Survival is up to 2.7 times higher at intermediate levels of micro-turbulence ($7.4 \times 10^{-9}\ m^2\ s^{-3}$ to $2.25 \times 10^{-8}\ m^2\ s^{-3}$ as an energy dissipation rate) than at lower or higher levels of turbulence. Cohort biomass, which is dominated by survival during the first week of feeding in the laboratory,

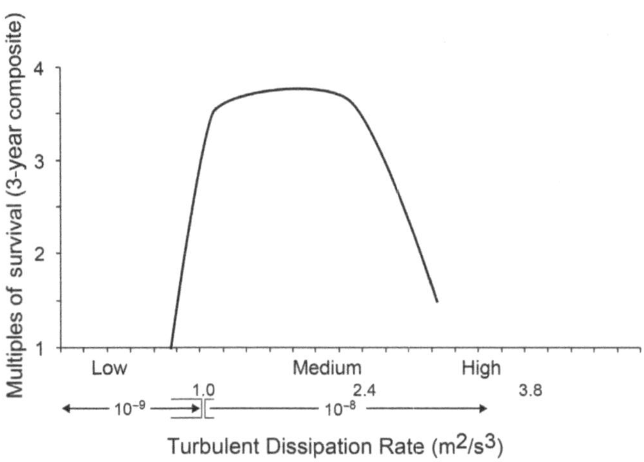

FIGURE 5.9 The relationship between micro-turbulence (estimated as the turbulent dissipation rate) and survival of YFT larvae during the first week of feeding in the laboratory. The survival curve is a smoothed, composite curve representing the mean survival estimated during four experiments over 3 years.

also exhibits a dome-shaped relationship with micro-turbulence. Growth, however, shows a moderately flat or weak U-shaped relationship with micro-turbulence level, at least under the low to intermediate prey conditions used in the laboratory. Growth rate is moderately high, but not necessarily maximum, over the optimal micro-turbulence range for larval survival. The apparent uncoupling of larval survival and growth under intermediate micro-turbulence conditions is likely the result of density effects on growth; higher survival at intermediate micro-turbulence levels produces a slight dampening effect on mean growth rate.

The effects of micro-turbulence on larval survival are significant and should be taken into account in rearing designs for YFT larvae. In most experimental investigations and rearing trials, our IATTC group routinely sets optimal (intermediate) levels of micro-turbulence in rearing tanks to maximize survival and cohort biomass. If maximum larval growth rates are the main goal of a rearing program (independent of maximum survival), it might be necessary to restrict other physical or biological factors (light intensity, prey abundance, larval density, water temperature) to only narrowly optimal levels to sustain maximum larval growth rates.

5.3.2.3 Effects of Ocean Acidification on Development, Survival, and Growth

The 5th Intergovernmental Panel on Climate Change (IPCC) assessment (Stocker et al., 2013) estimates a global average decline in ocean surface pH of 0.30–0.32 by 2100 due to increasing concentrations of dissolved carbon dioxide (pCO_2) from anthropogenic activities. Across regions of the Pacific Ocean, where YFT spawn and develop, mean surface water pH is predicted to decrease between 0.26–0.49 pH units by 2100 (Ilyina et al., 2013). Ocean acidification is a concern for its potential effects on early life stages of tuna in oceanic habitats and in aquaculture systems based in either coastal waters or hatcheries fed by oceanic seawater systems.

To advance our knowledge of the potential effects of ocean acidification on YFT early life stages, a laboratory study was conducted by multiple collaborating organizations at the Achotines laboratory in 2011. Two separate trials were conducted to test the impact of increased pCO_2 on eggs, yolksac larvae, and first-feeding larvae. Acidification levels tested ranged from present day to levels predicted to occur in some areas of the Pacific within the next 100 years (near future) to 300 years (long term). The study results were variable between trials, but did indicate the potential for significantly reduced survival and size of larvae and prolonged egg hatch times at acidification levels that are relevant to near future predicted levels (Bromhead et al., 2015).

The potential impacts of ocean acidification on early life stages are an important consideration in the future development of aquaculture of YFT.

If acidification does progress at predicted levels for the Pacific Ocean, it is unclear whether tuna possess the capacity to adapt to acidification through selection for more resistant individuals (Bromhead et al., 2015). It is also unclear whether resistant individual traits are heritable (Munday et al., 2012). To date, there is evidence that near future levels of ocean acidification can have significant negative effects on development, survival and growth of YFT eggs and larvae. These results should be assessed in the future as YFT aquaculture programs are developed.

5.3.3 Visual Sensitivity Through Ontogeny

Larval tuna are relatively near-sighted and possess pure-cone retinas with no dark-light adaptive ability, thus restricting them to photopic (light) vision (Margulies, 1997). Larval tuna are diurnal, visual planktivores during the early feeding stages. Visual acuities are somewhat advanced in larval tuna compared to other marine fish larvae, and the visual system exhibits rapid development of acuity, distance accommodation, and rudimentary rod development (providing increased sensitivity in low light) by the juvenile transformation stage (Margulies, 1997; Loew et al., 2002).

The spectral sensitivity of larval and early-juvenile YFT is different from that of adults. The single-cone retina of first-feeding larvae shows a wide distribution in individual cone absorbances, containing not only mixtures of the two adult cone pigments (with maximum sensitivity at 423 and 495 nm), but at least a third green-sensitive pigment with maximum sensitivity greater than 560 nm. With growth, the longer wavelength visual pigments disappear in the early juvenile stages and the spectral sensitivity converges to the adult condition (Loew et al., 2002). The extended range of larval YFT pigments appears to enhance the detection and capture success of zooplankton.

The visual morphology and sensitivity of larval YFT is an important consideration for rearing protocols. First-feeding larvae must have sufficient light intensity to feed, and do not appear to feed in the dark (Margulies et al., 2007b). Green water seems to be an essential component of the physical environment for successful rearing of YFT larvae, and correlates with the green-sensitive peak (one of three peaks) in the visual pigments of larvae. In rearing tanks, a green or bluish-green visual background maximizes the visual contrast of zooplankton prey when silhouetted against the down-welling light (McFarland and Munz, 1975). With development, older YFT larvae and early juveniles no longer require green water conditions for successful feeding, as the green sensitivity in their retinas disappears and their visual sensitivity shifts to bluish-green and violet wavelengths, and they become more adapted to feeding in low-light conditions (Loew et al., 2002).

5.3.4 Rearing Protocol Studies

5.3.4.1 Stocking and Transfer of Eggs and Yolksac Larvae

The protocols for collecting and incubating YFT eggs that have been developed over the years at the Achotines laboratory are described in Margulies et al. (2007a). YFT eggs are routinely incubated in 240- to 300-L conical fiberglass tanks that are plumbed with flow-through systems that allow water to enter through the bottom and exit near the top. Water flow rates routinely range from 12 to 18 exchanges/day, but lower rates of 6 exchanges/day have been used with no deleterious effects on eggs.

Optimal incubation, stocking, and transfer protocols for eggs and hatched larvae have been investigated. Incubation tanks are stocked at a maximum of 170 eggs/L in order to maintain the cleanest water conditions and to avoid crowding in the tank. A two-fold difference in stocking densities (150 and 300 eggs/L) results in no significant difference in the percentage hatch, although the possible effects of egg stocking density on survival at later stages of larval development have not yet been examined. However, our egg incubation protocols consistently yield high percentage hatch rates (>85%) and our preference continues to be to use a lower egg-stocking density.

Optimal transfer and stocking protocols for larval rearing tanks have been investigated. Transfer and stocking of rearing tanks can take place near the end of the egg stage, just prior to hatching, or post-hatch. It may be advantageous to stock the rearing tanks before hatching, while the egg chorion still provides added physical protection. Alternatively, waiting until after hatching is complete and transferring yolksac larvae allows time for the egg chorion to be shed, and therefore reduces the amount of biological detritus, which acts as a medium for bacterial growth, and nitrogenous waste, in rearing tanks. Additionally, the physical environment of the rearing tank (e.g., aeration and water-flow rates) does not have to be adjusted for two different life stages if transfer occurs post-hatch. The effect of transfer stage on the survival at the first-feeding stage has been investigated, and preliminary results have indicated no significant differences in survival rates between the two methods. The transfer of YFT eggs and/or hatched larvae at the Achotines laboratory routinely occurs at both life stages.

Optimal yolksac stocking density has also been investigated. The effects of density on growth and survival during feeding stages have been discussed previously, and possible trade-offs in growth in weight and cohort biomass should be balanced with the number of survivors, depending on when the fish will be harvested. In order to examine the effect of yolksac stocking density on survival of first-feeding larvae, newly hatched yolksac larvae were stocked in replicate rearing tanks at densities of 5, 10, 20, or 40/L. Survival estimates were similar over the range of stocking densities tested, and it appears that any potential negative effects of high stocking densities in the yolksac stage occur at >40 yolksac larvae/L. The physical rearing environment (i.e., water flow and quality) is likely to influence these results, and sufficient water circulation and exchange should be maintained to avoid compromised water quality.

5.3.4.2 Probiotic Trials

The use of probiotics in aquaculture has gained popularity in recent years (Gatesoupe, 1999; Vine et al., 2006; Wang et al., 2008, Zink et al., 2011). The introduction of beneficial bacteria has been shown to reduce pathogenic bacteria levels and improve water quality in the marine rearing environment (Skjermo and Vadstein, 1999; Verschuere et al., 2000). Probiotics can be introduced at multiple stages during larviculture. Experiments have been conducted at the Achotines laboratory in order to investigate the effects of probiotics on YFT during the egg and yolksac stages and during the first week of feeding. Many of these investigations were the result of the collaboration between the Early Life History group of the IATTC and the University of Miami's Aquaculture Program during an annual workshop held at the Achotines laboratory.

As previously discussed in this chapter, there is an increasing need for the development of successful shipping protocols of YFT eggs and/or larvae in aquaculture. The ability to maintain high water quality is important for the survival and viability of shipped YFT larvae. Zink et al. (2011) describe the beneficial effect of the addition of a probiotic product containing a mix of bacterial strains (*Bacillus* spp., EcoAqua, EcoMicrobials™) to shipping bag water containing yolksac larvae of YFT during shipping simulations carried out at the Achotines laboratory. The incorporation of probiotics resulted in significantly lower final concentrations of total ammonia nitrogen and unionized ammonia, along with significantly higher dissolved oxygen concentration (possibly due to stress reduction) in comparison with a control group (Zink et al., 2011). Survival after 24 h was high (>85%) in both the probiotic and control treatments, although the effect of water quality during transport on larval viability could manifest itself over a longer period of time, i.e., survival at first-feeding. Based on these results, it appears that the use of probiotics in live shipment protocols may improve water quality during transport, and should be further investigated.

Laboratory investigations of the effect of probiotics on the growth and survival of YFT larvae after the first week of feeding show promising, but mixed, results (IATTC and University of Miami, unpublished data). The same probiotic product (EcoAqua, EcoMicrobials™) was tested by treating eggs, and the resulting larvae were treated again upon stocking, at first feeding (3 dah), and again at 3-day intervals. Larvae in the probiotic treatment were fed rotifers and reared in green water with microalgae that had been probiotic-treated. Control treatments contained untreated eggs, larvae, rotifers, and microalgae. The results from these investigations indicate a possible increase in larval survival (2–4 × higher survival) and faster growth in length (up to 9% more rapid growth) after the first week of feeding in some replicate tanks, although the results have been variable, so these trials require further study.

The use of probiotics may also enhance rotifer production. The IATTC and the University of Miami have conducted one pilot experiment to investigate the effects of probiotics and antibiotics on rotifer production in batch cultures. The probiotic treatment was created by the addition of probiotics to mass cultures of the microalga *Nannocloropsis oculata* used to feed rotifers. The antibiotic treatment was created by administering an oxytetracyline (OTC) solution to rotifers during enrichment and a control treatment received neither probiotics nor OTC. Preliminary analysis of the results indicated that probiotic treatments may enhance rotifer production, but further investigation is required.

The use of probiotics in YFT larviculture is an area of study that shows potential benefits in multiple applications, and should be explored further in the development of YFT aquaculture programs.

5.3.4.3 Effects of Micro-Turbulence on Survival Prior to Feeding

Creating micro-turbulence in rearing tanks assists in keeping yolksac larvae distributed in the water column before they reach the first-feeding stage. As part of the protocols developed at the Achotines laboratory, a very light aeration is created (~2.5–3.0 cm/s horizontal velocity at the water surface) and maintained during the yolksac stage via air diffusers placed on the bottom of the rearing tanks. Micro-turbulence is then increased once the larvae initiate feeding to the optimal levels previously described for first-feeding larvae. Experimental investigations of optimal micro-turbulence levels for survival of yolksac larvae support these protocols, with low levels of aeration supporting significantly higher survival (10–12 × higher) than those used for first-feeding larvae.

5.3.4.4 Tank Color Comparative Trials

YFT larvae are visual feeders and have highly developed visual systems that include violet-, blue-, and green-sensitive pigments until the early-juvenile stage (Loew et al., 2002, Margulies et al., 2007b). In combination with green water in the rearing tank, tank color has the potential to affect the survival and growth of larvae by influencing their ability to visually detect prey items and avoid contact with the tank walls, which could cause physical damage. Various tank colors have been used at the Achotines laboratory to rear larvae, including white, black, green, and blue, with light blue being the preferred color at the Achotines laboratory, based on historical rearing production and experimental survival. The effect of light blue-interior versus dark green-interior tanks on growth and survival of YFT larvae after the first week of feeding has been compared experimentally, with no statistical differences in survival detected between the two treatments during the first week of feeding.

5.3.4.5 *Enrichment of Larval Prey*

YFT larvae are characterized by high growth rates, combined with high metabolic requirements. Successful rearing of marine fish larvae with these growth characteristics is enhanced by the nutritional enrichment of cultured prey items in order to boost the highly unsaturated fatty acid (HUFA) content of the prey to equivalent levels found in wild planktonic prey (Tocher et al., 2008). Enrichment of cultured rotifers and *Artemia* as prey for YFT larvae has been successful, using either AlgaMac 2000 or AlgaMac Enrich (Aquafauna BioMarine, Inc., Hawthorne, CA, USA). However, the benefits of using an enrichment product with a high fat content should be balanced with the effect of a high-fat diet, which results in reduced activity of rotifers, as they can become lethargic due to insufficient available carbohydrate. Proximate fat content near 50% in some enrichment products has yielded suboptimal results in larval feeding due to the lack of activity and batch crashes of rotifers after enrichment. We have found that proximate fat content of between 30% and 35% with relative DHA levels by dry weight of around 10% is optimal for YFT, supporting normal rotifer activity and exponential larval growth in weight and length.

5.4 LABORATORY STUDIES OF EARLY JUVENILES

The juvenile life stage of YFT (juvenile transformation to reproductive stage) is protracted, with the minimum estimate of its duration in captivity being 16–18 months. Successful full-life-cycle aquaculture of YFT will require a thorough understanding of the growth and nutrition requirements of juveniles. Although the focus of IATTC YFT studies at the Achotines laboratory has been on spawning, egg and larval studies, a natural extension of the research has been the opportunity to conduct rearing trials of early juvenile YFT (generally 25–100 dah).

Reared cohorts of early juveniles have been studied since 1997, utilizing the survivors from larval ecology experiments or aquaculture-focused rearing trials. Early juveniles are transferred from larval experimental tanks (700–1000 L volume) to larger fiberglass tanks (4000–10,000 L volume) at an age of 25–35 dah. Juveniles have also been reared from the yolksac stage in 4000–10,000 L volume tanks. Early juveniles are fed a diet of newly hatched YFT larvae (1–10/L) until 25–40 dah, and subsequently fed minced fish which is offered at a daily ration of 50–125% of body weight/day (during some feedings a portion of the minced fish is not eaten, but the majority is consumed). From 25 to 40 dah, the diet is often a mixture of newly hatched YFT larvae and minced fish. Artificial diets and weaning diets have not been extensively studied with early juveniles, but some limited trials have been conducted (see summary below).

5.4.1 Survival, Growth, and Nutrition

Survival and growth during the early juvenile stage appear to be most strongly affected by diet, and may be secondarily influenced by water temperature and fish density. At 30 dah, survival in the laboratory is generally 1–3% of the number of hatched larvae originally stocked, although survival of individual cohorts has been as high as 5–10%. Laboratory cohorts of juveniles are routinely reared to 60–65 dah, however as is described in further detail below, juveniles are now being reared beyond 115 dah.

Growth of early juveniles in the laboratory is rapid when they are fed a fish diet (newly hatched larvae and/or minced fish) (Figure 5.10). From 25 to 45 dah, fish grow very rapidly in length (Figure 5.7), exhibiting daily growth rates of 1.0–3.8 mm/day (average 2.5 mm/day). The instantaneous daily growth rate in length can approach 9% per day during this stage. Specific growth rates in dry weight are high, ranging from 25% to 45% per day (average 30–35% per day), which is similar to growth in weight during the late larval stage. Growth data from 50 to 100 dah are limited, but suggest equivalent or slightly slower daily growth compared to the 25–45 dah period.

Nutritional studies of early juveniles are limited, but can provide important insights into the physiology of this life stage and the development of appropriate rearing diets. Buentello et al. (2011) investigated the appearance and trends in activity of the main digestive enzymes present in early juveniles of YFT. They reported that early juveniles have a digestive system that is precocious in development, which allows consumption and efficient utilization of larvae of other marine fish species. The enzyme activities of early juvenile YFT are comparable to those of other fish species at much later developmental stages. A fish diet promotes very rapid somatic growth, and the high metabolic rates require a large provision of free amino acids. Arginine, in particular, is rapidly mobilized from the free amino acid pool to promote protein deposition and rapid growth. Protocols for live-food production in YFT hatcheries should be structured to maximize the free amino acid pool of live prey for larvae and early juveniles.

The greatest challenges in rearing early juvenile YFT in laboratory tanks are providing optimal physical conditions, maintaining good nutrition, and preventing the occurrence of secondary bacterial and fungal infections. Juveniles exhibit high activity levels, and are sensitive to sudden changes in light intensity, particularly at dusk and dawn. Tank lighting must be consistent, and changed gradually in the early morning and at twilight. Juveniles

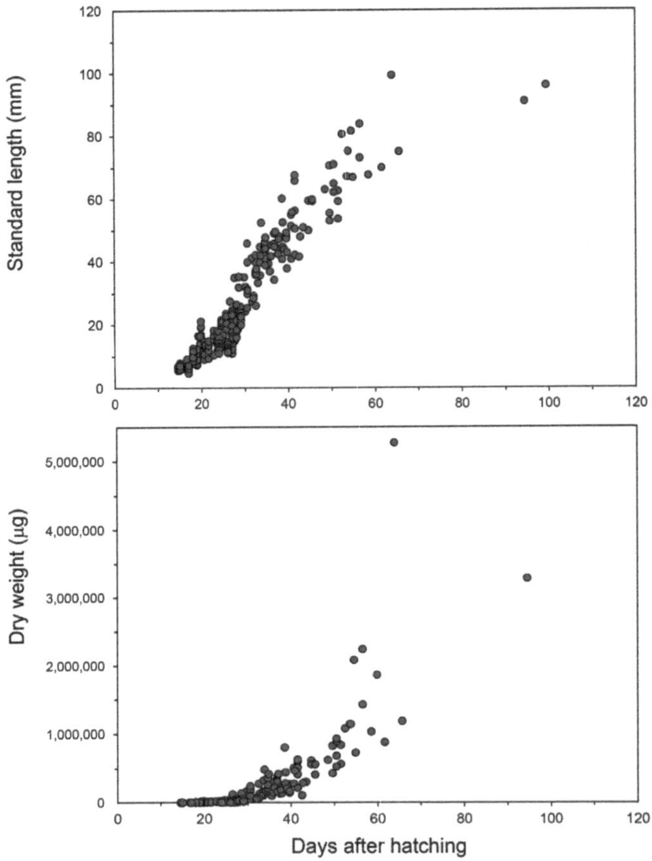

FIGURE 5.10 Growth in length (top panel) and dry weight (bottom panel) of YFT early-juveniles from 15 to 100 days after hatching (dah) in the laboratory.

often deteriorate in condition over a period of weeks or months in tanks. The minced fish diets offered to juveniles to date are probably deficient in nutritional quality, but the development of artificial feeds nutritionally matched to tuna digestion and nutritional requirements would significantly improve rearing protocols. In the laboratory, even incidental contact with tank walls can result in the development of secondary bacterial or fungal infections on the head and trunk. Our research group has explored the use of bubble curtains and strong currents to prevent contact of juveniles with tank walls, with only limited success to date. The use of sea cages for juvenile rearing may be essential for successful full-life-cycle production (see discussion below).

5.4.2 Feeding Studies with Artificial Foods and Weaning Diets

The focus of IATTC experimental studies of YFT has been on pre-juvenile life stages. However, some limited trials have been conducted investigating the effectiveness of artificial foods and weaning diets in supporting early juvenile growth and survival.

Over the past 10 years, the IATTC has tested several weaning diets with early juvenile YFT. The diets have included granulized or marumerized products, including Otohime (manufactured by Marubeni Nisshin Feed Co., Chuo, Japan) and diets manufactured by Skretting USA and INVE Aquaculture. In these limited trials, early juvenile YFT were reared in 12,000-L capacity fiberglass tanks and fed at intervals of 0.5 to 1 h (diets offered at each interval until feeding stopped, no daily rations estimated). To date, early juvenile YFT have first accepted an artificial diet at about 25 dah, with an exclusively artificial diet fed successfully from 25 to 115 dah. These results include trials conducted during 2015 by Kinki University, in collaboration with the IATTC and ARAP. Artificial feeds have also been provided to early juveniles in a mixed diet with minced fish from 30 to 93 dah. Survival with artificial diets has been moderate to poor, while growth has been slower than that produced under a minced-fish diet. Up to 115 dah, an exclusively artificial diet has produced daily growth rates in length of 1.0 to 2.3 mm/day, while a mixed artificial/minced-fish diet has produced growth rates of 1.1 to 2.3 mm/day up to 93 dah. In contrast, a minced-fish diet has resulted in growth rates of 2.0 to 3.8 mm/day from 25 to 100 dah (Figure 5.10).

The development of effective juvenile and weaning diets will be a key factor in the improvement of rearing success and the completion of full-life-cycle aquaculture of YFT. Juvenile diets must be attractive and palatable and should maximize nutritional factors such as free amino acids that promote protein synthesis and growth.

5.4.3 Future Studies: Sea Cage Rearing of Juveniles

Pioneering work on the rearing of early juvenile Pacific bluefin in sea cages has been conducted by researchers at Kinki University (Sawada et al., 2005; Masuma et al., 2008). Early juvenile survival and overall condition has been improved by transferring fish from walled laboratory tanks to sea cages. The IATTC Early Life History group has partnered with KU and the Autoridad de Recursos Acuáticos de Panama (ARAP) in a multi-year project (SATREPS), funded by JICA and JST to study the feasibility of rearing early juvenile YFT in sea cages in nearshore waters adjacent to the Achotines laboratory.

Preliminary rearing trials were conducted in late 2014 at the Achotines laboratory and more extensive sea cage trials were conducted during 2015. Early juvenile YFT were reared at the Achotines laboratory by KU and ARAP, supported by the IATTC, in 1000 to 7000-L capacity fiberglass tanks. The juveniles were transferred to a juvenile rearing cage (20 m diameter, 6 m depth) moored approximately 0.5 km offshore of the Achotines laboratory. At the time of transfer, the juveniles were 52 dah and ranged in length from 9 to 13 cm SL. This activity represents the first successful transfer worldwide of YFT early-juveniles to a sea cage. Rearing protocols for the juvenile YFT in the sea cages followed those developed by KU for PBFT. Due to vessel and staff limitations, the YFT juveniles were held in the sea cage for 26 days, and then transferred to an in-ground tank (8.5 m diameter, 3 m deep) at the Achotines laboratory. As of mid-August 2015, a small group of the juveniles had reached an age of 115 dah and averaged approximately 23 cm SL.

5.5 SUMMARY OF AQUACULTURE-RELATED RESEARCH ON YFT

The YFT exhibits strong potential as a candidate species for full-life-cycle aquaculture. YFT are highly fecund, spawn at near-daily intervals, reach sexual maturity in as little as 2 years, and grow rapidly during juvenile life stages. Broodstock YFT in Panama have proved amenable to capture, transport, stocking, and maintenance in land-based tanks. Captive spawning of YFT is near-daily in our tropical laboratory system. Courtship and spawning behaviors and the effects of physical and biological factors on spawning have been studied for over 18 years. Long-term captive spawning at the Achotines laboratory has provided the opportunity to conduct detailed studies of the egg, larval, and early-juvenile life stages of YFT. The early life stages of YFT are characterized by fast growth, high metabolic demands, and high mortality. Mortality during the yolksac and early-feeding stages is particularly high, even though first-feeding YFT are highly capable planktivores and exhibit a very high incidence of successful first feeding. Larval and early-juvenile growth is exponential in both length and weight, and the timing of the onset of piscivory during the post-flexion larval stage seems to promote fast growth and greater survival. YFT have been successfully reared in the laboratory up to 115 days after hatching. Juvenile rearing in sea cages may support higher survival than rearing in walled laboratory tanks. The development of effective weaning diets and artificial feeds for juveniles will provide a key contribution to successful full-life-cycle aquaculture of YFT.

ACKNOWLEDGMENTS

We appreciate the significant contributions of these former members of the IATTC Early Life History group: Robert Olson, Jenny Suter, Sharon Hunt, and Kevin Bentler. We thank the following staff of the Achotines laboratory for their many important contributions to the research conducted at the laboratory: Susana Cusatti, Luis Tejada, Diomedes Ballesteros, Heriberto Espinoza, Daniel Solis, Carlos Vergara, Lina Castillo, Danisin Dominguez, Dario Ramirez, Augustin Ortega, Daniel Perez, Yovani Ballesteros, and Daniel Mancilla. Charles Farwell of the Monterey Bay Aquarium provided advice and guidance on culture of tuna in captivity. We thank Akira Suda for his assistance in the early development of YFT studies at the Achotines laboratory, and Shukei Masuma for advice on tank design and tuna husbandry. We are grateful to the OFCF, JICA, and JST for invaluable financial support for YFT studies at the Achotines laboratory. We thank colleagues of the Fisheries Laboratory of KU, in particular Yoshifumi Sawada, for supporting collaborative studies of tuna. We acknowledge the support of the ARAP, and in particular, resident biologist Amado Cano. Patrick Tomlinson of the IATTC provided invaluable advice on growth curve models, and Christine Patnode skillfully prepared the final versions of the figures. William Bayliff provided helpful review comments. Finally, we would like to thank the Commissioners of the IATTC, former Directors, James Joseph and Robin Allen, current Director, Guillermo Compeán, and Chief Scientist, Rick Deriso, for their support of the research conducted at the Achotines laboratory.

REFERENCES

Agawa, Y., Iwaki, M., Komiya, T., Honryo, T., Tamura, K., Okada, T., et al., 2015. Identification of male sex-linked DNA sequence of the cultured PBFT *Thunnus orientalis*. Fish. Sci. 81, 113–121.

Bromhead, D., Scholey, V., Nicol, S., Margulies, D., Wexler, J., Stein, M., et al., 2015. The potential impact of ocean acidification upon eggs and larvae of yellowfin tuna (*Thunnus albacares*). Deep-Sea Res. II 113, 268–279.

Buentello, J.A., Pohlenz, C., Margulies, D., Scholey, V.P., Wexler, J.B., Tovar-Ramírez, D., et al., 2011. A preliminary study of digestive enzyme activities and amino acid composition of early juvenile yellowfin tuna (*Thunnus albacares*). Aquaculture 312 (1–4), 205–211.

Chow, S., Scholey, V.P., Nakazawa, A., Margulies, D., Wexler, J.B., Olson, R.J., et al., 2001. Direct evidence for Mendelian inheritance of the variations in the ribosomal protein gene introns in yellowfin tuna (*Thunnus albacares*). Mar. Biotechnol. 3, 22–26.

Davis, T.L.O., Clementson, L.A., 1989. Data report on the vertical and horizontal distribution of tuna larvae in the East Indian Ocean, January-February 1987. CSIRO Mar. Lab. Rep., 206.

Dower, J.F., Miller, T.J., Leggett, W.C., 1997. The role of microscale turbulence in the feeding ecology of larval fish. In: Blaxter, J.H.S., Southward, A.J. (Eds.), Advances in Marine Biology, vol. 31. Academic Press, San Diego, pp. 169–220.

Fukuda, H., Torisawa, S., Sawada, Y., Takagi, T., 2010. Developmental changes in behavioral and retinomotor responses of PBFT on exposure to sudden changes in illumination. Aquaculture 305 (1–4), 73–78.

Gatesoupe, F.J., 1999. The use of probiotics in aquaculture. Aquaculture 180, 147–165.

Honryo, T., Tanaka, T., Guillen, A., Wexler, J.B., Cano, A., Margulies, D., et al., 2014. Effect of water surface condition on survival, growth and swim bladder inflation of yellowfin tuna, *Thunnus albacares* (Temminck and Schlegel), larvae. Aquacult. Res., 1–9.

Ilyina, T., Six, K.D., Segschneider, J., Maier-Reimer, E., Li, H., Núñez-Riboni, I., 2013. Global ocean biogeochemistry model HAMOCC: model architecture and performance as component of the MPI-Earth System Model in different CMIP5 experimental realizations. J. Adv. Model. Earth Syst. 5, 287–315.

Ishibashi, Y., Inoue, K., Nakatsukasa, H., Ishitani, Y., Miyashita, S., Murata, O., 2005. Ontogeny of tolerance to hypoxia and oxygen consumption of larval and juvenile red sea bream, *Pagrus major*. Aquaculture 244, 331–340.

Ishibashi, Y., Kotaki, T., Yamada, Y., Ohta, H., 2007. Ontogenetic changes in tolerance to hypoxia and energy metabolism of larval and juvenile Japanese flounder *Paralichthys olivaceus*. J. Exp. Mar. Biol. Ecol. 352, 42–49.

Izquierdo, M.S., Fernández-Palacios, H., Tacon, A.G.J., 2001. Effect of broodstock nutrition on reproductive performance of fish. Aquaculture 197, 25–42.

Jenkins, G.P., Young, J.W., Davis, T.L.O., 1991. Density dependence of larval growth of a marine fish, the southern bluefin tuna, *Thunnus maccoyii*. Can. J. Fish. Aquat. Sci. 48, 1358–1363.

Kimura, S., Nakata, H., Margulies, D., Suter, J.M., Hunt, S.L., 2004. Effect of oceanic turbulence on the survival of yellowfin tuna larvae. Nippon Suisan Gakkaishi 70, 175–178 (In Japanese with English abstract).

Lang, K.L., Grimes, C.B., Shaw, R.F., 1994. Variations in the age and growth of yellowfin tuna larvae, *Thunnus albacares*, collected about the Mississippi River Plume. Envir. Biol. Fishes 39, 259–270.

Lauth, R.R., Olson, R.J., 1996. Distribution and abundance of larval scombridae in relation to the physical environment in the northwestern Panama Bight. Inter-Am. Trop. Tuna Comm. Bull. 21, 125–167.

Loew, E.R., McFarland, W.N., Margulies, D., 2002. Developmental changes in the visual pigments of the yellowfin tuna, *Thunnus albacares*. Mar. Fresh. Behav. Physiol. 35 (4), 235–246.

MacKenzie, B.R., Miller, T.J., Cyr, S., Peters, R.H., 1994. Evidence for a dome-shaped relationship between turbulence and larval fish ingestion rates. Limnol. Oceanogr. 39, 1790–1799.

Magnuson, J.J., 1963. Tuna behavior and physiology, a review. Methodological Paper No. 5. In: Proceedings of the World Scientific Meeting on the Biology of Tunas and Related Species. FAO Fisheries Reports 3. pp. 1057–1066.

Margulies, D., 1993. Assessment of the nutritional condition of larval and early juvenile tuna and Spanish mackerel (Pisces: Scombridae) in the Panama Bight. Mar. Biol. 115 (2), 317–330.

Margulies, D., 1997. Development of the visual system and inferred performance capabilities of larval and early juvenile scombrids. Mar. Fresh. Behav. Physiol. 30, 75–98.

Margulies, D., Wexler, J.B., Bentler, K.T., Suter, J.M., Masuma, S., Tezuka, N., et al., 2001. Food selection of yellowfin tuna, *Thunnus albacares*, larvae reared in the laboratory. Inter-Am. Trop. Tuna Comm. Bull. 22 (1), 9–51.

Margulies, D., Suter, J.M., Hunt, S.L., Olson, R.J., Scholey, V.P., Wexler, J.B., et al., 2007a. Spawning and early development of captive yellowfin tuna (*Thunnus albacares*). Fish. Bull. 105, 249–265.

Margulies, D., Scholey, V.P., Wexler, J.B., Olson, R.J., Suter, J.M., Hunt, S.L., 2007b. A review of IATTC research on the early life history and reproductive biology of scombrids conducted at the Achotines Laboratory from 1985 to 2005. Inter-Am. Trop. Tuna Comm. Spec. Rep. 16, 63.

Masuma, S., Miyashita, S., Yamamoto, H., Kumai, H., 2008. Status of bluefin tuna farming, broodstock management, breeding and fingerling production in Japan. Rev. Fish. Sci. 16, 385–390.

McFarland, W.N., Munz, F.W., 1975. Part III: The evolution of photopic visual pigments in fishes. Vision Res. 15 (10), 1071–1080.

Miller, D.C., Poucher, S.L., Coiro, L., 2002. Determination of lethal dissolved oxygen levels for selected marine and estuarine fishes, crustaceans, and a bivalve. Mar. Biol. 140, 287–296.

Miyashita, S., Hattori, N., Sawada, Y., Ishibashi, Y., Nakatsukasa, H., Okada, T., et al., 1999. Ontogenetic change in oxygen consumption of bluefin tuna, *Thunnus thynnus*. Suisanzoshoku 47, 269–275.

Munday, P.L., Mccormick, M.I., Meekan, M., Dixson, D.L., Watson, S., Chivers, D.P., et al., 2012. Selective mortality associated with variation in CO_2 tolerance in a marine fish. Ocean Acidif. 1, 1–6.

Nakamura, E.L., 1969. Visual acuity of yellowfin tuna, *Thunnus albacares*. In: Ben-Tuvia, A., Dickson W., (Eds.), Proceedings of the FAO Conference on Fish Behaviour in Relation to Fishing Techniques and Tatics. Bergen, Norway, pp. 19–27.

Niwa, Y., Nakazawa, A., Margulies, D., Scholey, V.P., Wexler, J.B., Chow, S., 2003. Genetic monitoring of spawning ecology of captive yellowfin tuna (*Thunnus albacares*) using mitochondrial DNA variation. Aquaculture 218, 387–395.

Olson, R.J., Scholey, V.P., 1990. Captive tunas in a tropical marine research laboratory: Growth of late-larval and early-juvenile black skipjack *Euthynnus lineatus*. Fishery Bull. 88 (4), 821–828.

Owen, R., 1997. Oceanographic atlas of habitats of larval tunas in the Pacific Ocean off the Azuero Peninsula, Panama. Inter-Am. Trop. Tuna Comm. Data Rep. 9, 32.

Partridge, G., Benetti, D.D., Stieglitz, J.D., Hutapea, J., McIntyre, A., Chen, B., et al., 2011. The effect of a 24-hour photoperiod on the survival, growth and swim bladder inflation of pre-flexion yellowfin tuna (*Thunnus albacares*) larvae. Aquaculture 318, 471–474.

Pelster, B., 1999. Environmental influences on the development of the cardiac system in fish and amphibians. Comp. Biochem. Physiol. A 124, 407–412.

Reglero, P., Balbín, R., Ortega, A., Alvarez-Berastegui, D., Gordoa, A., Torres, A.P., et al., 2013. First attempt to assess the viability of bluefin tuna spawning events in offshore cages located in an a priori favourable larval habitat. Sci. Mar. 77 (4), 585–594.

Rothschild, B.J., Osborn, T.R., 1988. Small-scale turbulence and plankton contact rates. J. Plankton Res. 10, 465–474.

Sawada, Y., Okada, T., Miyashita, S., Murata, O., Kumai, H., 2005. Completion of the PBFT *Thunnus orientalis* (Temminck et Schlegel) life cycle. Aquacult. Res. 36, 413–421.

Schaefer, K.M., 1998. Reproductive biology of yellowfin tuna (*Thunnus albacares*) in the eastern Pacific Ocean. Inter-Am. Trop. Tuna Comm. Bull. 21 (5), 205–272.

Schaefer, K.M., 2001. Reproductive biology of tunas. In: Block, B.A., Stevens, E.D. (Eds.), Fish Physiology, Tuna: Physiology, Ecology and Evolution, vol. 19. Academic Press, San Diego, pp. 225–270.

Scholey, V.P., 1997. Construction and operation of a laboratory or aquaculture facility in a developing country. Rev. Fish. Sci. 5 (3), 279–302.

Skjermo, J., Vadstein, O., 1999. Techniques for microbial control in the intensive rearing of marine larvae. Aquaculture 177, 333–343.

Stocker, T.F., Qin, D., Plattner, G.K., Tignor, M., Allen, S.K., Boschung, J., et al., 2013. IPCC, 2013: climate change 2013: the physical science basis. Contribution of working group I to the fifth assessment report of the intergovernmental panel on climate change.

Susca, V., Corriero, A., Bridges, C.R., De Metrio, G., 2001. Study of the sexual maturity of female bluefin tuna: purification and partial characterization of vitellogenin and its use in an enzyme-linked immunosorbent assay. J. Fish Biol. 58, 815–831.

Takagi, M., Chow, S., Okamura, T., Scholey, V.P., Nakazawa, A., Margulies, D., et al., 2003. Mendelian inheritance and variation of four microsatellite DNA markers in the yellowfin tuna *Thunnus albacares*. Fish. Sci. 69, 1306–1308.

Tamura, T., Wisby, W.J., 1963. The visual sense of pelagic fishes especially the visual axis and accommodation. Bull. Mar. Sci. 13, 433–448.

Tanaka, Y., Minami, H., Ishihi, Y., Kumon, K., Higuchi, K., Eba, T., et al., 2014. Differential growth rates related to initiation of piscivory by hatchery-reared larval PBFT *Thunnus orientalis*. Fish. Sci. 80 (6), 1205–1214.

Tocher, D.R., Bendiksen, E.A., Campbell, P.J., Bell, J.G., 2008. The role of phospholipids in nutrition and metabolism of teleost fish. Aquaculture 280, 21–34.

Uotani, I., Matsuzaki, K., Makino, Y., Noda, K., Inamura, O., Horikawa, M., 1981. Food habits of larvae of tunas and their related species in the area northeast of Australia. Bull. Jpn. Soc. Sci. Fish. 47, 1165–1172 (In Japanese with English abstract).

Verschuere, L., Rombaut, G., Sorgeloos, P., Versraete, W., 2000. Probiotic bacteria as biological control agents in aquaculture. Microbiol. Mol. Biol. Rev. 64, 655–671.

Vine, N.G., Leukes, W.D., Kaiser, H., 2006. Probiotics in marine larviculture. FEMS. Microbiol. Rev. 30, 404–427.

Wang, Y., Li, J., Lin, J., 2008. Probiotics in aquaculture: challenges and outlook. Aquaculture 281, 1–4.

Watanabe, T., Vassalb-Agius, R., 2003. Broodstock nutrition research on marine finfish in Japan. Aquaculture 227, 35–61.

Wexler, J.B., 1993. Validation of daily growth increments and estimation of growth rates of larval and early juvenile black skipjack, *Euthynnus lineatus*, using otoliths. Inter-Am. Trop. Tuna Comm. Bull. 20 (7), 399–440.

Wexler, J.B., Margulies, D., Masuma, S., Tezuka, N., Teruya, K., Oka, M., et al., 2001. Age validation and growth of yellowfin tuna, *Thunnus albacares*, larvae reared in the laboratory. Inter-Am. Trop. Tuna Comm. Bull. 22 (1), 52–71.

Wexler, J.B., Scholey, V.P., Olson, R.J., Margulies, D., Nakazawa, A., Suter, J.M., 2003. Tank culture of yellowfin tuna, *Thunnus albacares*: developing a spawning population for research purposes. Aquaculture 220, 327–353.

Wexler, J.B., Chow, S., Wakabayashi, T., Nohara, K., Margulies, D., 2007. Temporal variation in growth of yelowfin tuna (*Thunnus albacares*) larvae in the Panama Bight, 1990–97. Fishery Bull. 105, 1–18.

Wexler, J.B., Margulies, D., Scholey, V.P., 2011. Temperature and dissolved oxygen requirements for survival of yellowfin tuna, *Thunnus albacares*, larvae. J. Exp. Mar. Biol. Ecol. 404, 63–72.

Wexler, J.B., Margulies, D., Scholey, V.P., 2012. Corrigendum to "Temperature and dissolved oxygen requirements for survival of yellowfin tuna, *Thunnus albacares*, larvae" [J. Exp. Mar. Biol. Ecol. 404(1–2), 63–72] .J. Exp. Mar. Biol. Ecol. 430–431, 87.

Wild, A., 1986. Growth of yellowfin tuna, *Thunnus albacares*, in the eastern Pacific Ocean based on otolith increments. Inter-Am. Trop. Tuna Comm. Bull. 18, 421–482.

Zink, I.C., Benetti, D.D., Douillet, P.A., Margulies, D., Scholey, V.P., 2011. Improvement of water chemistry with *Bacillus* probiotics inclusion during simulated transport of yellowfin tuna yolk sac larvae. North Am. J. Aquacult. 73, 42–48.

Chapter 6

Tuna Aquaculture in Europe

Fernando de la Gándara[1], Aurelio Ortega[1] and Alejandro Buentello[2]
[1]*Centro Oceanografico de Murcia, Spanish Institute of Oceanography, Murcia, Spain,*
[2]*Ichthus Unlimited, LLC., West Des Moines, IA, USA*

6.1 INTRODUCTION

6.1.1 Mediterranean Tuna History

The Atlantic bluefin tuna (ABFT) (*Thunnus thynnus*) (L. 1758), is a highly migratory scombrid fish (Fromentin and Powers, 2005) that has been harvested for millennia by hominids and humans around the Mediterranean Sea. Evidence of tuna fisheries in this region extend back to 7000 BC (Mather et al., 1995; Doumenge, 1998) and the Phoenician trap fisheries occurring around 2000 BC are well-documented (Ravier and Fromentin, 2001). "*Thunnos*" are mentioned by Aristotle, Oppian, and Pliny the Elder and are also recorded in excavation logbooks of prehistoric sites such as the recently unearthed archeological evidence near Gibraltar in the Vanguard Cave, which points to the reliance of Neanderthal populations on seafood resources, including ABFT (Figure 6.1; Brown et al., 2011).

6.1.2 Origins of Tuna Fattening in Europe

Although the trap-capturing system can be traced back to the eighteenth and nineteenth centuries, it was not until 1979 that tuna farming per se started in Ceuta, in the vicinity of Gibraltar (Miyake et al., 2003). The methodology followed closely a Canadian model that had been implemented more than a decade earlier, in St. Mary's Bay, Nova Scotia. The European version started with the capture of large (>150 kg), lean, post-spawned ABFT, on their way out to the Atlantic Ocean, using the traditional Mediterranean trapping system. These fish were then transferred to large rectangular cages ($\sim 70 \times 40$ m) equipped with a permanent mooring system. Generally, fish would be captured from July to August, fattened for four months, and sold to

D.D. Benetti, G.J. Partridge, A. Buentello (Eds): Advances in Tuna Aquaculture.
DOI: http://dx.doi.org/10.1016/B978-0-12-411459-3.00005-9

FIGURE 6.1 Neanderthal utilization of tuna resources near Gibraltar (Cort, 2007) and Spanish trap "Almadraba" ≈ Italian "Tonnara" (XVIII century).

the Japanese market in December, although the fattening season is sometimes extended into February (Ottolenghi, 2008). These caged and fed ABFT rapidly added fat and value, then were sold as prime ABFT in Japan.

Although smaller sized fish (≤120 kg) normally have lower fat and a flesh (loin) color less intense than the desirable cherry-red, smaller animals subjected to the fattening process as described above were also able to increase in value and the practice of selling these fish at local markets therefore shifted in the 1980s toward the better paid sashimi markets (Miyake et al., 2003). While these smaller tuna did not fetch top prices, they were profitable enough to warrant their fattening and exportation to Japanese markets.

Modern ABFT ranching, modeled after the Australian system (Clarke, 2005), did not start in Europe until 1996, and was apparently imported by Croatian nationals with Australian experience. From 1996 to 2001, ABFT ranching increased more than 20-fold in the Mediterranean, with operations beginning in many different regions. Operations were established first off Murcia, Spain (Miyake et al., 2003), and also in the Adriatic Sea off the Croatian coast (Katavic et al., 2003b). By 1997, Japan was buying 7,700 MT/year of ABFT fattened in the Mediterranean Sea and today over 96% of the ABFT farmed in the Mediterranean are sold at the Japanese market.

ABFT ranching expanded to Malta in 2000 (Peric, 2003a), and to Italy in 2001, with a farm being established in the Castellammare region (western Sicily). This farm had an average annual production of 777 MT (Santulli, 2005). Turkish operations followed shortly after, fueled by private-sector investment, and by 2002 that country was producing about 2,000 MT/year (Colak et al., 2003). In 2003, Tunisia entered the ABFT-ranching industry, with establishment of six independent companies that use the majority of Tunisia's ABFT quota to produce around 2,400 MT/year (Hattour, 2005). More recently, other Mediterranean countries such as Morocco, Libya, Greece, and Cyprus, as well as Portugal on the Atlantic coast, have added their efforts to the tuna farming activity in Europe (Tudela and Garcia, 2004; Ottolenghi, 2008).

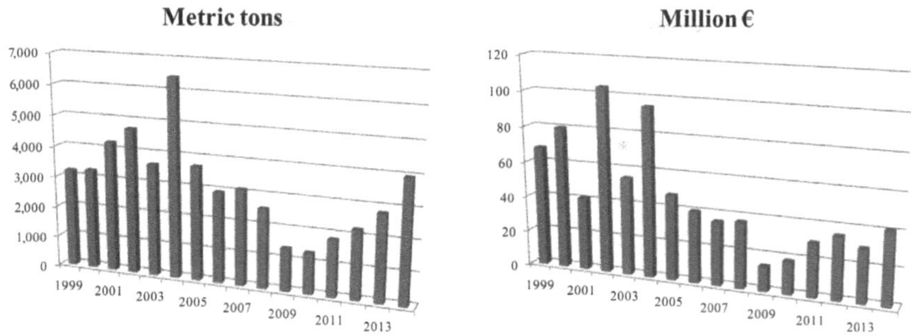

FIGURE 6.2 Annual production of ABFT in Spain (Spanish Ministry of Agriculture and Fisheries, 2015).

6.2 PRODUCTION BY GEOGRAPHIC AREA

In Europe, the vast majority of ABFT ranchers are located in the Mediterranean Sea with a small number located on the Atlantic coast (Portugal). Leading producing countries are Spain, Italy, Tunisia, Malta, Cyprus, Greece, Croatia, and Turkey, although, in recent years Malta, Tunisia, Croatia, Italy, and Turkey have reported the most significant increases in farmed ABFT production. In 2007, Spain had 11 facilities registered as tuna ranches with nine of those located near the Murcia coastline, one in Cataluña, one in Tarragona, and the last one in Andalucía (Almería). Figure 6.2 illustrates the total production and annual income generated by ABFT ranching activities in Spain.

6.3 ASSESSMENT OF EASTERN ATLANTIC AND MEDITERRANEAN TUNA STOCKS

The International Commission for the Conservation of Atlantic Tunas (ICCAT) is the international organization responsible for the quantitative estimation of the fishery status and conservation of the ABFT (and other scombrids) in the Atlantic Ocean and its adjacent seas. Tuna ranchers—utilizing mostly the Mediterranean and Black Sea ABFT sub-populations (Figure 6.3)—and/or ranching advocates argue that farming operations take pressure off wild populations because the tuna fishermen secure higher prices for their catch and therefore need to fish less intensively. However, ABFT populations are considered severely over-fished and critics of the industry respond that ranching operations actually add pressure to the stocks, steadily increasing their rate of depletion.

In reality, since 1996–1997, there has been an explosive increase in ABFT farming activities in the Mediterranean and the vast majority of purse-seine catches now are being moved into cages for fattening (Miyake et al., 2010). As an illustration, declared ABFT catches in the Eastern

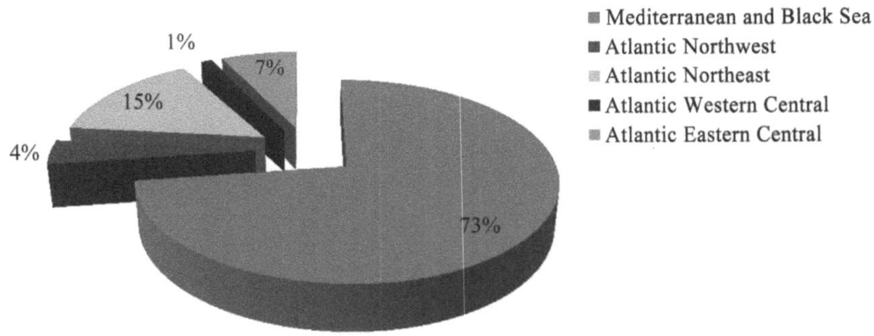

FIGURE 6.3 2004 ABFT global catch by geographic area. *Modified from Ottolenghi (2008).*

Atlantic peaked in 1996 at more than 50,000 MT (Anon., 2006, 2007), but since then have declined with the Total Allowed Capture (TAC), which was only 25,500 MT for 2010. Nearly all of this was exported overseas (ICCAT, 2014).

For the Eastern Atlantic stock—which includes the ABFT Mediterranean population—ICCAT's Standing Committee on Research and Statistics found strong declines in the spawning stock biomass since 1993, as well as an increase in fishing mortality rates (Tudela, 2001). Even before 1993, natural populations were struggling and, to avoid the collapse of the fishery, ICCAT implemented in 1990 a TAC (or quota system) and the TAC has been progressively reduced since its inception. As an example, the Mediterranean TAC established in 2001 was 29,500 MT/year from which 18,589 MT were allocated to EU countries and 876 MT to Croatia (member since July 2013).

As the ABFT monitoring agency, ICCAT has seen its sampling efforts shifted from collections made just after capture at sea, to having to procure data at the end of the fattening period, often at the processing plant. Several difficulties came into play after the adoption of the quota system and with the advent of ABFT farming. Therefore, ICCAT has implemented new sampling strategies—including cage sampling—for captured ABFT. European Commission Regulation 1543/00, also called "EC Data Collection Program," is supposedly a more efficient sampling procedure that was designed to resolve several difficulties in obtaining reliable size data from the Mediterranean ABFT purse-seining fleet (Di Natale et al., 2006).

These sampling efforts, together with ABFT stock assessments and evaluations of fishing intensity, make it clear that:

- The increased demand of ABFT for tuna ranching has had major impacts on the wild stock having contributed significantly to wild-stock depletion (Fromentin and Powers, 2005; Mac Kenzie et al., 2009).
- Tuna ranching, in its present form, negatively affects the marine ecosystem (Tudela and Garcia, 2004; Forrestal et al., 2012).

- There is a very high degree of uncertainty in the models used to evaluate the ABFT stocks (Bard, 2013), including the unaccounted migration between Eastern and Western stocks.

ICCAT's Scientific Committee asserted that future catches in excess of 33,000 MT/year would not be sustainable and only a level of ~25,000 MT/year would stand a chance to stop the progressive decline in ABFT biomass (ICCAT, 2014). Moreover, this same committee expressed concerns about the excessive intensity of fishing on both large and small fish, indicating that the later not only exacerbates over-fishing but also reduces the long-term potential yield from ABFT resources.

Declines in ABFT are not new, and the attempts which began in 1981 to stop the decline, including the TAC system, have been amended several times because the stocks did not recover as predicted. A 15-year recovery plan was adopted by ICCAT in Dubrovnik, Croatia, in 2006, for the Eastern ABFT stock (ICCAT, 2006b). The plan was further revised in Marrakech, Morocco, in 2008, to include multiple levels of monitoring, control, and surveillance and was transposed into EU law in April 2009 (Fromentin, 2009). This plan enforces three major rules:

1. TACs of 29,500, 28,500, 27,500, and 25,500 MT/year for 2007, 2008, 2009, and 2010, respectively
2. Extended closed fishing seasons:
   a. June 1st to December 31st for large long-liners over the whole area except the area delimited by West of 10°W and North of 42°N
   b. July 1st to December 31st for purse seiners over the whole area
   c. November 15 to May 15 for bait boats and pelagic trawlers over the whole area
3. Minimum size of 30 kg (with an 8% tolerance on the by-catch of fish less than 30 kg but no less than 10 kg), with the exception of bait boats and pelagic trawlers catches in the Eastern Atlantic and catches for farming purposes in the Adriatic Sea for which the minimum size is set at 8 kg.

Anon (2008) considered that even if the implementation of this plan were to be perfect, and assuming future recruitment is at least as good as it was in the 1990s, and also presuming this recruitment would be unaffected by recent reductions in the spawning biomass, there would be only about 50% probability of rebuilding the stock by 2023, under the current regulations. Stock rebuilding is even more unlikely because the premises that support this plan are unattainable.

There are, however, some benefits stemming from this plan. According to Cort and Martinez (2010), at least 840,000 juvenile (≤30 kg) ABFT/year have been spared from fishing mortality in the western Mediterranean since the prohibition was implemented in 2007. These juveniles are now being recruited to the spawning population, as evidenced by fish being caught by traditional tuna traps ("almadrabas" in Spanish) on the Atlantic side of the Strait of Gibraltar and at the Bay of Biscay, by the bait-boat fishery. The implementation of this multi-annual recovery plan comes at a price, borne by tuna farmers who sense an inequitable restriction of their ranching activities.

Ultimately, these are reactionary attempts to prevent the collapse of the ABFT fishery. A better way to alleviate fishing pressure on the ABFT stock and aid in its recovery would be "domestication" of the ABFT through further development of closed-cycle aquaculture. This goal finds its inspiration in the pioneering efforts to close the life cycle of the Pacific bluefin tuna (PBFT, *Thunnus orientalis*) in captivity, which is today a reality (Sawada et al., 2005) and is thoroughly described in Chapter 8 of this book. The significant strides made in Europe on ABFT domestication are the focus of a later section in this chapter.

6.4 TUNA FARMING METHODOLOGIES

6.4.1 Capture and Towing

For farming purposes, schools of ABFT are caught in various regions of the Mediterranean Sea from May to July by purse-seining, the fishing method that allows for far greater survival than all others tried to date. After capture, fish are transferred into towing cages—either circular, with a 50 m diameter or hexagonal with 22 m per side—and transported at very slow speed ($\sim$1 knot) to prevent the collapse of the cage (Ottolenghi, 2008), thus avoiding fish stress and mortalities. Due to the improvement in transportation techniques, towing-related mortalities decreased in Spanish operations from 21% in 1995 to $\sim$4% in 2000 (Mylonas et al., 2010). However, towing mortalities were still $\sim$10% in Turkey (Oray and Karakulak, 2003) and between 10% and 15% in Croatia (Katavic et al., 2003a), highlighting variability in practices from country to country. Because of the travel distance between capture locations and fattening facilities, transport may take several weeks. This is an expensive component of the overall operation, as daily towing costs may reach €3000 (Mylonas et al., 2010). Therefore, Mediterranean tuna farmers often relocate permanent cages closer to fishing grounds. This, together with the enhancement of the capture, transport, and transfer techniques, has resulted in both cost reduction and decreases in mortality rates to $\leq$2% (Ottolenghi, 2008).

6.4.2 Transfer to Permanent Cages

To be able to service the daily husbandry needs of these valuable fish, fattening cages are typically attached to a mooring grid and are located in close proximity to shore. In Europe, fattening cages of a more permanent nature are most often cylindrical (10–120 m diameter × 25–35 m deep; Belmonte and de la Gándara, 2008), but a few are rectangular (Hattour, 2005). Once the towing cage arrives at the fattening location, fish are carefully transferred to the permanent cages and an effort is made to assess the biomass. As from the purse seine (capture) to the towing cages (transport), the transfer process from the towing to the fattening cage involves the gentle encouragement of ABFT to swim from one cage to the other using divers and seine-nets, with the two cages often being sewn together. A target density in these cages is typically ≤ 6 kg fish/m^3 (Farwell, 2003; Belmonte and de la Gándara, 2008). Animals are then allowed a 10–15 day acclimation period before feeding starts (Hattour, 2005).

6.4.3 Feeding

The fattening period in the Mediterranean generally runs from July to December but can be extended into February if necessary (Belmonte et al., 2007). During this period, ABFT are fed a variety of defrosted, previously frozen baitfish—typically imported—including sardinella (*Sardinella aurita*), pilchard (*Sardina pilchardus*), herring (*Clupea harengus*), mackerel (*Scomber scombrus*), horse mackerel (*Trachurus* sp.), chub mackerel (*Scomber japonicus*), bogue (*Boops boops*), and some cephalopods (Vita et al., 2004). For the year 2004 alone, an estimated 225,000 MT of baitfish was utilized by Mediterranean tuna farming operations (Huntington, 2008).

Feeding practices vary greatly between countries and farms, with the availability of feed items often dictating the ration's composition. In general, fish are offered 1–3 daily meals (Ottolenghi, 2008), but this can climb up to 6 times per day, depending on fish size, water temperature, and feeding responses. To minimize wastage and environmental impacts, most Mediterranean farms use SCUBA divers to monitor feeding activities and to signal when fish are satiated (Sanz-Lazaro and Marin, 2008), but overfeeding is still a common occurrence (Aguado and Garcia-Garcia, 2005b). If an *ad libitum* regime is not targeted, daily feed inputs are provided at set amounts which vary from 2% to 10% of the estimated biomass (Farwell, 2003; Ottolenghi et al., 2004; Ottolenghi, 2008). In general, the highest feeding rates occur in summertime and between 1 and 2 months after stocking, when the farmed ABFT are smallest (Norita, 2003).

Several methods are used in Europe to dispense forage fish, including thawing and pumping it with water through a hose that reaches the center of the cage, or offering frozen (buoyant) blocks in a central feeding ring. As the forage fish thaws, it sinks; then, tuna are able to consume it. Feeding occurs in hierarchical order, with larger animals feeding first closer to the water surface and smaller tuna swimming lower in the water column until dominant fish lose interest (Belmonte and de la Gándara, 2008).

6.4.4 Biomass Estimation and Growth Rates

As it is practiced today, tuna farming in Europe involves capture, confinement, daily feeding at a level close to satiation and subsequent modification of some ABFT morphometric relationships before harvest. Because the accurate measurement of fish lengths and weights as the animals are transferred into the fattening cages is not possible, the computation of feed conversion rates (FCR) is today a rough approximation at best. These ranges from 15–20 to 1, for animals below 30 kg (forage fish wet weight: tuna wet-weight gain; Farwell, 2003; Katavic et al., 2003b), and although this could be considered extremely high as compared to FCRs for other cultured marine fish, it is only about half of the 40:1 FCRs calculated for ABFT over 150 kg (De la Gándara and Ortega, 2013). This is both because tunas, like other fishes, tend to grow more slowly at larger sizes and because they change body composition, increasing fat. The slow growth of larger ABFT was documented by Norita (2003) in which smaller animals increased 40–50% in 8 months; whereas, larger tuna increased only 10–30% in body weight in the same period. These growth rates were recently confirmed for ABFT in Spain where 40–50 kg fish grew 30–40% in the 6–7 month period they remained in the fattening cage, while 150–200 kg animals gained only 12–15% in the same time interval (De la Gándara and Ortega, 2013). Under similar conditions, in a tuna farm off Izmir, Turkey, ~60 kg ABFT grew between 25% and 35% during the fattening season (Percin and Konyalioglu, 2008).

Aguado and Garcia-Garcia (2003a) pioneered the assessment of growth in captive ABFT using indirect, noninvasive methods (image analysis) to estimate size/weight, and also testing/optimizing several biometric equations to compute ABFT weight under farming conditions. These authors documented extremely high growth rates for ABFT as compared to other marine fish. Contrasting wild versus farmed ABFT, Aguado and Garcia-Garcia (2005a) studied the relationship between fork length (FL), weight, and condition index, finding that farmed animals ≥ 180 cm became significantly heavier than their wild counterparts of the same size and their condition index also increased significantly. However, these wild versus farmed differences were not apparent for smaller ABFT, indicating that smaller animals are not as affected by the imposed feeding regimes in tuna farming. Seemingly, the higher metabolic and growth rates of smaller fish are sufficient to prevent ABFT from becoming heavier than what is considered normal for wild fish. In contrast, larger animals which are prevented from their natural migratory activity continue to feed vigorously, storing surplus calories as fat. Similar findings were described by Tzoumas et al. (2010), for ABFT farmed off the coast of Greece. However, the inflexion point documented in Greece was 98 cm and differences in weight (farmed vs wild fish) ranged from 0.6 to 227 kg for the size class (in FL) 105–295 cm, respectively.

To add to the complexity in size/weight and performance estimations, size-at-capture can be quite heterogeneous. Although Galaz (2011) indicated that the 200–240 cm range is perhaps the most common in Mediterranean tuna farming, mixed-size catches are also frequent. While computations of growth for a mixed-size batch range from 20% to 35% in a typical farming season, a school of more homogeneously sized ABFT, consisting of 120–150 cm (FL) animals, can grow up to 80% in the same time period. The presence of large and small fish in the same batch renders weight determinations inaccurate; and, thus, alternative assessment methods need to be developed. The growing/fattening period also varies greatly between Mediterranean regions, making comparisons between fattened (typically 6–7 months), farmed (smaller fish reared in cages for up to 18 months), and wild ABFT specimens problematic (Milatou and Megalofonou, 2014). Croatia is one such Mediterranean country which often grows smaller ABFT for longer periods, along the eastern Adriatic Sea. In collaboration with industry partners, Katavic et al. (2009) described the growth progression of advanced ABFT juveniles (~2 years old, average weight ~10 kg/fish) which reached 45 and 90 kg within 18 and 30 months, respectively.

Based on the innovative work of Aguado and Garcia-Garcia (2003a) in Murcia, Spain, the use of stereo-video cameras for estimating size and weight of farmed ABFT has extended to other regions in that country (e.g., Ricardo Fuentes Group [RFG], Cartagena) as well as to other Mediterranean countries including Italy (Marpesca, Vibo, Valentia; Costa et al., 2009; Anon, 2010), Malta (Anon, 2011), and Greece (Bluefin Tuna Hellas, Ionian Sea, Astakos; Ramfos et al., 2011). In the case of Malta, the Australian AQ1 system (Hobart, Tasmania, Australia) was used to determine relatively accurate estimates of fish size and quantification of individual numbers during cage stocking. This is of significant relevance under ICCAT's recommendations which compel all tuna farmers to improve reporting practices by documenting the total number of fish stocked, as well as the size composition of the particular stock (ICCAT, 2006a).

Complementing this technological application, Puig et al. (2012) and Espinosa et al. (2012) combined the use of acoustic and imaging approaches to expand data collection efforts throughout the farming cycle. By integrating information from top (surface) and bottom (cage-bottom) transponders, these authors were able to gather, in a non-intrusive manner, information on the tridimensional fish-use of a cage—as previously reported by De la Gándara et al. (2005) and Sainz-Pardo et al. (2010)—and also incorporating quantification of fish numbers and approximate specimen size in fulfillment of ICCAT's recommendation.

By the cumulative effects of the many strategies described above, farm mortalities have been progressively reduced from 15% to 20% in 2003 (Katavic et al., 2003b) to almost negligible under normal conditions nowadays. However, sharp environmental variations always have the potential for catastrophic damage to captive tuna; and, under conditions of extreme weather, it is not uncommon for farmers to report complete loses of their tuna stock (Mylonas et al., 2010).

6.5 PRELIMINARY EUROPEAN RESEARCH EFFORTS IN SUPPORT OF TUNA FARMING

The burgeoning ABFT fattening activities in Mediterranean waters has also provided access for European researchers to study various aspects of tuna physiology and in doing so, contribute to sustainability and permanence of this industry. In addition, insight has been gained into practical aspects on live-tuna handling such as the fact that for these pelagic animals, even non-abrasive netting materials like soft cotton twine promote mucus losses and dermal scraping (Ticina et al., 2004), which, together with general holding stress, resulted in complete (100%) mortality at day 6, when 40 kg ABFT were transported from oceanic net pens to land-based facilities (Garcia-Gomez et al., 2003). Efforts have been made to reduce handling stress through the use of traditional anesthetics in ABFT only to find inconclusive results (Garcia-Gomez et al., 2003; Diaz et al., 2005; Anon, 2007; Bridges et al., 2009). The anesthesia, numbing, and immobilization of ABFT brood-fish is desirable because this would make possible the accurate application of hormonal therapies aimed at maturation induction and the procurement of length–weight measurement, ovarian biopsies, and archival/identification tags (Bridges et al., 2009).

6.5.1 ABFT Ethology

Although studies on tuna behavior in the wild have been undertaken using archival tags (Kitagawa et al., 2004), and experimental ultrasonic telemetry (Newlands et al., 2004), captive tunas afford the unique opportunity to derive detailed ethograms, including the identification of anomalous behaviors as affected by anthropogenic stressors (e.g., service boats and SCUBA divers), which could eventually influence their overall performance in fattening cages. The use of digital video acquisition systems allowed Nucci et al. (2007, 2010), to conduct noninvasive, behavioral studies on ABFT in captivity in a farm located in the southern Tyrrhenian Sea (about 0.5 nm offshore Bivona, Sicily, Italy). A preliminary analysis of changes in daily distributions and swimming patterns (klinokinesis and orthokinesis) indicated that the prolonged presence of divers may disrupt the normal metabolic expenditure in ABFT, thus negatively affecting its culture efficiency. As mentioned above, echo sounders have also been used to pinpoint the tuna position relative to the cage and behavioral approaches have correlated this with feeding, courtship, and reproductive processes (De la Gándara et al., 2005; Sainz-Pardo et al., 2010).

6.5.2 Proximate Composition

Research by Roncarati et al. (2008), documented equivalences existing between tissue lipid content, as assessed by established chemical methods (e.g., A.O.A.C., 2005), and values obtained using an automated device termed a "fat meter." However, muscle tissue fatty acid profiles from ABFT captured with traditional Italian set nets ("tonnara"), long line ("palangri"), or fattened in net pens and slaughtered with the Japanese technique differ significantly depending upon the fish's origin. In a different study, Sanchis (2012) suggested that the biosynthetic pathway for long-chain polyunsaturated fatty acids (LC-PUFA) was perhaps most active in ABFT brain tissue, based on gene expression evidence collected from two genes coding for enzymes involved in rate-controlling steps in this metabolic pathway (6fad and Elov5), with samples collected from other tissues expressing these genes at much lower levels. The same author analyzed the fatty acid profile evolution over a complete fattening cycle, finding relatively constant fatty acid compositions in baitfish-fed ABFT.

Furthermore, Sanchis (2012) also documented a comparatively homogeneous distribution of docosahexaenoic acid (DHA) in muscle samples taken from various locations. In contrast, Grigorakis et al. (2010), also studying fatty acid profiles from muscle tissue collected in different body regions, determined that the proximal (cephalic) and abdominal (ventral) regions contain significantly higher lipid content with concomitant lower moisture and protein content, as other ABFT body regions. Findings from this study indicate that dorsal loins of ABFT have the highest concentration of PUFA n-3 fatty acids and the lowest monounsaturated fatty acids. The belly flaps contained the exact opposite fatty acid profile.

Some Croatian scientists have centered their attention on the assessment of ABFT blood biochemical parameters as indicators of the overall metabolic equilibrium in farmed tuna and tried to relate these parameters to the ABFT's tissue proximate composition (Popovic et al., 2008, 2012; Jelavic et al., 2012). Changes in proximate composition brought about by farming activities also in Croatia were investigated by Yerlikaya et al. (2009), observing increases in fat and ash contents, but protein content remained relatively unchanged during this period.

Focusing sampling efforts on ABFT fattening operations in southeastern Spain, Aguado and Garcia-Garcia (2003b, 2005b) evaluated the macronutrient composition of baitfish and other feed items used in tuna farming and—in perhaps the first study to do so with replicate net pens—more accurately determined ABFT feed utilization parameters including growth, digestibility, and waste production. These studies are discussed in further detail below.

6.5.3 Incipient Reports on Farmed Tuna Diseases

As an integral part of tuna farming technologies, management of fish diseases is of paramount importance for both grow out and hatchery operations.

In addition to the obvious advantage afforded by increased survival to production efficiency, farmers who can document the health status of their product have a competitive advantage with customers and regulators, who recognize the benefits of a safer product. Successful fish health management begins with prevention of disease rather than treatment. Therefore, any sound tuna mariculture program needs to incorporate management practices which are designed to prevent diseases.

Although low stocking densities are prevalent in the tuna mariculture industry at present, and thus infectious diseases or parasites have yet to become a serious concern, the majority of aquatic food animal production systems undergo intensification as technology evolves, and experience indicates that intensification will certainly increase the risk of infectious disease outbreaks in ABFT farming.

For the most part, catastrophic losses in tuna farms have been confined to events of rapid water quality deterioration associated with excessive run-off after rainfall (Mylonas et al., 2010). The abrupt increase in water turbidity and eutrophication is something that this pelagic fish, accustomed to oligotrophic, blue-water environments, does not tolerate well. Noninfectious diseases related to nutritional factors have been reported due to the rancidity of the lipid fraction of tuna feeds (Roberts and Agius, 2008). The condition, diagnosed as pan-steatitis, seen in other farmed fish species, was reverted once the diet was modified.

A mortality event probably caused by a combination of stress and a secondary bacterial infection after parasitological infestation of plerocercoids (cestode *Hepatoxylon trichiuri* embedded in the gastric mucosa) was reported by Mladineo (2006). An outbreak of *Photobacterium damselae*, subspecies piscicida, was identified as the causative agent for substantial mortalities of ABFT reared in the Adriatic Sea (Mladineo et al., 2006). This adds to the list of reported bacterial and parasitic infestations of farmed scombrids (Sanchez Serrano and Caceres Martinez, 2003; Nowak et al., 2006; Muñoz et al., 2007; Mladineo et al., 2008, 2011; Rodriguez Llanos, 2012; Toksen et al., 2012; Vardic Smrzlic et al., 2012).

The incidence of lipomas, hepatocarcinomas, and other structural abnormalities in various ABFT organs has been reviewed by Peric (2003b), Marino et al. (2006), Roberts and Agius (2008), Corriero et al. (2013), Diler et al. (2013), and Passantino et al. (2013). For an expanded discussion on this topic, the reader is referred to Chapter 11 written by Balli et al.

Interestingly, a thorough assessment of the potential transmission of viral diseases using forage fish as vector to reach farmed ABFT yielded negative results in all 180 fish evaluated by Peñalver et al. (2007). However, horizontal transmission of pathogenic organisms is a real and latent threat to farmed tuna and to other species occupying the surrounding environment. Therefore, every effort should be made to avoid the environmental catastrophe documented by Ward et al. (2001), in Australia, as a consequence of pilchard importation for tuna farming.

6.5.4 Environmental Impact of Tuna Cages

The relevance of the mass-balance approach used by Aguado and Garcia-Garcia (2003a) to ascertain ABFT nitrogenous and phosphoric waste lies in the fact that these elements have major environmental effects in promoting eutrophic conditions in specific areas, thus limiting its carrying capacity for ABFT ranching and/or the natural health of the marine ecosystem. Aguado and Garcia-Garcia (2003a) demonstrated that ABFT growth rates in captivity are exceptionally high in contrast to adult specimens of other cultured marine fish species; but, the metabolic cost of this growth, inherent to the ABFT's unique physiology and behavior, also resulted in very poor feed conversion ratios with waste outputs being correspondingly high. For example, dissolved nitrogen and particulate phosphorous waste was significantly higher for ABFT (mg of nutrient/kg fish/day) than for other species cultured in the Mediterranean. In a subsequent publication, Aguado et al. (2004), evaluated the total nitrogen and total phosphorous digestibility under routine ABFT farming conditions with only one daily ration of mackerel, herring, pilchard, and gilt sardine fed to a level approaching satiation. Body weight of sampled tuna ranged between 150 and 350 kg. Feces were collected either by dissection of the distal intestine immediately after slaughtering or obtained directly from the cage environment via a manual sieve handled by SCUBA divers. Acid insoluble ash was used as inert marker. The computed apparent digestibility coefficients (ADCs) for total nitrogen were significantly lower when the fecal samples were collected directly from the intestine than from the water (628.2 vs 937.4 g/kg, respectively). Likewise, ADCs for total phosphorus were lower in intestinal samples than in fecal samples (481.4 and 661.9, respectively). While nitrogen digestibility coefficients were found to be similar in ABFT than those obtained for other marine carnivorous fish, those for phosphorus were lower and this could be related to the comparatively high dietary phosphorus provided to ABFT by batfish and other feed items. These results further underline the urgent need to develop suitable formulated and environmentally friendly diets for ABFT.

Also assessing the nutrient footprint in offshore tuna operations, Vita et al. (2004) used sediment traps to measure the particulate organic waste output. Findings from this work indicate that tuna farming has particulate outputs that differ both qualitatively and quantitatively from those produced by mariculture operations of sparid fish (gilthead sea bream, *Sparus aurata*), also in the Mediterranean. Tuna farms generate 5.8 mg N/kg fish/day and 9.2 mg P/kg fish/day and the settling rates approximate 14-fold those of cage-free environments in the same area. While the environmental footprint of gilthead sea bream farms was confirmed to be substantial, Vezzuli et al. (2008) produced contrasting results for ABFT farming focusing on organic waste. These authors found only negligible differences between farms and control sites, for most parameters tested. In agreement, Aksu et al. (2010) also found no significant differences in the water column—dissolved nutrients including nitrite, nitrate, ammonium, and phosphate or chlorophyll α—or sediments (organic carbon) between sampling stations located directly underneath the ABFT cage or controls outside the mooring grid at Gerence Bay (Eastern Aegean Sea, Turkey). Trends were detected, however, toward increased nutrient loads at the cage station and the authors indicated that the specific location of this farming site was far from the coastline (deeper water), with strong currents, which may have influenced the results.

Looking into shifting population assemblages at the benthic environment as a consequence of tuna farming, Sanz-Lazaro and Marin (2008) found this parameter to be quite sensitive and thus able to detect changes effected by slight alterations in the chemical environment. These authors also highlighted the importance of using toxicology approaches to better understand tuna-related environmental modifications in farming conditions. Using stable isotopes to trace waste deposition in the Gulf of Castellammare (NW of Sicily, Italy), Vizzini et al. (2007), and subsequently Vizzini and Mazzola (2012), examined dispersal patterns and fate of ABFT waste. Whereas only a slight accumulation was observable in the sediment below cages, waste was isotopically shown to contribute to the diet of demersal and bentho-pelagic fish collected around the cages. Among the multiple fates for waste products, dilution into the water column and its hydrodynamic dispersion were important in preventing organic matter accumulation and marker isotopes were detected up to 1 km away from the cages. While the ultimate carrying capacity of a marine environment is dependent on a variety of factors, these studies should provide insight and viable solutions for mitigating nutrient-related environmental effects of tuna aquaculture.

6.5.5 Yield and Dressing Percentage

It is important for those marketing ABFT to determine how much meat a farmed tuna provides. The kilograms of marketable meat a tuna farmer secures per animal will be dependent upon the dressing percentage (DP) and the carcass cutting yields. Ticina et al. (2005) studied the relationship between ABFT "live" weight (round weight) and the gilled and gutted product (G&G). A total of 165 fish from two different farming sites and four separate cages were analyzed immediately upon slaughter. Live weights of these fish ranged from 19.4 to 81.7 kg and on average 11.2% of this was lost after G&G processing. Therefore, a conversion factor of 1.112 can be used on the known G&G product weight to estimate the round weight of farmed tuna. Conversely, the DP is the percent of the live animal that ends up as carcass. Various factors will affect this indicator including how much the fish has eaten before it is slaughtered and how thorough the bleeding was. Because farmed ABFT swim in pristine waters, there are seldom foreign materials that would negatively correlate with DP; thus, this statistic compares favorably with terrestrial species, and even other marine fish. The amount of fat and muscle tissue will positively affect DP, with larger (heavier) or fatter tuna, typically having higher DPs than smaller/leaner fish.

6.5.6 Harvesting and Slaughtering Methods, Flesh Quality, and Welfare Issues

Flesh quality of tuna is very much dependent on pre-slaughter handling (Buentello et al., 2008); and so, all pre-slaughter processes including the round-up are planned around these overriding considerations (Lines and Spence, 2011). Due to the large size and high individual value of tuna, it is commercially feasible to take considerable precautions to protect the welfare of individual tuna.

The objective of ABFT slaughter is to ensure that death occurs swiftly to prevent lactic acid formation and accumulation in the muscle, because this imparts a metallic taste to tuna flesh when eaten raw as sushi or sashimi. Lactic acid accumulation is referred to as "burnt flesh syndrome" (Buentello et al., 2008; Galaz et al., 2001) or "yake niku" in Japanese, and it causes a steep drop in market price or even refusal by Japanese purchasers (Mylonas et al., 2010; De la Gándara and Ortega, 2013). With the highest lactate dehydrogenase activity found in the animal kingdom (5700 units/g; Guppy and Hochachka, 1978), tuna is a lactate-producing machine. As an illustration, normal circulating lactate levels in humans range from 0.5 to 1 mmol/L, with hyperlactatemia and potentially fatal lactic acidosis occurring between 2 and 5 mmol/L (Kimmoun et al., 2015). In contrast, roundup-stressed PBFT accumulate a colossal 20.9 mmol/L in the blood before slaughter, highlighting the enormous aerobic/anaerobic range of tunas as well as the need to implement nutritional, managerial, and even pharmacological strategies to prevent burnt flesh syndrome in farmed tuna (Buentello et al., 2008).

With the beginning of European tuna farming in the late 1990s, new slaughtering methods were implemented and described by several authors, including Soto et al. (2006), Belmonte and de la Gándara (2008), Anon (2009b), and Mylonas et al. (2010). Pre-slaughter procedures may be improved such that fish excitement or stress are minimized, ensuring both appropriate fish welfare standards and high-quality tuna flesh for selective markets (Buentello et al., 2008).

Tuna harvesting may take several forms, but all are oriented at reducing the volume of water in which fish swims to allow for swift captures (reduced stress). Some may use specially constructed seining nets with knotless interlacing (inside or out the fattening cage), barbless hooks, or other means of capture. Fish size often dictates the subsequent slaughtering method. As the human palate can easily detect lactate accumulation in tuna, less stressed animals regularly fetch better prices, adding to the importance of expediting the process, thus reducing suffering and improving flesh quality (Ugolini et al., 2005; Messina and Santulli, 2008; De la Gándara and Ortega, 2013).

Prevalent in Europe is the harvest strategy which involves lifting up the net pen bottom, detaching and pulling one of the pen walls or using a seining net—also referred to as "culling net"—to reduce water volume and isolate the targeted fish to be harvested from the rest of the stock. Alternatively, targeted animals may be herded into a smaller "slaughtering cage" ("copo" in Spanish; 300–400 m^2 surface area), which is fastened to either the culling net or the fattening cage.

Adapted from bovine slaughter houses, electro-narcosis was attempted as a stunning/killing option that used an electric charge which, in theory, could efficiently and safely stun large-sized tuna. Electro-narcosis transponders were fitted to spearing harpoons (Soto et al., 2006) but, efficient electrical current adjustments were difficult to attain in the highly conductive and variable sea water environment. The use of this method was discontinued in the EU as a result of excessive charges breaking the tuna spine and damaging adjacent tissues (reduced marketability), or suboptimal current failing to instantaneously stun/kill fish.

Although the use of firearms is considered as a high-risk operation in slaughter plants for terrestrial animals, trained marksmen standing on a service boat or a platform next to the slaughter cage shoot tuna on the head using a shotgun loaded with single-bullet cartridges (slugs). Typically, a single shot suffices to inflict immediate death, but sometimes subsequent shots are necessary. Gun shooting allows for the fast slaughtering of a large number of tuna in a short period of time, once the target animals (perhaps 30–70 fish) are separated from the standing stock and rapidly ($\leq$3 min) brought to the surface by the hydraulic pull of the pen's bottom (De la Gándara and Ortega, 2013).

A variant of the Hawaiian harpoon called "lupara" (2–3 m long jab stick)—more suitable for selective harvest because it allows singling out chosen pieces—becomes impractical for harvesting large numbers of tuna. This time-consuming but selective method delivers a blow to the brain and, when properly applied, it results in the immediate immobilization and death of the specimen. If the animal is not immobilized, another diver is charged with the delivery of a coup de grâce or grace shot (Anon, 2009b; De la Gándara and Ortega, 2013).

Irrespective of the slaughtering method, bleeding is always conducted by quickly severing the lateral blood vessels and sometimes the gills with pointed sharp knives in order to prevent metabolites such as lactic acid, histamines, and cortisol from reaching the flesh (Mylonas et al., 2010). This is followed by a coring step which consists in driving a hollow metal rod

(~2–3 cm in diameter) through the skull and into the tuna's brain. Slaughtering personnel aim for a specific cranial area covered by a soft translucent tissue over the pineal gland in the frontal part of the cranium called "pineal window." At this point, a procedure called pithing is performed to completely destroy the spinal cord in the neural canal. This procedure, first applied in tuna in Japan (called Tanigushi or Ikijime 活け締め, indistinctively), uses a long metal or nylon semi-flexible wire to destroy the spinal cord, reducing reflex tailing/flapping, and is oriented at disrupting nerve connections such that any remnant metabolism is brought to a complete halt and is followed by G&G and rapid cooling, often using slush ice which is capable of a very rapid reduction in the tuna's core temperature to 0°C (Mylonas et al., 2010).

For smaller sized animals (≤100 kg) spiking rather than coring is used, which is a similar procedure except that the metal rod is thin (~0.5 cm in diameter), solid and pointy (Anon, 2009b).

The various slaughtering methods described above may result in differences in quality/appearance of the end product. The Tsukiji and Nagoya markets in Japan are the main destination for ABFT harvested off the southeast coast of Spain and both adamantly reject lactate-affected fish. That the grading system of traditional Japanese experts is subjective can be evidenced using published data on human eye aging (Pokorny et al., 1987). Over time, substantial structural modifications occur in optical lenses, resulting in the loss of flexibility and increases in optical density. Lenses also become yellow with age, but chromatic adaptation prevents awareness of these changes. The effects are most noticeable when performing critical color matching or comparing color matches with those of other observers. Very often older observers tend to report that the objects are significantly more red than those reported by younger observers (Werner, 1996), adding subjectivity to the Japanese tuna grading system. To resolve some of these issues, Mateo et al. (2006) evaluated an automated image analysis system which was able to objectively define not only color hues for tuna meat at origin, but also to correlate the effects of various slaughtering options on color and other quality indicators.

The interpretation of welfare in commonly farmed fish has yet to be resolved. Therefore, welfare standards for relatively new cultured fish such as ABFT are necessarily tentative, and using the same technologies and rearing systems already in use for other fish species may not be adequate to ensure ABFT welfare (Martins et al., 2012). These concerns prompted the European Commission to request a scientific opinion from the Panel on Animal Health and Welfare about the main systems used to stun and kill farmed tuna in the EU (Anon, 2009b). The panel used a semi-quantitative risk assessment approach to evaluate the potential of incurring poor welfare conditions during the harvest-slaughtering process. Although the availability of data on indicators of poor welfare—physiological or otherwise—was very limited, the Panel identified crowding as the most important hazard during the pre-slaughter period (Anon, 2009b). To parameterize this indicator, crowding was defined as 0.2 fish/$m^3 \approx$ 20 kg fish/$m^3 \approx$ 300 fish of 150 kg, each, in 2,000 m^3 of water, which is the fish density utilized for the coring/spiking method. The crowding necessary for shotgun slaughter is ~20-fold higher (4 fish/$m^3 \approx$ 150 kg fish/$m^3 \approx$ 40 fish of 150 kg, each/40 m^3), obviously undermining tuna welfare; whereas, the lupara method was more benign, needing only 0.07 fish/$m^3 \approx$ 10 kg fish/$m^3 \approx$ 20 fish of 150 kg, each/300 m^3 of water. Assumptions about method-related welfare are true only if a single shot (lupara or shotgun) is sufficient for the kill. Another variant scoring higher in welfare metrics was spiking small fish underwater but this method is seldom used and gaffing and hoisting before spiking leads to severe pain, distress, and a significantly compromised tuna welfare (Anon, 2009b; EFSA, 2009).

6.6 COOPERATIVE EUROPEAN RESEARCH EFFORTS IN SUPPORT OF TUNA FARMING

6.6.1 Close-Cycle Cultivation of Tuna in Europe

Recognizing that tunas are the most valuable fishery resource worldwide, worth over US$6 billion per year (FAO, 2005a,b); and, that, with the exception of skipjack tuna (*Katsuwonus pelamis*), global tuna stocks are either fully exploited, over-exploited, or depleted (Anon, 2009a, 2011), the EU saw in the captive propagation of the ABFT a way to alleviate the overpowering pressure on the Mediterranean wild ABFT fishery. Closing the life cycle of the PBFT has already been attained in Japan, using the technology described by Sawada et al. (2005), with an account provided in Chapter 8 of this book. The Japanese success with PBFT raised European interest in developing similar spawning and larval rearing protocols to support the ABFT farming industry (Mylonas et al., 2010; Ottolenghi, 2008; De Metrio et al., 2010b; De la Gándara, 2012; De la Gándara and Ortega, 2013). Such protocols, in theory, would close the life cycle for captive ABFT, enabling new technologies for successful aquacultural production with minimal impact on wild stocks. Captive, domesticated ABFT could be fed scientifically formulated and environmentally friendly diets, following successful EU models for species such as Atlantic salmon (*Salmo salar*), European sea bass (*Dicentrarchus labrax*), turbot (*Psetta maxima*), and gilthead sea bream (*S. aurata*).

6.6.1.1 DOTT

In the late 1990s, the Israeli Oceanographic and Limnological Research Center-National Center for Mariculture (IOLR-NCM, Dr Hillel Gordin) proposed the formation of a research group charged with the development of closed-cycle mariculture technologies—stocking cages with laboratory produced juveniles (F2 and beyond)—for ABFT under the initiative termed "Domestication of *Thunnus thynnus*" (DOTT; Anon, 2003). Funding was secured from the EU to organize the 1st International Symposium on ABFT Domestication, held in Cartagena, Spain, in 2000. Together with state-of-the-technology descriptions of this industry as it existed around the world, several of the constraints then limiting its full potential were thoroughly discussed. Working groups were matched with areas of expertise including genetics, reproduction, larval rearing, nutrition, disease prevention, husbandry, aquaculture engineering, fish processing as well as socioeconomic and environmental aspects surrounding ABFT ranching. At the end of the first DOTT meeting, the following recommendations were offered:

1. Closed-cycle mariculture technology is a promising alternative (to sourcing of wild fish) to deliver high-quality ABFT to international markets in a sustainable way. If available at a commercial level, this technology would stimulate aquacultural tuna production as well as promote remediation of wild stocks for commercial and recreational fishing.
2. The undertaking is bigger and more costly than what any single research institution is able to afford and thus, cooperative work between ABFT farmers, the fishing industry, academic institutions, and policy makers emerge as the most feasible alternative to master close-cycle tuna mariculture in the short term.
3. The DOTT group recognized that successful and sustainable ABFT aquaculture was only possible if its practices were also environmentally and socially sound.
4. As the multinational undertaking was complex and expensive, a concomitant funding platform would also require actions of a similar magnitude and international expanse.
5. Among several recommendations, DOTT proposed the establishment of a center for ABFT domestication which would include a future land-based facility for holding mature animals.
6. A "Liaison Committee" was solicited from participating nations to facilitate the interactions among the various participants and end users.

6.6.1.2 REPRODOTT

Also emerging from the first DOTT symposium, a collaborative research proposal was structured and submitted to the EU for funding support. A 3-year grant (2003–2005) was secured to evaluate the feasibility of ABFT reproduction in captivity. Research efforts were coordinated by the Spanish Institute of Oceanography (IEO, for its Spanish acronym) with nine additional research partners from seven Mediterranean countries (Garcia-Gomez et al., 2005). The overall objective was to increase understanding on the ABFT reproductive physiology through the attainment of four research objectives:

1. Evaluate endocrine and reproductive status of migrating ABFT.
2. Document whether or not ABFT of reproductive size are able to attain sexual maturity and reproduction in captivity.
3. Obtain fertilized/viable ABFT eggs as well as sperm samples to evaluate the feasibility of artificial fertilization.
4. Develop suitable handling techniques for reproductive size ABFT.

To pursue these research objectives, nine lines of investigation were organized into specific working packages (WP), namely, WP1, coordination and reporting; WP2, development of procedures for ABFT handling; WP3, husbandry of captive broodfish; WP4, sample collection; WP5, histological evaluation of gonads; WP6, determination of sexual maturity (analyses of brain and pituitary tissue as well as blood steroid hormones); WP7, determination of stress in both captive and wild ABFT; WP8, spawning induction (hormonal); and WP9, characterization of gametes.

Chapter 7 in this book presents an expanded account of results on brood stock management, reproduction, and spawning. In summary, the main results of the REPRODOTT project were:

1. Over 120, 80–120 kg ABFT brood stock were captured and confined to experimental cages, contributing to the improvement of handling techniques.
2. Over 800 wild ABFT from different Mediterranean regions were sampled from April to July during the 3 years of the project.
3. Histological evaluation of collected gonads was conducted, assays for circulating sexual and maturity hormones (steroids) and other indicators (vitellogenin) were completed.
4. Complete cDNA sequences for ABFT gonadotropin hormone subunit β (GtH-β) and gonadotropin-releasing hormone (GnRH) were obtained.
5. ELISA assays for native GnRH forms of ABFT were optimized and validated
6. A new ELISA assay specific for ABFT GnRHa was developed.
7. Levels of cortisol, lactate, and catecholamines were evaluated as indicators of stress in captive and free-roaming ABFT.
8. A novel GnRHa-delivery system was develop and tested for inducing ABFT to spawn.
9. ABFT larvae were obtained after a procedure of artificial fertilization collecting sperm and eggs from hormonally induced broodfish, after slaughter.

6.6.1.3 SELFDOTT

Interest in the "domestication" of the ABFT continued in the EU and in 2007, the European Commission launched the 7th Framework Program targeting this topic. The DOTT consortium was successful again in securing funding support and coordination for the new 3-year, €3 million project entitled "From Capture to Self-Sustained Aquaculture and Domestication of Bluefin Tuna, *Thunnus thynnus*" (SELFDOTT, 2008–2011). This effort was entrusted again to IEO leadership. Building on REPRODOTT results, SELFDOTT aimed at developing larval rearing technology and also at addressing the need for formulated diets for ABFT. Other members of the SELFDOTT consortium included the Hellenic Centre for Marine Research (HCMR, Greece); the Institut Français d'Exploitation de la Mer (IFREMER, France); the Heinrich-Heine University of Düsseldorf (UD, Germany); IOLR-NCM (Israel); the University of Cádiz, (UC, Spain); the University of Bari (UB, Italy); the Malta Aquaculture Research Centre (MARC, Malta); the Centre National pour la Recherche Scientifique (CNRS, France); and the Université de Montpellier II, (UM, France). Reproductive ABFT were maintained at RFG (Spain) and Malta Fish Farming LTD (MFFL, Malta) facilities, which together with Skretting's Aquaculture Research Center (SARC)—a global fish feed manufacturing company—were industry partners in the consortium (Anon., 2012b).

SELFDOTT research objectives were as follows:

1. Enhance knowledge of ABFT reproductive biology derived from the REPRODOTT project. To this end, two broodfish populations were established at El Gorguel (Cartagena, southeast Spain) and Marsaxlokk bay (Malta).
2. Promote successful development of fertilized eggs and master ABFT larviculture. To attain this objective, larval rearing experiments were conducted at four different laboratories in various Mediterranean regions, namely, IEO (Mazarron, southeastern Spain), IFREMER (Palavas, France), HCMR (Crete, Greece), and the Eilat laboratory (part of IOLR-NCM, Israel).
3. Develop suitable ABFT diets, defined as both efficient and environmentally friendly. To meet this objective, wild-caught juvenile ABFT were offered several experimental diets developed by SARC while being held at dedicated cages of Tuna Graso and Caladeros del Mediterraneo, both companies belonging to RFG.

An independent regional initiative in the Italian province of Apulia (ALLOTUNA) also contributed funding support which complemented research efforts by SELFDOTT, with some Italian members participating in both (De Metrio et al., 2010b).

6.6.1.4 ABFT Reproductive Biology

Addressing the first objective, ABFT captured during the 2007 and 2008 purse-seining seasons were transported to oceanic cages in Spain and Malta and were fed forage fish/squid rations shown to support reproductive function during REPRODOTT. ABFT were hormonally implanted and eggs were collected via the various devices described in the egg-collection section, below. Reproduction efforts in Spanish waters took place at the RFG facilities located in El Gorguel Bay, where 38 ABFT (wild-caught in the Balearic Sea) were stocked in cylindrical cages (25 m diameter × 20 m deep). A second batch of 25 reproductive-sized fish (~80 kg each) caught in the same location was donated by the Spanish Fisheries Authority to the IEO, as a quota excess from 2009 became available. This second batch was stocked in a different cage in the same area, also belonging to RFG. Over a 4-year rearing period, fish in both batches adapted well to captivity, with maximum mortality reaching only 5%.

As a direct result of the REPRODOTT initiative, a hormonal-implant technology described by Mylonas et al. (2007) was used on several captive ABFT groups in the Mediterranean, successfully inducing massive spawning events (De Metrio et al., 2010a; Caggiano et al., 2011). However, Gordoa et al. (2009), Gordoa (2010), and Gordoa and Carrerras (2014) have repeatedly reported that spontaneous spawning events also occur in non-implanted ABFT shortly after being captured from the wild in towing cages, and also after 1 year of being held in captivity. These reports are similar to those of Deguara (2011), reporting spontaneous spawning in a commercial facility off Malta. Moreover, that mature ABFT release and fertilize eggs at the peak of their spawning season (June–July, for southern Spain) with or without implants have been thoroughly documented (Table 6.1; De la Gándara et al., 2009, 2011b).

TABLE 6.1 ABFT Spawning Events in Cartagena, Spain (De la Gándara et al., 2009, 2011b)

Year	GnRHa Implant	Spawn	1×10^6 Eggs	Project
2008	Yes	No		SELFDOTT
2009	Yes	Yes	140	SELFDOTT
2010	No	Yes	60	SELFDOTT
2011	No	Yes	160	SELFDOTT
2012	No	Yes	*	TCAR[a]
2013	No	Yes	*	TCAR
2014	No	Yes	*	TCAR

* After 2012 eggs have been collected on as needed basis and have not been systematically quantified.
[a]*Técnicas de Cultivo del Atún Rojo – TCAR ≈ ABFT culture techniques.*

Water temperature throughout the cage's vertical extent in the water column was monitored with data loggers (HOBO® pendant sensors) attached to the net at 3 m (surface), 10 m (mid-water), and 17 m (cage bottom). Loggers were programmed to record temperature every 10 min.

The effect of brood stock diets on growth and proximate composition of ABFT gonadal tissue reared in experimental cages in Malta was described by Vassallo-Agius et al. (2011), and represents the first attempt to develop suitable diets for ABFT broodfish. The use of squid and paprika powder, proven beneficial to improve spawning performance in other marine fish (Vassallo-Agius et al., 2001), was tested also on ABFT brood stock. Animals were fed to a level approaching satiety 6 days per week for 3 months prior to the normal ABFT spawning period (May–July). The incorporation of nutrients including fatty acids, carotenoids, amino acids, vitamin C, and vitamin E was evaluated in gonadic tissue. A control group of fish was fed in a similar manner as for regular fattening described above. Fish fed with the squid/paprika supplements had a feed intake two-fold higher than that of the control group. Growth and average body weight per fish, together with condition factor, FL, and gonadosomatic index, were significantly higher for fish fed the experimental ration. Contrasting these levels with those found in samples from wild fish indicated that supplementation of astaxanthin, DHA, total amino acids, and vitamin C in the diet offered to recently caught wild ABFT may result in an improved reproductive performance.

Leptin is a hormone synthesized by adipose cells which helps regulate energy balance by inhibiting hunger and is thought to play a role also in sexual maturity in fish (Trombley and Schmitz, 2013). As part of the analytical procedures for objective 1 in the SELFDOTT initiative, Yanowski et al. (2011a,b) examined expression patterns of cloned ABFT leptin during the reproductive season and found clear correlations between regressed gonads, energy depletion, and an increased transcription of the ABFT Lep gene in the hypothalamic tissue.

6.6.1.5 Egg Collection, Shipment, Development, and Larviculture of ABFT

6.6.1.5.1 Egg Collectors

The initial design of egg collector used in 2008 was a curtain-type egg collector consisting of two panels (each 100 m long × 2 m deep) sewn together to form a cylinder. The upper part was a solid PVC sheet (tarpaulin) which surrounded the inside perimeter of the cage from 1 m above the surface to 1 m underwater; whereas, the bottom sheet was made of 500 μm mesh netting material which surrounded the same perimeter from 1 to 3 m underwater such that a continuous wall surrounded the upper part of the spawning cage (Figure 6.4).

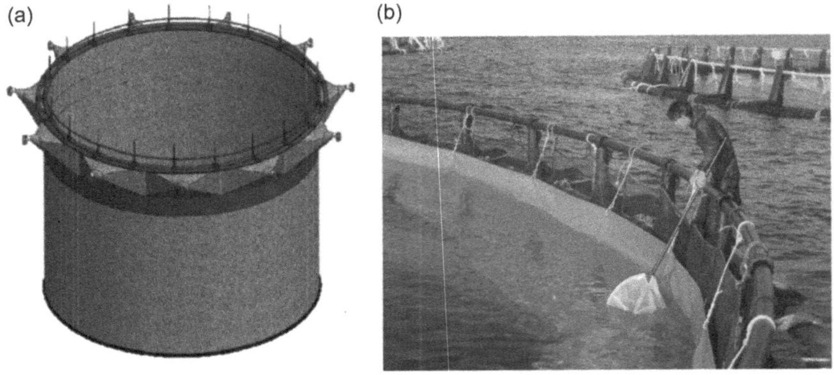

FIGURE 6.4 Egg collectors: (a) conceptual diagram of the initial device and (b) diver gathering eggs from an improved collector made of PVC tarp.

Improvements were necessary because some eggs—despite being buoyant—were caught in transverse currents and transported under the curtain and out the cage. In 2009, the new collector increased the vertical extent of the curtain, now surrounding the entire cage perimeter from 50 cm above water to 6 m deep. In addition, the lower part of the curtain (500 μm mesh) was fitted with 11 funneling cones evenly distributed around the perimeter. Each cone ended in a cylindrical PVC egg collector. To resolve the potential danger of having opportunistic fish eating the fertilized eggs, a screen covered each cone entry, but allowed eggs to pass through. In 2010, further improvements were implemented by replacing cylindrical PVC collectors with soft mesh tubular collectors which improved access to secure eggs. The improvements performed well, allowing for the continued use of this egg collector in 2011.

Initially, a water temperature of 24°C—considered optimal for ABFT spawning—occasioned the deployment of egg collectors. However, monitoring agencies in 2010 documented fish spawning at 20°C in the Balearic Sea; thus, the decision was made for an early deployment (June 14–16, 2010), even though water temperature had not yet reached 24°C. Calm seas and lack of currents on those dates caused the majority of fertilized eggs to reach the water surface and a dip net (500 μm) was used to skim the eggs directly from the perimeter curtain with only a minor fraction of the total egg count (60 million; Table 6.1) actually reaching the collectors, as these require at least a weak current for transport to the collection tube. Based on this and other performance issues, it is clear that significant room for improvement exists in designing egg collectors that perform well under various climate conditions.

6.6.1.5.2 Egg Shipment

Spawned eggs (140 million in 2009) were sent to IEO, packed, and immediately dispatched to hatcheries at the various collaborating institutions around

TABLE 6.2 Shipments of ABFT Fertilized Eggs in 2009 for Larval Rearing (De la Gándara et al., 2010)

Year	Institution	Number of Eggs	Project
2009	IEO	346,000	SELFDOTT
	IFREMER	540,000	SELFDOTT
	HCMR	386,000	SELFDOTT
	MARC	450,000	SELFDOTT
	Panittica Pugliese SPA	836,000	ALLOTUNA

the Mediterranean Sea (IEO, IFREMER, MARC, HCMR, and Panittica Pugliese SPA; Table 6.2) to conduct larval rearing trials. Critical issues that contributed to satisfactory egg viability after shipment included pre-packing temperature, insulation of transport boxes, and minimizing transit time (De la Gándara et al., 2010). Eggs originated in southeastern Spain closest to the IEO hatchery (Mazarron). This allowed for rapid delivery and straightforward incubation. Average hatching rate at this facility exceeded 90%. No differences in hatching rates were observed between eggs collected in 2009 (from fish treated with hormonal implants) and those collected in 2010 (non-hormonally treated) (De la Gándara et al., 2011a).

6.6.1.5.3 Larval Rearing

Three strategies were used for larval rearing at destination hatcheries, namely, mesocosm (at HCMR, MARC, and IEO), pseudo-green water (MARC, HCMR, and IEO), and clear water (IFREMER and IEO) in several larval trials (Table 6.3, Papandroulakis et al., 2010).

As illustrated in Table 6.3, either the presence (coculture or mesocosm) or the addition (pseudo-green water) of eutrophic water appears advantageous to ABFT for early development, yielding superior growth, swim bladder inflation, and survival, compared with the clear-water method (Ortega et al., 2009, 2011; Anon, 2010; De la Gándara et al., 2010).

Supplementation of yolk-sac larvae of other marine fish as a transitional feed from *Artemia nauplii* also promotes growth and survival of ABFT larvae, highlighting the importance of suitable nutrition during early larval stages for this rapidly growing fish. Thus far, ABFT larvae necessitate the supplementation of live fish larvae of a surrogate species of marine fish (e.g., *S. aurata*) as a feed item to successfully transition into the juvenile phase. Performance of ABFT juveniles after weaning is illustrated in Figure 6.5. Maximum sizes attained in these preliminary trials (Ortega et al., 2014) ranged from 8 to 14 cm in total length. Other key factors weighing

TABLE 6.3 ABFT Larval Rearing Trials at Participating Hatcheries (Anon., 2012a; Ortega et al., 2011)

Year	Institution	Strategy	Age of Longest Surviving Fish (dah)	Larval Size g - cm	Swim Bladder Inflation (%)
2009	IEO	Mesocosm	17		–
		Pseudo-green	73	30 – 14	58
		Clear water	10		5
	HCMR	Mesocosm	60		50
		Pseudo-green	60		50
	IFREMER	Clear water	20		1.2
	MARC	Mesocosm	24		–
2010	IEO	Pseudo-green	110		–
2011	IEO	Pseudo-green	Some still alive		–

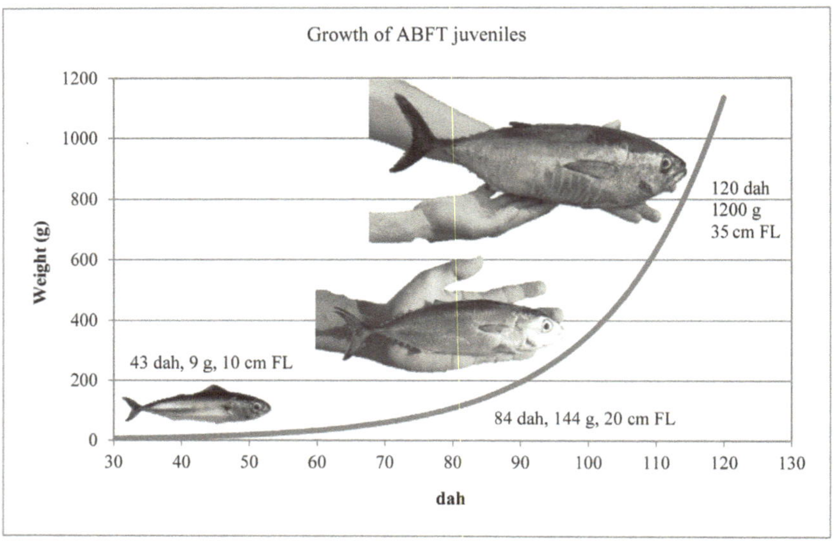

FIGURE 6.5 Growth of juvenile ABFT after weaning from live feeds.

heavily in the larval rearing outcome were light intensity, lighting regime, hydrodynamic control, cleaning of tank surfaces, sequence of feeding items (and/or prey), and nutritional quality of these. Although significant progress was made—only one fish survived to 70 days after hatching (dah) in the first SELFDOTT attempt (2009); whereas, several thousand 30-day-old ABFT juveniles from reproduction and larviculture trials were stocked into oceanic cages in 2011 (Ortega et al., 2011)—further refinements in all of these factors will likely further improve results.

In general, ABFT hatched 26–34 h after spawning, depending on water temperature. Also controlled by this factor, exogenous feeding started 2–3 dah. Feeding protocols were based on enriched (Origreen – Skretting) rotifers (*Brachionus plicatilis*), followed by enriched *Artemia* and recently hatched gilthead sea bream (*S. aurata*) yolk-sac larvae. At 22–25 dah in larval rearing tanks, ABFT juveniles (0.1–0.3 g/fish) were counted (~2% survival from hatch) and transferred to weaning tanks where they were offered an artificial SARC diet or minced raw fish (*Ammodytes* sp). Weaning lasted 7–10 days, with survival during this period ranging from 20% to 50%. As indicated above, by 2011, ~3,000 juvenile ABFT (3–5 g) survived to be stocked in oceanic net pens. Today, some of these fish are still alive and growing (~30 kg each). The moment these animals spawn, the life cycle of ABFT will be closed for the first time, evidenced by production of an F2 generation from laboratory-reared individuals (Ortega et al., 2011).

6.6.1.6 *Early Ontogeny and Digestive Enzymes*

The larval rearing trials conducted at various hatcheries in Spain, France, Malta, Greece, and Israel also afforded the opportunity for a thorough description of the early ontogeny of this species. Cahu et al. (2010) and Mazurais et al. (2014) provide accounts of the onset of activities of selected digestive and antioxidant enzymes in ABFT larvae reared under clear-water (intensive) or mesocosm systems. Both groups reported that the expression of genes coding for the targeted enzymes followed normal patterns. Although the activities of selected digestive enzymes were somewhat elevated as compared to other marine fish, this alone could not explain the high rate of mortalities experienced in the preliminary trials.

The development of the lateral line system in ABFT larvae was characterized by Ghysen et al. (2010, 2012); whereas, Ortega and Mourente (2010) measured lipid profiles of ABFT eggs and larvae deprived of food and compared these profiles with those of other scombrids (Atlantic bonito, *Sarda sarda*), also maintained at IEO facilities. Further, Sanchis (2012) described the effects of developmental stages and dietary intake on fatty acid composition of ABFT larvae and juveniles during the nutritional progression from enriched rotifers → *Artemia* → sea bream larvae → weaning diets. The expression of fatty acyl desaturase and elongase genes and the evolution of the DHA:eicosapentanoic acid (EPA) ratio during development of unfed ABFT larvae have been described by Morais et al. (2011).

In a recent publication, Yufera et al. (2014) provided a histological description of the anatomical development of ABFT during the first month after hatching, emphasizing the development of the gastrointestinal tract and of other organs required to establish the hunting behavior characteristic of this species (e.g., eyes). Complementary to this, the ontogeny of the ABFT visual and neuroendocrine systems has been characterized by Besseau et al. (2011), who observed that at hatching (0 dah) the retina had not yet differentiated and identification of the pineal gland was not possible at this stage, nor at 1 dah. However, by 3 dah the fully developed retina and pineal organ were both observable.

6.6.1.7 *Microbiota*

A description of the bacterial populations (microbiota) present at various developmental stages in ABFT larvae was provided by Makridis et al. (2011). Anon (2012a) further indicated that high variability in bacterial populations is present in different groups of ABFT yolk-sac larvae. Total bacterial counts increased exponentially as the larvae grew, with the genus *Vibrio* becoming dominant, possibly contributing to larval mortality. In this research, bacterial isolates collected at different time points during the rearing process were used for homology searches, making possible the establishment of dendrograms (tree-like diagrams) which describe the phylogenetic similarities of the various bacterial groups present in ABFT larvae. In addition, Gatesoupe et al. (2013) conducted a spatiotemporal study of bacterial communities associated with ABFT larvae in three different Mediterranean hatcheries. Concurrent with Anon (2012b), Gatesoupe et al. (2013) also found high variability in bacterial community profiles before the onset of exogenous feeding, but were not able to demonstrate an ontogenic fluctuation in microbial populations related to ABFT larval development.

6.6.1.8 *Bone Development*

Because of suboptimal larval rearing protocols, skeletal abnormalities often affect ABFT larvae, so much so that several researchers studying ABFT bone development under the SELFDOTT initiative concluded that a significant portion of the mortality experienced during the first few larval rearing attempts could in fact be attributed to skeletal malformations (Boglione et al., 2012, 2013a,b; Libert et al., 2013; Cloutier et al., 2013). De la Gándara et al. (2013) also identified a good correlation between the timing of the notochord flexion and water temperature. This is relevant because, together with the ontogenic development of eyes, gut, jaws, and teeth, the notochord flexion enables the ABFT larvae to engage in early piscivory.

6.6.1.9 Growing Laboratory Produced ABFT in Oceanic Net-Pens

At the conclusion of the SELFDOTT project (November 2011), a stakeholders' meeting was held to review results and issue recommendations that were later transcribed by Anon (2012b). Several members of the consortium have continued the lines of research initially explored by this initiative, including IEO and RFG. Today, more than 7000 laboratory-reared ABFT thrive in various oceanic cages belonging to this group. These juvenile cages are 25 m in diameter and 12 m deep.

When initially transferred to the cages, juvenile ABFT (3–4 g; 6–8 cm total length) were fed a sand-eel diet (*Gymnammodytes cicerellus*), progressing to sardine (*Sardina pilchardus*), mackerel (*Scomber scombrus*), and anchovy (*Engraulis encrasicolus*), as they grew. The feeding regime has also progressed from feeding to apparent satiation in ≥ 8 meals per day to only 1 or 2 daily meals.

The main cause of mortality at this stage is due to collisions with the net and the resulting skin abrasion. This is because in tunas, the fight-or-flight response against perceived danger is biased to the latter in these superb swimmers. Larger-size cages may contribute to higher survival at this stage. During the first month after stocking, mortalities climbed to 60% and reached 75% by the time animals attained 1 kg ($\sim$4 months). After peaking at month 5—associated with a net-related incident (failed net exchange)—mortality progressively decreased and from month 5 to 12 was only 15%. Growth rates were very high with animals stocked at 5 g reaching 500 g in a period of 2 months. Specific growth rates averaged 8.1%/day and achieved a remarkable 9,900% percent weight gain (WG) in 8 weeks. For comparison, red drum *Sciaenops ocellatus*, a relatively fast growing fish starting at 5 g achieves $\sim$1,000% in WG in the same time period (Wurts and Stickney, 1993).

In spite of growth deceleration with increasing size, most juvenile ABFT reached 2 kg by month 6. Figure 6.6 illustrates growth of juvenile ABFT in oceanic net pens, together with a temperature profile during the recorded period (Ortega et al., 2014). Inferences derived from this figure indicate that growth rates in juvenile ABFT (5–2,000 g) are strongly influenced by water temperature, which also exerts rigid control over the animal's feed intake. Feed consumption starts at 16°C, in a desultory manner, but both feed intake and growth increase substantially at 18°C. This underlines the importance of formulated diets to promote insights about the ABFT's specific physiology which may lead not only to good performance and reduced costs but also to a lessened dependence on wild fisheries for forage fish, potentially also lowering vertical and horizontal disease transmission.

6.6.1.10 Formulated Diets for ABFT

To address the last SELFDOTT objective of formulating diets for ABFT, wild juveniles ($\leq$1 year) were captured in southeastern Spain (Cabo de

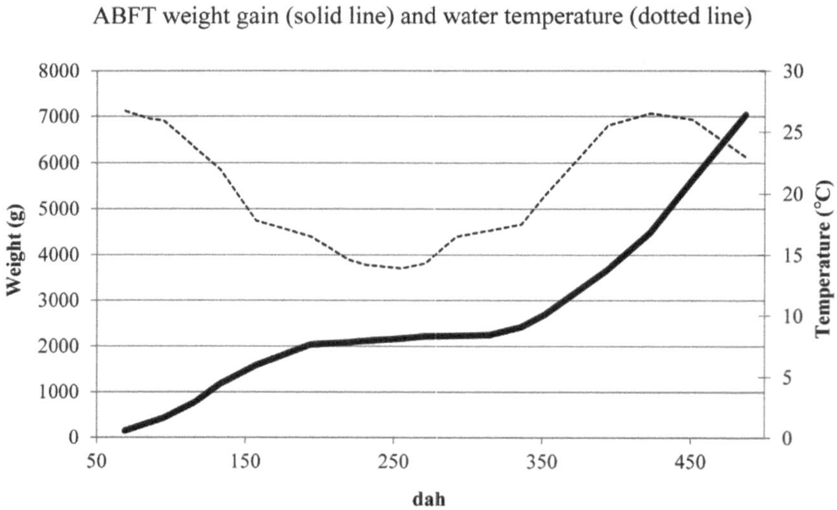

FIGURE 6.6 Growth of juvenile ABFT in oceanic cages and temperature profiles in southeastern Spain.

Palos, Murcia) by the hook-and-line method (barbless hook) and were acclimated to the cage environment. Plastic lures ("currican" in Spanish) trawled at a constant speed of 6 knots were utilized. Handling only by line and lure, captured animals were unhooked into a 1,200 L cylindrical plastic tank (1.5 m in diameter) and transported either to net pens or land-based facilities. At the time of capture, water temperature was 19.5 ± 0.5°C. Similar tanks were used for land transport in a service van. Transport tanks were fitted with a bottom baffle with holes to house air stones for oxygen delivery and a small pump to provide circulation and promote swimming activity in the captured fish. At their destination, fish were transferred to a cylindrical net-pen (25 m diameter × 20 m deep). Although hooks were barbless, 50% of the transport mortality was due to bleeding. Often, a fish hooked by the gills will bleed to death. A lesser problem was abrasion against the tank. Of those fish arriving alive at the cage, 75% survived the first month in captivity. Handling was accomplished with a plastic tarpaulin which proved much less abrasive than traditional netting. Once in the cage, fish recovered quickly from transport stress. As for land transportation, a complete account can be found in Anon (2012a). Briefly, 22 ABFT juveniles were transported to IEO land-based facilities from October to November 2010. Overall survival at the end of the process was 100% and by December 31 had dropped to only 78%.

Transport casualties were used for whole-body composition and stomach content analyses to be used in baseline determination of nutritional requirements. A preliminary analysis of carcass protein and fat content, n-3 fatty acids, and amino acid concentrations revealed that juvenile ABFT may have nutritional requirements that are distinctly different from those of other Mediterranean marine fish, being higher for most indicators. Amino acid profiles of tuna carcasses and stomach contents indicated that higher concentrations of histidine, methionine, lysine, leucine, and aspartic acid may be necessary to satisfy the nutritional requirements of juvenile ABFT.

Although an important component of the overall SELFDOTT initiative, nutritional work was somewhat limited in lieu of higher priority reproduction and larval rearing techniques. A special diet named Aurora incorporated higher than normal dietary amino acids and crude protein concentrations, lower crude lipid, higher dietary n-3 fatty acid concentrations, and higher contents of vitamins C and E than conventional diets for marine fish. Aurora was tested on 105 young ABFT with an average initial weight of 850 g (De la Gándara, 2012). Performance of these fish was compared against 85 ABFT of 750 g average initial weight fed with round sardinella, *S. aurita*. The Aurora diet was only marginally accepted by fish and severe mortalities in the net pen under this regime caused the experiment to be terminated ahead of schedule. All other formulated diets tested as part of the last SELFDOTT objective were absolutely refused by the experimental fish. Although these first attempts to wean wild-caught juvenile tuna from a forage fish diet to a formulated feed have provided some useful preliminary information, clearly more research is necessary before suitable balanced diets are able to meet the metabolic needs of juvenile ABFT. A comprehensive review of the most updated information regarding tuna nutrition and the development of formulated diets is provided in Chapter 12 of this book.

At the conclusion of the SELFDOTT project, the IEO, in collaboration with industrial partners (RFG and SARC), conducted complementary research oriented at improving the production of ABFT juveniles (project TCAR, for its Spanish acronym) and contrasting larval development in the natural environment or under laboratory conditions (project ATAME, funded by the Spanish National Science Program).

6.6.1.11 TRANSDOTT

Building on results from the REPRODOTT and SELFDOTT projects, the European Commission approved funding support in 2012 for a new project called TRANSDOTT (www.transdott.eu), to pursue a top-down approach and apply the newly established knowledge base to further ABFT aquaculture (Bridges, 2012). This initiative aimed at translating this technology into commercial applications and, to this end, established partnerships with one global enterprise (SARC), four small/medium enterprises (Futuna Blue S.L. Spain, ARDAG Red Sea Mariculture Ltd Israel, MFFL and Panittica

Pugliese SPA, Italy) and three Public Research Institutions (Heinrich-Heine University of Düsseldorf, MARC and IOLR-NCM).

Fertilized ABFT eggs produced in Malta were provided in June 2012, 2013, and 2014 to a number of hatcheries in Malta, Spain, Israel, and Italy, to conduct larviculture of ABFT at a commercial scale. This has provided validation for the larval rearing protocols generated during the previous projects and juveniles have been transferred to oceanic cages. Weaning diets provided by SARC have been used with relative success and now the economic feasibility of conducting these activities in support of the sustainable ABFT aquaculture is being evaluated.

6.6.2 Newly Built Land-Based Facility

To resolve issues of climate variability and to reduce uncertainty in ABFT egg production as affected by a constantly shifting aquatic environment, the IEO has recently completed (April 2015) construction of a new land-based facility, with formal inauguration planned for June 2015 (Figure 6.7). Construction funds were provided the Spanish Ministry of Science and Innovation, the Murcia Regional Community and the IEO, with co-funding from the regional FEDER funds (operational program, knowledge-based economy, 2007–2013). The goal for this facility is to provide fertilized ABFT eggs, in a predictable manner, in quantity and quality necessary to service the ABFT aquaculture industry, academic research, and conservation efforts, thus fostering an ecologically sound ABFT aquaculture industry.

The building area of the new facility totals 2660 m^2, with 1960 m^2 assigned to tank space, 300 m^2 to wet and dry laboratories and office space and, 400 m^2 to water depuration and recirculation equipment. Two broodfish

FIGURE 6.7 Newly built, land-based ABFT aquaculture facility in Murcia, Spain. (a) General lay out, (b) water proofing concrete tanks, (c) paint pattern for visual background, (d) roofing, (e) filled broodfish tank, (f) and (g) recirculation systems and filtration, and (h) protein skimmers and cooling units.

tanks are 22 and 20 m in diameter and 9 m deep (volume of 3500 and 2500 m^3, respectively). Tanks for juvenile fish—planned for quarantine, acclimation, and grow out purposes—are 14 and 8 m in diameter and 3 m deep (volume of 900 and 150 m^3, respectively) amounting to 7000 m^3 of tank volume for the entire facility.

Broodfish tanks have separate recirculation systems and can be independently controlled for photoperiod and water temperature (heating or cooling). Water depuration equipment includes filtering systems (physical, biological and chemical filtration), protein skimmers, and sterilization (ultraviolet light). The laboratory/office area includes a storage room, cold and freezer rooms, two laboratories, and four small offices. The complete facility is fitted with closed-circuit cameras (including underwater video), various sensors, and automated alarm systems for the continuous monitoring of culture conditions and fish behavior.

This facility will allow for manipulation of various environmental factors (particularly photo-thermal manipulations and nutrient availability) to initiate gonadal recrudescence and to accomplish and extend ABFT spawning at predicted times. This extension in the ABFT reproductive period will increase egg and juvenile availability, thus promoting commercial and scientific productivity (increased number of experiments) in support of the establishment of the ABFT aquaculture industry (Anon., 2012a).

6.7 FUTURE PROSPECTS

In 2014, the IEO and the Spanish company Fortuna Mare Ltd. signed a research contract for the improvement of ABFT larval rearing by using *Acartia tonsa*, a marine calanoid copepod, as complementary live feed for tuna larvae. Preliminary work started in 2013, and by 2014 feeding regimes using the standard enriched rotifers or *A. tonsa* had been compared and contrasted. Results indicated that copepod usage improved ABFT larvae growth and survival by 15% and 200%, respectively, and the mixture of copepods and rotifers also demonstrated improvements from the rotifer regime alone. A second prospective line of research, pursued in collaboration with the University of Cadiz (Spain) and the University of Stirling (Scotland, UK), uses gene expression to assess essential fatty acid requirements of ABFT during the early ontogeny, thus helping to prevent deficiency pathologies and to optimize growth. This collaborative work will require the generation of detailed knowledge on the biochemical and molecular basis of essential fatty acid requirements and metabolism of ABFT. These efforts will likely provide practical solutions to resolve technical issues of ABFT larviculture making significant contributions to the continuous improvement of the ABFT culture technology.

REFERENCES

Aguado, F., Garcia-Garcia, B., 2003a. Estima del crecimiento y de la producción de desechos mediante un modelo bioenergético en Atún Rojo (*Thunnus thynnus thynnus* Linnaeus, 1758), sometido a engrase intensivo en jaulas flotantes en mar abierto. CIVA 2003. http://www.civa2003.org, pp. 271–283. (In Spanish)

Aguado, F., Garcia-Garcia, B., 2003b. Macronutrient composition of food for bluefin tuna (*Thunnus thynnus thynnus*) fattening. Cah. Options Medit̀err. 60, 15–16.

Aguado, F., Garcia-Garcia, B., 2005a. Changes in some morphometric relationships in Atlantic bluefin tuna (*Thunnus thynnus thynnus* Linnaeus, 1758) as a result of fattening process. Aquaculture 249, 303–309.

Aguado, F., Garcia-Garcia, B., 2005b. Growth, food intake and feed conversion rates in captive Atlantic bluefin tuna (*Thunnus thynnus* Linnaeus, 1758) under fattening conditions. Aquacultural Res. 36, 610–614.

Aguado, F., Martinez, F.J., Garcia-Garcia, B., 2004. In vivo total nitrogen and total phosphorous digestibility in Atlantic Bluefin Tuna (*Thunnus thynnus thynnus* Linnaeus, 1758) under industrially intensive fattening conditions in Southeast Spain Mediterranean coastal waters. Aquacultural Nutr. 10, 413–419.

Aksu, M., Kaymakci-Basaran, A., Egemen, O., 2010. Long-term monitoring of the impact of a capture-based bluefin tuna aquaculture on water column nutrient levels in the Eastern Aegean Sea, Turkey. Environ. Monit. Assess. 171, 681–688.

Anon., 2003. Domestication of *Thunnus thynnus*—DOTT. Proceeding of the first International Symposium held at the University of Cartagena, Spain, 3–8 February 2002. Cah. Options Medit̀err. 60, 198 pp.

Anon., 2006. The plunder of bluefin tuna in the Mediterranean and East Atlantic in 2004 and 2005. Uncovering the real story. WWF-World Wide Fund for Nature, 101 pp.

Anon., 2007. REPRO-DOTT Final Report. Reproduction of the bluefin tuna in captivity - feasibility study for the domestication of *Thunnus thynnus*. Contract number: Q5RS-2002-0153. 220 pp. http://hdl.handle.net/10508/1010

Anon., 2008. Eastern Atlantic and Mediterranean bluefin tuna: moving in a good direction and how long will it take to get there. Collect. Vol. Sci. Pap. ICCAT 62 (4), 1291–1297.

Anon., 2009a. SELFDOTT REPORT 2008. pp. 139 <http://hdl.handle.net/10508/355>.

Anon., 2009b. Species-specific welfare aspects of the main systems of stunning and killing of farmed tuna. EFSA J. 1072, 1–53.

Anon., 2010. SELFDOTT REPORT 2009. In: De la Gándara, F., Mylonas, C.C., Covès, D., Bridges, C.R. (Eds.), pp. 279 <http://hdl.handle.net/10508/356>.

Anon., 2011. Malta's report on a Pilot study carried out in 2011 to estimate the number and weight of bluefin tuna at the point of capture and caging by stereoscopic camera. SCRS 2011/191, 1–10.

Anon., 2012a. SELFDOTT Report 2010–2011. In: De la Gándara, F., Mylonas, C.C., Covès, D., Bridges, C.R., (Eds.), 488 pp. <http://hdl.handle.net/10508/1118>.

Anon., 2012b. SELFDOTT Final report (Abstract). In: De la Gándara, F., Mylonas, C.C., Covès, D., Bridges, C.R., (Eds.), 46 pp. <http://hdl.handle.net/10508/1119>.

A.O.A.C., 2005. Official Methods of Analysis of AOAC International, 12th ed Association of Official Analytical Chemist, Washington, DC, USA (17th ed.).

Bard, D., 2013. The story of Atlantic bluefin tuna. Fact sheet. The PEW Charitable Trusts. http://www.pewtrusts.org/en/research-and-analysis/fact-sheets/2013/10/08/the-story-of-atlantic-bluefin-tuna

Belmonte, A., de la Gándara, F., 2008. El Cultivo del atún rojo *Thunnus thynnus*. Fundación Observatorio Español de Acuicultura. Consejo Superior de Investigaciones Científicas. Ministerio de Medio Ambiente y Medio Rural y Marino, Madrid (Spain) (In Spanish).

Belmonte, A., Ortega, A., De la Gándara, F., 2007. Cultivo de túnidos. Actas del XI Congreso Nacional de Acuicultura, Vigo (In Spanish).

Besseau, L., Fromentin, S., Coves, D., Escande, M.L., Fuentes, M., De la Gándara, F., et al., 2011. Ontogeny of the visual and neuroendocrine systems in the Atlantic bluefin tuna, *Thunnus thynnus*. Proceeding of the EAS2011, Rhodes, Greece, pp. 120–121.

Boglione, C., Marroncini, M., Prestinicola, L., Caggiano, M., Catudella, S., 2012. Normal and anomalous morphogenesis of splanchnocranium, vertebral column and fins in reared larvae and post larvae of Atlantic bluefin tuna *Thunnus thynnus*. Proceedings of the AQUA 2012, WAS, Prague, Czech Republic. p. 1158.

Boglione, C., Gavaia, C., Koumoundouros, G., Gisbert, E., Moren, M., Fontagne, S., et al., 2013a. Skeletal anomalies in reared European fish larvae and juveniles. Part 1: normal and anomalous skeletogenic processes. Rev. Aquacultural 5 (Suppl. 1), S99–S120.

Boglione, C., Gisbert, E., Gavaia, C., Witten, P.E., Moren, M., Fontagne, S., et al., 2013b. Skeletal anomalies in reared European fish larvae and juveniles. Part 2: main typologies, occurrences and causative factors. Rev. Aquacultural 5 (Suppl. 1), S121–S167.

Bridges, C.R., 2012. European contributions to sustainable ABFT aquaculture - past, present and future. Broodstock management, induction and larval rearing. In: Miyashita, S., Takii, K., Biswas, A. (Eds.), Proceedings of The Final Global COE Program International Symposium of Kinki University in Kushimoto, Kushimoto (Japan), Kinki University (Japan), 50-57.

Bridges, C.R., Borutta, F., Krohn, O., Schulz, S., 2009. Broodstock monitoring, handling and induction techniques – recent advances in Atlantic bluefin tuna (*Thunnus thynnus*) reproduction. In: Allan, G., Booth, M., Mair, G., Clarke, S., Biswas, A. (Eds.), Proceedings of the Second Global COE Program Symposium of Kinki University, Adelaide, Australia, pp. 20–24.

Brown, K., Fa, D.A., Finalyson, G., Finalyson, C., 2011. Small Game and Marine Resource Exploitation by Neanderthals: The Evidence from Gibraltar. Trekking the Shore: Changing Coastlines and the Antiquity of Coastal Settlement, Interdisciplinary Contributions to Archaeology. Springer Science + Business Media.

Buentello, J.A., Pohlenz, C., Neill, W.H., Gatlin III, D.M., Ascencio, F.J., 2008. Physiological indicators for tuna cultured in sea cages: a preliminary approach to prevention of the burnt-flesh syndrome. World Aquacultural 2008, Busan, South Korea.

Caggiano, M., Campana, M., Moscato, M., Bridges, C.R., Mylonas, C.C., Deflorio, M., et al., 2011. Atlantic bluefin tuna *Thunnus thynnus* larval and juvenile rearing: three consecutive years of experiments. Proceedings of the World Aquaculture 2011. Natal, Brasil.

Cahu, C.L., Papandroulakis, N., Ortega, A., Coves, D., Le Gall, M.M., Desbruyeres, E., et al., 2010. Onset of selected digestive and antioxidant enzymes in bluefin tuna (*Thunnus thynnus*) larvae reared in intensive or mesocosm system. In: Proceedings of the Aquaculture Europe 2010, 5–8 October, Porto, Portugal, pp. 260–261.

Clarke, S.M., 2005. Farming of southern bluefin tuna *Thunnus maccoyii*: an Australian perspective. Proceedings of the WAS Bali 2005, Bali, Indonesia, 125 pp.

Cloutier, O., Cloutier, R., Libert, E., Ortega, A., De la Gándara, F., 2013. *Condrogénesis y Malformaciones Craneales de Larvas y Juveniles de atún rojo (*Thunnus thynnus*) Criados en Cautividad: identificación de una Fuente Importante de Mortalidad de las Larvas.*. Actas del XIV Congreso Nacional de Acuicultura, Gijón (Spain) (In Spanish).

Colak, A., Colak, S., Guven, E., 2003. Primary application on bluefin tuna (*Thunnus thynnus*) fattening in Turkey. Proceedings of the Workshop on Farming, Management and Conservation of Bluefin Tuna. Istambul, Turkey, 5–7 April, pp. 1–11.

Corriero, A., Medina, A., Mylonas, C.C., Bridges, C.R., Santamaria, N., Deflorio, M., et al., 2013. Proliferation and apoptosis of male germ cells in captive Atlantic bluefin tuna (*Thunnus thynnus* L.) treated with gonadotropin-releasing hormone agonist (GnRHa). Anim. Reprod. Sci. 116 (3–4), 346–357.

Cort, J.L., 2007. El enigma del atún rojo reproductor del Atlántico Nororiental. Bedia Artes Gráficas, S.C. Santander (In Spanish).

Cort, J.L., Martinez, D., 2010. Possible effects of the bluefin tuna (*Thunnus thynnus*) recovery plan in some Spanish fisheries. Collect. Vol. Sci. Pap. ICCAT 65 (3), 868–874.

Costa, C., Scardi, M., Vitalini, V., Cataudellas, S., 2009. A dual camera system for counting and sizing Northern Bluefin Tuna (*Thunnus thynnus*; Linnaeus, 1758) stock, during transfer to aquaculture cages, with a semi-automatic Artificial Neural Network tool. Aquaculture 291 (3–4), 161–167.

De la Gándara, F., 2012. Avances en la domesticación del atún rojo, 2012. Investigación y Ciencia, Julio (In Spanish).

De la Gándara, F., Ortega, A., 2013. Cultivo de escómbridos: el atún rojo y el bonito Atlántico. In: Diversificación de especies en la piscicultura marina española. Publicaciones Científicas y Tecnológicas de la Fundación Observatorio Español de Acuicultura nº 9. Ministerio de Agricultura, Alimentación y Medio Ambiente. Madrid, capítulo 7. pp. 283–320. (In Spanish)

De la Gándara, F., Miquel, J., Iglesias, M., Belmonte, A., Ayora, E., Garcia-Gomez, A., 2005. Using an echosounder system to study the vertical movements of captive bluefin tuna (*Thunnus thynnus*) in floating cages. European Aquaculture Society Special Publication 35, pp. 171–172. <http://hdl.handle.net/10508/1009>.

De la Gándara, F., Ortega, A., Belmonte, A., Mylonas, C.C., Coves, D., Bridges, C.R., et al., 2009. Reproducción del atún rojo *Thunnus thynnus* en jaulas flotantes. Obtención de puestas masivas mediante implantes hormonales. Libro de resúmenes del XII Congreso Nacional de Acuicultura. Madrid Noviembre de 2009. pp. 542–543. http://hdl.handle.net/10508/1076 (In Spanish)

De la Gándara, F., Mylonas, C.C., Coves, D., Ortega, A., Bridges, C.R., Belmonte, A., et al., 2010. Seedling production of Atlantic bluefin tuna (ABFT) *Thunnus thynnus*. The SELFDOTT project. Joint International Symposium of Kinki University and Setouchi Town on The 40th Anniversary of Pacific Bluefin Tuna Aquaculture, 2010. Amami, Japan, pp. 45–52.

De la Gándara, F., Ortega, A., Belmonte, A., Mylonas, C.C., 2011a Spontaneous spawning of Atlantic bluefin tuna *Thunnus thynnus* kept in captivity. In: Proceedings of the EAS2011, Rhodes, Greece, pp. 249–250. http://hdl.handle.net/10508/375

De la Gándara, F., Ortega, A., Belmonte, A., Mylonas, C.C., Prieto, J.R., Viguri, J., 2011b Obtención de puestas masivas de huevos de atún rojo (*Thunnus thynnus*) en cautividad, durante 3 años consecutivos. Actas del XIII Congreso Nacional de Acuicultura, Barcelona, Spain, P-151–2. http://hdl.handle.net/10508/450 (In Spanish)

De la Gándara, F., Ortega, A., Blanco, E., Viguri, J., Reglero, P., 2013. La flexión de la notocorda en larvas de atún rojo, *Thunnus thynnus* (L, 1758) cultivadas a diferentes temperaturas. In: Actas del XIV Congreso Nacional de Acuicultura. Gijón, Spain, pp. 180–181. http://hdl.handle.net/10508/1559 (In Spanish)

De Metrio, G., Bridges, C.R., Mylonas, C.C., Caggiano, M., Deflorio, M., Santamaria, N., et al., 2010a. Spawning induction and large-scale collection of fertilized eggs in captive Atlantic bluefin tuna (*Thunnus thynnus* L.) and the first larval rearing efforts. J. Appl. Ichthyol. 26, 596–599.

De Metrio, G., Caggiano, M., Deflorio, M., Mylonas, C.C., Bridges, C.R., Santamaria, N., et al., 2010b. Reproducing the Atlantic bluefin tuna in captivity: the Italian experience. Collect. Vol. Sci. Pap. ICCAT 65 (3), 864–867.

Deguara, S., 2011. Natural spawning of captive Atlantic bluefin tuna, *Thunnus thynnus* L., in farm cages in Malta. SCRS SCRS/2011/, pp. 1–4.

Di Natale, A., Mangano, A., Vasallo-Agius, R., 2006. Bluefin tuna (*Thunnus thynnus* L.) size composition in cages from the Tyrrhenian Sea and the strait of Sicily in 2004. Collect. Vol. Sci. Pap. ICCAT 59 (3), 851–857.

Diaz, J., Garcia-Gomez, A., Bridges, C.R., Belmonte, A., Medina, A., Mylonas, C.C., et al., 2005. Development of handling procedures for captive bluefin tuna *Thunnus thynnus*. In: Proceedings of the WAS Bali 2005. Bali, Indonesia, 166 pp.

Diler, O., Ozen, M.R., Didinen, B.I., Cicek, N.L., Ozvarol, Y., 2013. Lipidiozis in captive reared bluefin tuna (*Thunnus thynnus* L. 1758) from the western Mediterranean region in Turkey. J. Fisher. Sci. 7 (3), 287–291.

Doumenge, F., 1998. L'histoire des pêches thonières. Collect. Vol. Sci. Pap. ICCAT 50 (2), 753–803 (In French).

EFSA - European Food Safety Authority, 2009. Species-specific welfare aspects of the main systems of stunning and killing of farmed tuna. EFSA J. 1072, 1–53.

Espinosa, V., Puig, V., Soliveres, E., Estruch, V., Andreu, G., Atienza, V., et al., 2012. Bluefin tuna counting and biomass estimation during transfers through the combined use of optical and acoustical techniques. SCRS SCRS/2012/133, pp. 1–10.

FAO, Food and Agriculture Organization of the United Nations, 2005a. Fishery statistics: capture production. FAO Year book, Fish. Stat., 96/1.

FAO, Food and Agriculture Organization of the United Nations, 2005b. Report of the third meeting of the ad hoc GFCM/ICCAT working group on sustainable bluefin tuna farming/fattening practices in the Mediterranean. FAO Fisheries Report 779, pp. 1–108.

Farwell, C., 2003. Management of captive tuna: Collection and transportation, holding facilities, nutrition, growth, and water quality. Cah. Options Medit̀err. 60, 65–68.

Forrestal, F., Coll, M., Die, D.J., Christensen, V., 2012. Ecosystem effects of bluefin tuna *Thunnus thynnus thynnus* aquaculture in the NW Mediterranean Sea. Mar. Ecol. Prog. Ser. 456, 215–231.

Fromentin, J.M., 2009. The key importance of the underlying stock-recruitment assumption when evaluating the potential of management regulations of Atlantic bluefin tuna. Collect. Vol. Sci. Pap. ICCAT 64 (2), 558–567.

Fromentin, J.M., Powers, J.E., 2005. Atlantic bluefin tuna: population dynamics, ecology, fisheries and management. Fish Fisheries 6, 281–306.

Galaz, J.M., 2011. Eleven years—1995–2005—of experience on growth of bluefin tuna (*Thunnus thynnus*) in farms. SCRS SCRS/2011/160, pp. 1–13.

Galaz, J.M., Norita, T., Albaladejo, J., 2001. Incidencia de la aparición de B.T.S. Burnt Tuna Syndrome en el cultivo y engorde de atún rojo *Thunnus thynnus thynnus*. In: Actas del VIII Congreso Nacional de Acuicultura. Santander, Dirección General de Pesca y Alimentación. Gobierno de Cantabria, pp. 35–36 (In Spanish).

Garcia-Gomez, A., De la Gándara, F., Raja, T., Atunes de Mazarron, S.L., 2003. First experience on adult bluefin tuna (BFT), *Thunnus thynnus*: Transportation from rearing cages to inland facilities. Cah. Options Medit̀err. 60, 77–79.

Garcia-Gomez, A., Diaz, M.V., De la Gándara, F., De la Serna, J.M., Belmonte, A., Ayora, E., et al., 2005. Posibilidades de reproducción del atún rojo, *Thunnus thynnus*, en cautividad. Actas del X Congreso Nacional de Acuicultura. Universidad Politécnica de Valencia, Gandía (Spain) (In Spanish).

Gatesoupe, F.J., Coves, D., Ortega, A., Papandroulakis, N., Vadstein, O., De la Gándara, F., 2013. A spatiotemporal study of bacterial community profiles associated with Atlantic bluefin tuna larvae, *Thunnus thynnus* L., in three Mediterranean hatcheries. Aquacultural Res. 44, 1511–1523.

Ghysen, A., Schuster, K., Coves, D., De la Gándara, F., Papandroulakis, N., Ortega, A., 2010. Development of the posterior lateral line system in *Thunnus thynnus*, the atlantic blue-fin tuna, and in its close relative Sarda sarda. Int. J. Dev. Biol. 54, 1317–1322.

Ghysen, A., Dambly-Chaudiere, C., Coves, D., De la Gándara, F., Ortega, A., 2012. Developmental origin of a major difference in sensory patterning between zebrafish and bluefin tuna. Evol. Dev. 14 (2), 204–211.

Gordoa, A., 2010. The Atlantic bluefin tuna: study of the temporal pattern of spawning in the western Mediterranean region and reproductive capacity in captivity. Collect. Vol. Sci. Pap. ICCAT 65 (3), 837–847.

Gordoa, A., Carreras, G., 2014. Determination of temporal spawning patterns and hatching time in response to temperature of Atlantic Bluefin tuna (*Thunnus thynnus*) in the Western Mediterranean. PLoS ONE 9 (3), e90691.

Gordoa, A., Olivar, M.P., Arevalo, R., Viñas, J., Moli, B., Illas, X., 2009. Determination of Atlantic bluefin tuna (*Thunnus thynnus*) spawning time within a transport cage in the western Mediterranean. ICES J. Mar. Sci. August 2006, 1–6.

Grigorakis, K., Negas, I., Fououlaki, E., Panayiotidou, M., Rigos, G., Vassilaki, A., 2010. Body composition and fatty acid profiles from different fillet parts of cultured bluefin tuna *Thunnus thynnus*. In: Proceedings of Aquaculture Europe 2010. Porto, Portugal, pp. 890–891.

Guppy, M., Hochachka, P.W., 1978. Controlling the highest lactate dehydrogenase activity known in nature. Am. J. Physiol. (RICP) 3 (2), R136–R140.

Hattour, A., 2005. Concernant l'activite d'engraissement du thon rouge dans les eaux Tunisiennes. Collect. Vol. Sci. Pap. ICCAT 58 (2), 606–614 (In French).

Huntington, T., 2008. Use of wild fish and other aquatic organisms as feed in aquaculture-a review of practices and implications in Europe: key issues to be addressed. In: Report of the FAO Expert Workshop on the Use of Wild Fish and/or Other Aquatic Species as Feed in Aquaculture and Its Implications to Food Security and Poverty Alleviation. Kochi, India, 16–18 November 2007. FAO Fisheries Report 867, pp. 209–268.

ICCAT, International Commission for the Conservation of Atlantic Tunas, 2006a Recommendation by ICCAT on Bluefin Tuna farming. Rec. [05–04]. pp. 7–11.

ICCAT, International Commission for the Conservation of Atlantic Tunas, 2006b. Recommendation by ICCAT to establish a multi-annual recovery plan for bluefin tuna in the eastern Atlantic and Mediterranean. Rec. [06–05]. 14 pp.

ICCAT, International Commission for the Conservation of Atlantic Tunas, 2014. SCRC Report for biennial period, 2012–13. 349 pp.

Jelavic, K.M., Stepanowska, K., Grubisic, L., Segvic-Bubic, T., Katavic, I., 2012. Reduced feeding effects to the blood and muscle chemistry of farmed juvenile bluefin tuna in the Adriatic Sea. Aquacultural Res. 43 (2), 317–320.

Katavic, I., Ticina, V., Franicevic, V., 2003a. Bluefin tuna (*Thunnus thynnus* L.) farming on the Croatian coast of the Adriatic Sea: Present stage and future plans. Cah. Options Meditèrr. 60, 101–106.

Katavic, I., Ticina, V., Franicevic, V., 2003b. Rearing of small bluefin tunas (*Thunnus thynnus* L.) in the Adriatic Sea Preliminary study. Cah. Options Meditèrr. 60, 95–99.

Katavic, I., Grubiscic, L., Ticina, V., Jelavic, K.M., Franicevic, M., Skakelja, N., 2009. Growth performances of the bluefin tuna (*Thunnus thynnus*) farmed in the Croatian waters of Eastern Adriatic. SCRS SCRS/2009/120, pp. 1–8.

Kimmoun, A., Novy, E., Auchet, T., Ducrocq, N., Levy, B., 2015. Hemodynamic consequences of severe lactic acidosis in shock states: from bench to bedside. Critical Care 9-19 (1), 175.

Kitagawa, T., Kiruma, S., Nakata, H., Yamada, H., 2004. Diving behavior of immature, feeding Pacific bluefin tuna (*Thunnus thynnus orientalis)* in relation to season and area: the East China Sea and the Kuroshio-Oyashio transition region. Fish. Oceanogr. 13, 161–180.

Libert, E., Cloutier, R., Ortega, A., De la Gándara, F., 2013. Malformaciones postcraneales y esqueletogénesis en larvas y juveniles de atún rojo *Thunnus thynnus* (L, 1758) criados en cautividad. In: Actas del XIV Congreso Nacional de Acuicultura. Gijón, Spain, pp. 260–261. (In Spanish)

Lines, J.A., Spence, J., 2011. Safeguarding the welfare of farmed fish at harvest. Fish Physiol. Biochem.. Available from: http://dx.doi.org/10.1007/s10695-011-9561-5.

Mac Kenzie, B.R., Mosegaard, H., Rosenberg, A.A., 2009. Impending collapse of bluefin tuna in the northeast Atlantic and Mediterranean. Conserv. Lett. 2, 25–34.

Makridis, P., Sarropoulou, E., Papandroulakis, N., Divanach, P., 2011. Microbiota in bonito (*Sarda sarda*) and blue fin tuna (*Thunnus thynnus*) larvae. In: Proceedings of the World Aquaculture 2011. Natal, Brasil, p. 178.

Marino, F., Monaco, S., Salvaggio, Al, Macri, B., 2006. Lipoma in a farmed northern bluefin tuna, *Thunnus thynnus* (L.). J. Fish Dis. 29, 697–699.

Martins, C.I.M., Galhardo, L., Noble, C., Damsgard, B., Spedicato, M.T., Zupa, W., et al., 2012. Behavioural indicators of welfare in farmed fish. Fish Physiol. Biochem. 38, 17–41.

Mateo, A., Soto, F., Villarejo, J.A., Roca-Dorda, J., De la Gándara, F., Garcia-Gomez, A., 2006. Quality analysis of tuna meat using an automated color inspection system. Aquacultural Eng. 35, 1–13.

Mather, F.J., Mason, J.M., Jones, A.C., 1995. Historical Document: Life History and Fisheries of Atlantic Bluefin Tuna. NOAA Technical Memorandum NMFS-SEFSC 370, 165 pp.

Mazurais, D., Coves, D., Papandroulakis, N., Ortega, A., Desbruyeres, E., Huelvan, C., et al., 2014. Gene expression pattern of digestive and antioxidant enzymes during the larval development of reared Atlantic bluefin tuna (ABFT), *Thunnus thynnus* L. Aquaculture Res.1–27, ARE-OA-13-Feb-132.R1.

Messina, C., Santulli, A., 2008. Effect of slaughtering methods on stress and quality of caged Bluefin Tuna (*Thunnus thynnus*). 16 IFOAM, pp. 29–31.

Milatou, N., Megalofonou, P., 2014. Age structure and growth of bluefin tuna (*Thunnus thynnus*, L.) in the capture-based aquaculture in the Mediterranean Sea. Aquaculture 424-425, 35–44.

Miyake, P.E., De la Serna, J.M., Di Natale, A., Farrugia, A., Katavic, I., Miyabe, N., et al., 2003. General review of bluefin tuna farming in the Mediterranean area. Collect. Vol. Sci. Pap. ICCAT 55 (1), 114–124.

Miyake, M.P., Guillotreau, P., Sun, C.H., Ishimura, G., 2010. Recent developments in the tuna industry: stocks, fisheries, management, processing, trade and markets. FAO Fisheries and Aquaculture Technical Paper No. 543. FAO, Rome.

Mladineo, I., 2006. *Hepatoxylon trichiuri* (Cestoda: Trypanorhyncha) plerocercoids in cage-reared northern bluefin tuna, *Thunnus thynnus* (Osteichthyes: Scombridae). Acta Adriatica 47 (1), 79–83.

Mladineo, I., Miletic, I., Bocina, I., 2006. *Photobacterium damselae* subsp. piscicida outbreak in cage-reared Atlantic bluefin tuna (*Thunnus thynnus*). J. Aquat. Anim. Health 18 (1), 51–54.

Mladineo, I., Zilic, J., Cankovic, M., 2008. Health survey of Atlantic Bluefin Tuna, *Thunnus thynnus* (Linnaeus, 1758), reared in Adriatic cages from 2003 to 2006. J. World Aquaculture Soc. 39 (2), 281–289.

Mladineo, I., Segvic, T., Petric, C., 2011. Do captive conditions favor shedding of parasites in the reared Atlantic bluefin tuna (*Thunnus thynnus*)? Parasitol. Int. 60, 25–33.

Morais, S., Pratoomyot, J., Taggart, J.B., Bron, J.E., Guy, D.R., Bell, J.G., et al., 2011. Expression of fatty acyl desaturase and elongase genes, and evolution of DHA:EPA ratio during development of unfed larvae of Atlantic bluefin tuna (*Thunnus thynnus* L.). Aquaculture 313, 129–139.

Muñoz, P., Meseguer, J., Peñalver, J., Martinez-Carrasco, C., Ruiz de Ybañez, R., Del RIos, L., et al., 2007. An epidemiological study of *Cardicola* spp. infection in Mediterranean red tuna. In: Proceedings of 13th International Conference of fish and shellfish diseases. Grado, Italy, 17–21 September 2007. 304 pp.

Mylonas, C.C., De la Gándara, F., Corriero, A., Belmonte, A., 2010. Atlantic bluefin tuna (*Thunnus Thynnus*) farming and fattening in the Mediterranean Sea. Rev. Fish. Sci. 18 (3), 266–280.

Newlands, N.K., Lutcavage, M.E., Pitcher, T.J., 2004. Analysis of foraging movements of Atlantic bluefin tuna (*Thunnus thynnus*) individuals switch between two modes of search behavior. Popul. Ecol. 46, 39–53.

Norita, T., 2003. Feeding of bluefin tuna: experiences in Japan and Spain. Cah. Options Meditèrr. 60, 153–156.

Nowak, B., Mladineo, I., Aiken, H., Bott, N., Hayward, C., 2006. Results of health surveys of two species of farmed tuna: southern bluefin tuna (*Thunnus maccoyii*) in Australia and northern bluefin tuna (*Thunnus thynnus*) in the Mediterranean. Bull. Eur. Ass. Fish Pathol. 26 (1), 38–42.

Nucci, M.E., Costa, C., Scardi, M., Catudella, S., 2007. Use of a digital video recording system for the study of the Mediterranean bluefin tuna (*Thunnus thynnus*, Linneo 1758) behavior under culturing conditions. In: Proceedings of the European Aquaculture Society Meeting, Istanbul, Turkey. 393 pp.

Nucci, M.E., Costa, C., Scardi, M., Catudella, S., 2010. Preliminary observations on bluefin tuna (*Thunnus thynnus*, Linnaeus 1758) behavior in captivity. J. Appl. Ichthyol. 26, 95–98.

Oray, I.K., Karakulak, F.S., 2003. Possibilities for the domestication of Bluefin tuna in the Eastern Mediterranean Sea. Cah. Options Meditèrr. 60, 157–158.

Ortega, A., Mourente, G., 2010. Comparison of the lipid profiles from wild caught eggs and unfed larvae of two scombroid fish: northern bluefin tuna (*Thunnus thynnus* L., 1758) and Atlantic bonito (*Sarda sarda* Bloch, 1793). Fish Physiol. Biochem. 36 (3), 461–471.

Ortega, A., Viguri, J., Belmonte, A., De la Gándara, F., 2009. Resultados preliminares de los 30 primeros días de cultivo larvario de atún rojo *Thunnus thynnus*, en el IEO. Libro de resúmenes del XII Congreso Nacional de Acuicultura. Madrid Noviembre de 2009, pp. 584–585 (In Spanish).

Ortega, A., Seoka, M., Belmonte, A., Prieto, J.R., Viguri, J., De la Gándara, F., 2011. Cultivo larvario de atún rojo (*Thunnus thynnus*) en el Centro Oceanográfico de Murcia. In: Actas del XIII Congreso Nacional de Acuicultura, Barcelona, Spain. <http://hdl.handle.net/10508/449> (In Spanish).

Ortega, A., J.R., Viguri, J.R. Prieto, A. Belmonte, D. Martínez, M. Velázquez, et al.. 2014. First results on ongrowing of hatchery reared Atlantic bluefin tuna, Thunnus thynnus, kept in sea cages. In: Proceedings of Aquaculture Europe 14, San Sebastián, Spain, 14–17 October 2014. pp. 931–932. <http://hdl.handle.net/10508/2757>.

Ottolenghi, F., 2008. Capture-based aquaculture of bluefin tuna. In: Lovatelli, A. (Ed.), Capture-Based Aquaculture. Global Overview. FAO Fisheries Technical Paper. FAO, Rome, pp. 169–182.

Ottolenghi, F., Silvestri, C., Giordano, P., Lovatelli, A., New, M.B., 2004. Capture-based aquaculture. The Fattening of Eels, Groupers, Tunas and Yellowtails. FAO, Rome.

Papandroulakis, N., Ortega, A., Coves, D., Vasallo-Agius, R., Tandler, A., Stefanakis, S., et al., 2010. First results of the Atlantic bluefin tuna (*Thunnus thynnus*) larval rearing in Europe in the frame of the SELF-DOTT project. In: Proceedings of the European Aquaculture Society, Porto, Portugal. pp. 968–969.

Passantino, L., Santamaria, N., Zupa, R., Pousis, C., Garofalo, R., Cianciotta, A., et al., 2013. Liver melanomacrophage centres as indicators of Atlantic bluefin tuna, *Thunnus thynnus* L. well-being. J. Fish Dis., 1–10. Available from: http://dx.doi.org/10.1111/jfd.12102.

Peñalver, J., Maria-Dolores, E., Tafalla, C., Diaz, R., Bermudez, L., Gomez, O., 2007. Valoración del riesgo de transmisión de enfermedades víricas a través de carnada usada en la alimentación del atún rojo (*Thunnus thynnus*). In: Actas del XI Congreso Nacional de Acuicultura. Vigo, Spain. pp. 1097–1100 (In Spanish).

Percin, F., Konyalioglu, S., 2008. Serum biochemical profiles of captive and wild northern bluefin tuna (*Thunnus thynnus* L. 1758) in the Eastern Mediterranean. Aquaculture Res. 39, 945–953.

Peric, Z., 2003a. Malta national program towards farming of the Bluefin tuna present and future. Cah. Options Medit̀err. 60, 167.

Peric, Z., 2003b. Morphological and histological changes of the parenchimatous organs of Bluefin tuna, *Thynnus thynnus* (Linnaeus, 1758). Cah. Options Medit̀err. 60, 165–166.

Pokorny, J., Smith, V.C., Lutze, M., 1987. Aging of the human lens. Appl. Opt. 26, 1437–1440.

Popovic, N.T., Srebocan, E., Coz-Rakovac, R., Hacmanjek, C., Strunjak-Perovic, I., Jadan, M., 2008. Blood biochemistry of captive Atlantic bluefin tuna *Thunnus thynnus* farmed in the Adriatic Sea. J. Appl. Ichthyol. 24, 614–616.

Popovic, N.T., Kozacinski, L., Strunjak-Perovic, I., Coz-Rakovac, R., Jadan, M., Cvrtila-Fleck, Z., et al., 2012. Fatty acid and proximate composition of bluefin tuna (*Thunnus thynnus*) muscle with regard to plasma lipids. Aquaculture Res. 43 (5), 722–729.

Puig, V., Espinosa, V., Soliveres, E., Ortega, A., Belmonte, A., De la Gándara, F., 2012. Biomass estimation of Bluefin tuna in sea cages by the combined use of acoustic and optical techniques. Collect. Vol. Sci. Pap. ICCAT 68 (1), 284–290.

Ramfos, A. Tzoumas, A., Alaveras, P., Katselis, G., 2011. Pilot study of a dual camera system for the determination of size of Atlantic bluefin tuna (*Thunnus thynnus*, L. 1758) during caging. SCRS 173, pp. 1–6.

Ravier, C., Fromentin, J.M., 2001. Long-term fluctuations in the eastern Atlantic and Mediterranean bluefin tuna population. ICES J. Mar. Sci. 58, 1299–1317.

Roberts, R.J., Agius, C., 2008. Pan-steatitis in farmed northern bluefin tuna, *Thunnus thynnus* (L.), in the eastern Adriatic. J. Fish Dis. 31, 83–88.

Rodriguez Llanos, J., 2012. Parásitos del atún rojo del Atlántico (*Thunnus thynnus*) en granjas de engorde del Mediterráneo Español. Patologías y uso como biomarcadores. Thesis. Universidad Politécnica de Valencia (In Spanish).

Roncarati, A., Felici, Al., Dees, A., Zalno, E., Forlini, L., Melotti, P., 2008. Quality traits of meat in wild and cage-fattened bluefin tuna *Thunnus thynnus* comparison between evaluation methods applied in Europe and in Japan. In: Proceedings of the World Aquaculture Society, Busan, Korea, 627 pp.

Sainz-Pardo, S., Espinosa, V., De la Gándara, F., 2010. Dorsal TS measurement analysis and tuna behavior in sea cages. In: European Underwater Acoustics Conference (ECUA 2010). Istambul, Turkey, pp. 49–53.

Sanchez Serrano, S., Caceres Martinez, J., 2003. Parásitos del atún aleta azul del norte *Thunnus thynnus orientalis* y su importancia potencial en maricultura. Boletín del Programa Nacional de Sanidad Acuícola y la Red de Diagnóstico 2 (22), 2–4 (In Spanish).

Sanchis, P.J., 2012. Fatty acid assimilation and LC-PUFA biosynthesis gene expression during different life stages of Atlantic Northern Bluefin Tuna (*Thunnus thynnus*). Nutrition Department, Institute of Aquaculture, University of Stirling, 44 pp.

Santulli, A., 2005. Environmental impact concerns related to bluefin tuna farming in the Mediterranean Sea. In: Proceedings of the World Aquaculture Society, Bali, Indonesia, 552 pp.

Sanz-Lazaro, C., Marin, A., 2008. Assessment of finfish aquaculture impact on the benthic communities in the Mediterranean Sea. Dyn. Biochem. Process Biotechnol. Mol. Biol. 2 (Special Issue 1), 21–32.

Sawada, Y., Okada, T., Miyashita, S., Murata, O., Kumai, H., 2005. Completion of the Pacific bluefin tuna *Thunnus orientalis* (Temminck et Schlegel) life cycle. Aquaculture Res. 36, 413–421.

Soto, F., Villarejo, J.A., Mateo, A., Roca-Dorda, J., De la Gándara, F., Garcia-Gomez, A., 2006. Preliminary experiences in the development of bluefin tuna *Thunnus Thynnus* (L., 1758) electroslaughtering techniques in rearing cages. Aquacultural Eng. 34 (2), 83–91.

Spanish Ministry of Agriculture and Fisheries, 2015. <http://www.magrama.gob.es/app/jacumar/datos_produccion/Documentos/Especies/PV_ATUN_ROJO.pdf> (In Spanish).

Ticina, V., Grubisic, L., Katavic, I., Miletic, I., Jeftimijades, I., Ticina, V.E., et al., 2004. Some difficulties in sampling and tagging live bluefin tuna in the growth-out floating cages. Collect. Vol. Sci. Pap. ICCAT 56 (3), 1218–1221.

Ticina, V., Grubisic, L., Katavic, I., Franicevic, V., Ticina, V.E., Ocic, M., 2005. Conversion of farmed bluefin tuna product weight to live weight of the fish. Collect. Vol. Sci. Pap. ICCAT 58 (4), 1271–1274.

Toksen, E., Nemli, E., Degirmenci, U., Karacalar, U., 2012. Parasitic diseases and their controls in sustainable development of aquqculture of bluefin tuna (*Thunnus thynnus*). In: Third International Symposium on Sustainable Development. Sarajevo, Croacia, pp. 41–50.

Trombley, S., Schmitz, M., 2013. Leptin in fish: possible role in sexual maturation in male Atlantic salmon. Fish Physiol. Biochem. 39 (1), 103–106.

Tudela, S., 2001. Tuna farming in the Mediterranean: some elements for a WWW positioning. WWW Rep., 10.

Tudela, S., Garcia, R., 2004. Tuna farming in the Mediterranean: the bluefin tuna stock at stake. WWW Rep., 35.

Tzoumas, A., Ramfos, A., De Metrio, G., Corriero, A., Spinos, E., Vavassis, C., et al., 2010. Weight growth of Atlantic bluefin tuna (*Thunnus thynnus*, l. 1758) as a result of a 6-7 months fattening process in the central Mediterranean. Collect. Vol. Sci. Pap. ICCAT 65 (3), 787–800.

Ugolini, R., Poli, B.M., Parisi, G., Lupi, P., Mecatti, M., Zampacavallo, G., et al., 2005. Harvesting/slaughtering stress effect on meat quality of reared bluefin tuna *Thunnus thynnus*. In: Proceedings of the World Aquaculture Society, Bali, Indonesia, 663 pp.

Vardic Smrzlic, I., Valic, D., Kapetanovic, D., Kurtovic, B., Teskeredzic, E., 2012. Molecular characterisation of Anisakidae larvae from fish in Adriatic Sea. Parasitol. Res., 1–7. Available from: http://dx.doi.org/10.1007/s00436-012-3094-4.

Vassallo-Agius, R., Watanabe, T., Satoh, S., Kiron, V., Imaizumi, H., Yamazaki, T., et al., 2001. Supplementation of paprika as a carotenoid source in soft-dry pellets for broodstock yellowtail *Seriola quinqueradiata* (Temminck & Schlegel). Aquaculture Res. 32S, 263.

Vassallo-Agius, R., Mylonas, C.C., Sharman, A., Negas, J., Corriero, A., Bridges, C.R., et al., 2011. Effect of broodstock diets on growth and chemical components of gonads of Atlantic bluefin tuna *Thunnus thynnus* (L.) reared in cages. In: Proceedings of the European Aquaculture Society, Rhodes, Greece, pp. 1143–1144.

Vezzuli, L., Moreno, M., Marin, V., Pezzati, E., Bartoli, M., Fabiano, M., 2008. Organic waste impact of capture-based Atlantic bluefin tuna aquaculture at an exposed site in the Mediterranean Sea. Estuarine Coast. Shelf Sci. 78, 369–384.

Vita, R., Marin, A., Jimenez-Brinquis, B., Cesar, A., Marin-Guirao, L., Borredat, M., 2004. Aquaculture of bluefin tuna in the mediterranean: evaluation of organic particulate wastes. Aquaculture Res. 35 (14), 1384–1387.

Vizzini, S., Mazzola, A., 2012. Tracking multiple pathways of waste from a northern bluefin tuna farm in a marine-coastal area. Mar. Environ. Res. 77, 103–111.

Vizzini, S., Savona, B., Micciche, L., Savona, A., Mazzola. A., 2007. Nitrogen stable isotopes as tracers of biodeposition from a tuna farm in the western Mediterranean. In: Proceedings of the European Aquaculture Society, Istanbul, Turkey, pp. 591–592.

Ward, T.M., Hoedt, F., McLeay, L., Dimmlich, W.F., Kinloch, M., Jackson, G., et al., 2001. Effects of the 1995 and 1998 mass mortality events on the spawning biomass of sardine, *Sardinops sagax*, in South Australian waters. ICES J. Mar. Sci. 58, 865–875.

Werner, J.S., 1996. Visual problems of the retina during ageing: Compensation mechanisms and color constancy across the life span. Prog. Retin. Eye Res. 15 (2), 621–645.

Wurts, W.A., Stickney, R.R., 1993. Growth rates of juvenile red drum *Sciaenops ocellatus* reared on commercial salmon feed in fresh and salt water. J. World Aquaculture Soc. 24 (3), 422–424.

Yanowski, E., Mylonas, C.C., Corriero, A., Bridges, C.R., Vassallo-Agius, R., De la Gándara, F., et al., 2011a. *Atlantic Bluefin Tuna (*Thunnus thynnus*) Leptin-Cloning and Expression Patterns During the Reproductive Season*. European Aquaculture Society, Rhodes, Greece.

Yanowski, E., Mylonas, C.C., Corriero, A., Bridges, C.R., Vassallo-Agius, R., De la Gándara, F., et al., 2011b. Molecular characterization and expression patterns of Atlantic bluefin tuna (*Tunnus thynnus*) leptin during the reproductive season. Indian J. Sci. Technol. 4 ((S8) 9th ISRPF Issue), 104–105.

Yerlikaya, P., Gokoglu, N., Topuz, O.K., Gokoglu, M., 2009. Changes in the proximate composition of bluefin tuna (*Thunnus thynnus*) reared in the cages located on the Gulf of Antalya (Turkey's Western Mediterranean coast) during the fattening period. Aquaculture Res. 40, 1731–1734.

Yufera, M., Ortiz-Delgado, J.B., Hoffman, T., Siguero, I., Urup, B., Sarasquete, C., 2014. Organogenesis of digestive system, visual system and other structures in Atlantic bluefin tuna (*Thunnus thynnus*) larvae reared with copepods in mesocosm system. Aquaculture, 1–34. Available from: http://dx.doi.org/10.1016/j.aquaculture.2014.01.031.

Chapter 7

Reproduction, Broodstock Management, and Spawning in Captive Atlantic Bluefin Tuna

Y. Zohar[1], C.C. Mylonas[2], H. Rosenfeld[3], Fernando de la Gándara[4] and Aldo Corriero[5]

[1]*Department of Marine Biotechnology, Institute of Marine and Environmental Technology, University of Maryland Baltimore County, Baltimore, MD, USA,* [2]*Institute of Marine Biology, Biotechnology and Aquaculture, Hellenic Center for Marine Research, Aqualabs, Iraklion, Crete, Greece,* [3]*National Center for Mariculture, Israel Oceanographic and Limonological Research Institute, Eilat, Israel,* [4]*Centro Oceanografico de Murcia, Spanish Institute of Oceanography, Murcia, Spain,* [5]*Department of Emergency and Organ Transplantation, Section of Veterinary Medicine and Animal Production, University of Bari Aldo Moro, Bari, Italy*

7.1 INTRODUCTION

The Atlantic (or Northern) bluefin tuna (*Thunnus thynnus*, ABFT) is one of the tuna species with the greatest commercial interest for fisheries (Ottolenghi, 2008). It is one of three species of bluefin tuna (with Pacific bluefin tuna, *Thunnus orientalis*, and southern bluefin tuna, *Thunnus maccoyii*). The ABFT can be found in the Western and Eastern Atlantic Ocean, as well as throughout the Mediterranean Sea (ICCAT, 2008). The ABFT is one of the largest fishes, attaining a body size of up to 700 kg. It can reach swimming speeds of 90 km/h and it is one of the most wide-ranging fish, with long trans-oceanic migrations (Cort and Liorzou, 1991; Safina, 1995). An event documented from ancient times by the Greek philosopher Aristotle (Aristotle, 1994), the ABFT migrates seasonally over long distances between the temperate waters of the Atlantic Ocean where it feeds and the warmer subtropical waters of the Gulf of Mexico and the Mediterranean Sea where it spawns.

ABFT farming and fattening in the Mediterranean Sea (Mylonas et al., 2010a) started in the mid-1990s in Croatia and spread quickly throughout the region, fueled by the insatiable demand of the Japanese sushi/sashimi market and the very high prices offered for fresh or frozen fish at the Tsukiji Market

D.D. Benetti, G.J. Partridge, A. Buentello (Eds): Advances in Tuna Aquaculture.
DOI: http://dx.doi.org/10.1016/B978-0-12-411459-3.00006-0

in Tokyo (aka the Tokyo Fish Market). The International Commission for the Conservation of Atlantic Tunas (ICCAT), the international body responsible for the management and allocation of fishing quotas (Total Allowable Catches, TACs) for the ABFT (Block et al., 2005; ICCAT, 2005; Rooker et al., 2007; Schaefer, 2001), reduced significantly their TACs in the last decade, in an effort to prevent the collapse of the wild stocks due to heavy overfishing. In parallel, studies were undertaken to examine the potential to produce ABFT entirely in captivity so that aquaculture can eventually become the main source for the seafood market and gradually replace the overexploited ABFT fishery, leading to the preservation of this important marine species.

As for any fish species, one of the prerequisites for domestication and the establishment of a sustainable aquaculture industry is the capacity to control reproductive processes in captivity, and to acquire high-quality gametes (i.e., eggs and sperm) and seeds (i.e., fertilized eggs and larvae) for grow-out to the marketable product (Mylonas et al., 2010b). This allows (i) the reliable, predictable, and consistent production of seeds to be provided to hatchery and nursery operations; (ii) the manipulation of the reproductive season through photo-thermal manipulations, in order to extend the period of seed production; and (iii) the improvement of desirable traits through selective breeding programs.

Achieving the above spawning-related goals for ABFT necessitates a basic understanding of its reproductive biology and endocrinology in the wild and in captivity, which has been studied during the last decade by several coordinated groups supported primarily by European Union multidisciplinary and multi-institutional efforts (e.g., Corriero et al., 2007; Mylonas et al., 2007; Rosenfeld et al., 2012). This chapter reviews the current knowledge on the reproductive biology and broodstock management of ABFT, and describes the current approaches for the production of high-quality fertilized eggs.

7.2 GAMETOGENESIS AND ENDOCRINE CYCLES

7.2.1 Gonad Structure

ABFT gonads are paired, elongated organs located in the abdominal cavity and suspended by a mesogonad (mesovarium or mesotestis) (Figure 7.1). The testes consist of two structurally different regions, an outer proliferative region and an inner spermatozoa storage zone. In the proliferative region, germ cells associated with somatic cells (Sertoli cells) develop in cysts (spermatocysts), which constitute the germinal epithelium lining the seminiferous lobules from the periphery of the testis, beneath the connective membrane (tunica albuginea), to a well-developed inner system of ducts (efferent ducts). In the lumen of the efferent ducts, spermatozoa accumulate and move toward the main sperm duct (or vas deferens).

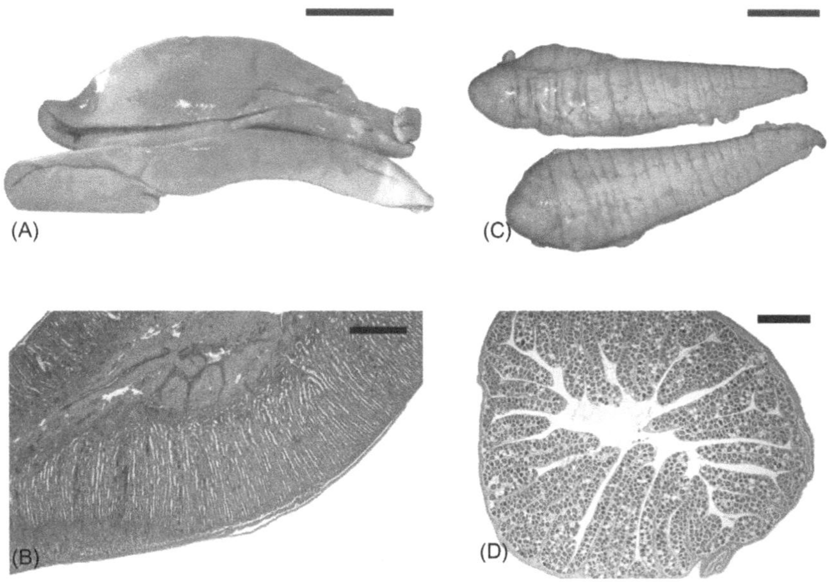

FIGURE 7.1 (A) Mature testes from an adult Atlantic bluefin tuna. (B) Histological section of the testis of a juvenile Atlantic bluefin tuna showing seminiferous lobules converging from the testis periphery to the main sperm duct in the center. Haematoxylin-eosin staining. (C) Ripening ovaries from an adult Atlantic bluefin tuna. (D) Histological section from the ovary of a juvenile Atlantic bluefin tuna. Ovigerous lamellae converging toward the lumen contain only oocytes at early developmental stages. Haematoxylin-eosin staining. Magnification bars: in (A) and (C) = 10 cm; in (B) and (D) = 500 μm.

Ovaries are hollow organs with a round cross section that join caudally in a common oviduct, which opens to the exterior in the urogenital pore. Ovaries of adult individuals consist of a thick muscle wall and ovigerous lamellae, containing oogonia and oocytes at different stages of development, embedded in a mass of connective tissue. A single oocyte, surrounded by a single layer of granulosa cells and a double layer of thecal cells (theca externa and theca interna), constitutes an ovarian follicle.

7.2.2 Spermatogenesis

Male germ cells develop within spermatocysts and each spermatocyst is comprised of a clone (group of isogenic cells) derived from a single spermatogonium that is associated with somatic (Sertoli) cells (Figure 7.2). As in other tuna species, the ABFT testis is characterized by the occurrence of spermatogonia along the majority of the seminiferous lobules and it belongs to the unrestricted spermatogonial type according to Grier et al. (1980) classification or the lobular type according to Billard's (1986) classification.

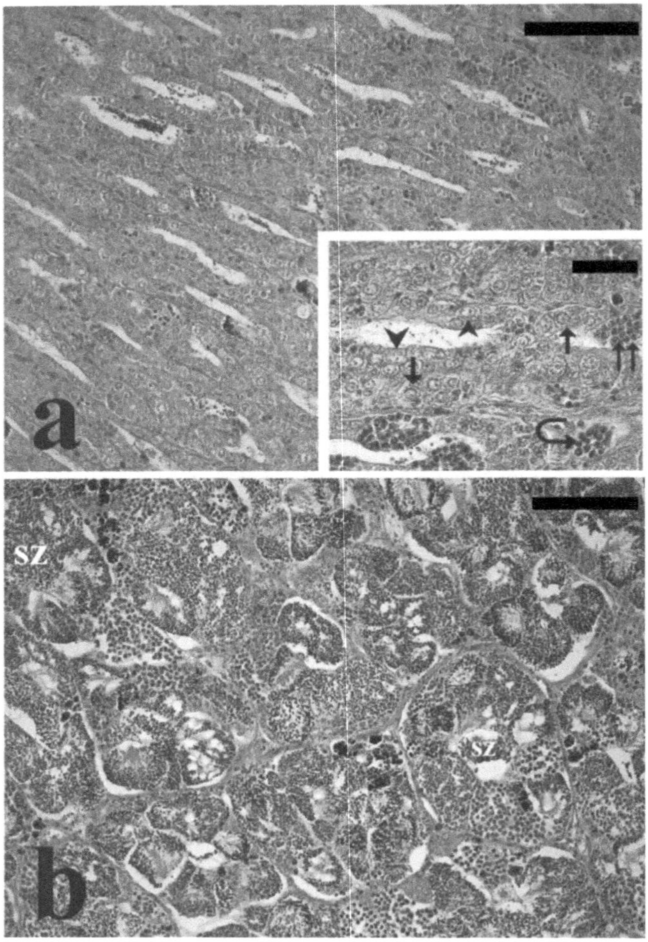

FIGURE 7.2 Micrograph of the testes of adult Atlantic bluefin tuna in early (A) and advanced (B) spermatogenesis. Haematoxylin-eosin staining. Magnification bars = 100 μm; bar in inset of (A) = 20 μm. Arrowhead: undifferentiated spermatogonium; arrow: type A spermatogonial cyst; curved arrow: type B spermatogonial cyst; double arrow: spermatocyte cyst; sz: spermatozoa released in the lumen of seminiferous lobules. *Reproduced from Zupa et al. (2013).*

Abascal et al. (2004) provided an ultrastructural description of ABFT male germ cells at different developmental stages. Spermatogonia A are ovoid cells (~10 μm in diameter), with a nucleus showing diffuse chromatin and a central nucleolus; their cytoplasm contains free ribosomes, mitochondria, endoplasmic reticulum cisternae, and "nuages" ("clouds" in French) of perinuclear electron-dense material that could possibly represent mRNA molecules. Spermatogonia B (~10 μm in diameter) are grouped in small clusters and show a nucleus containing patchy chromatin. Primary spermatocytes (~5 μm in diameter) are clustered cells with a heterochromatic nucleus and cytoplasm containing polysomes, mitochondria, and the diplosome. Secondary spermatocytes (3–4 μm in diameter), presumably short-lived germ cells rarely found in histological samples, have a nucleus with diffuse chromatin-forming electron-dense patches. In early spermatids (2–3 μm in diameter), the nucleus shows a dense chromatin and some electron-lucent areas. The chromatin becomes more homogeneous in mid-spermatids and condenses in a gross granular pattern in late spermatids (2 μm in diameter). Spermiogenesis involves flagellum elongation, cytoplasm reduction, and mitochondria rearrangement around the proximal axonema region. Rotation of the nucleus does not occur during spermiogenesis (characteristic of type-II teleost sperm).

7.2.3 Oogenesis

Female ABFT germ cells have been described in terms of morphology (Corriero et al., 2003), histochemical characteristics (Sarasquete et al., 2002), and ultrastructure (Abascal and Medina, 2005). The ABFT ovary shows an asynchronous pattern of oocyte development (i.e., presence of all oocyte stages in the same specimen during the spawning season) (Figure 7.3), which leads to the release of mature oocytes in multiple batches. The number of oocytes released per spawning season (absolute fecundity) reflects the number of oocytes released per spawning event (batch fecundity) and the number of spawning events per spawning season, which may vary according to the size of the fish, and environmental or individual conditions. Studies of wild ABFT (Medina et al., 2002) estimated that batch fecundity is around 93,000 eggs per kg body weight, which means that a 100 kg female can produce around 9 million eggs in a single spawn. As far as how many times (i.e., days, batches) each female can spawn, either from the wild fishery or from captive-reared broodstock, is not known at this time. Based on the presence of post-ovulatory follicles in spawning ABFT in the wild, it has been estimated that the spawning frequency is 1.2 days (Medina et al., 2002). This means that a large proportion of the population spawns every day, but it is not known for how many days this can continue.

Although different classification criteria have been used, oogonia and the following five oocyte developmental stages have been described for ABFT: early oocytes; primary oocyte growth; lipid stage; vitellogenesis; post-vitellogenesis; and oocyte maturation. Oogonia (8–15 μm in diameter) are found in small clusters at the periphery of the ovigerous lamellae. Their cytoplasm is weakly basophilic and possesses many ribosomes, round mitochondria, Golgi complexes, and a poorly developed endoplasmic reticulum. The nuclear envelope shows numerous pores and cloudy material (nuages) can be observed around it. Early oocytes (15–20 μm in diameter) are

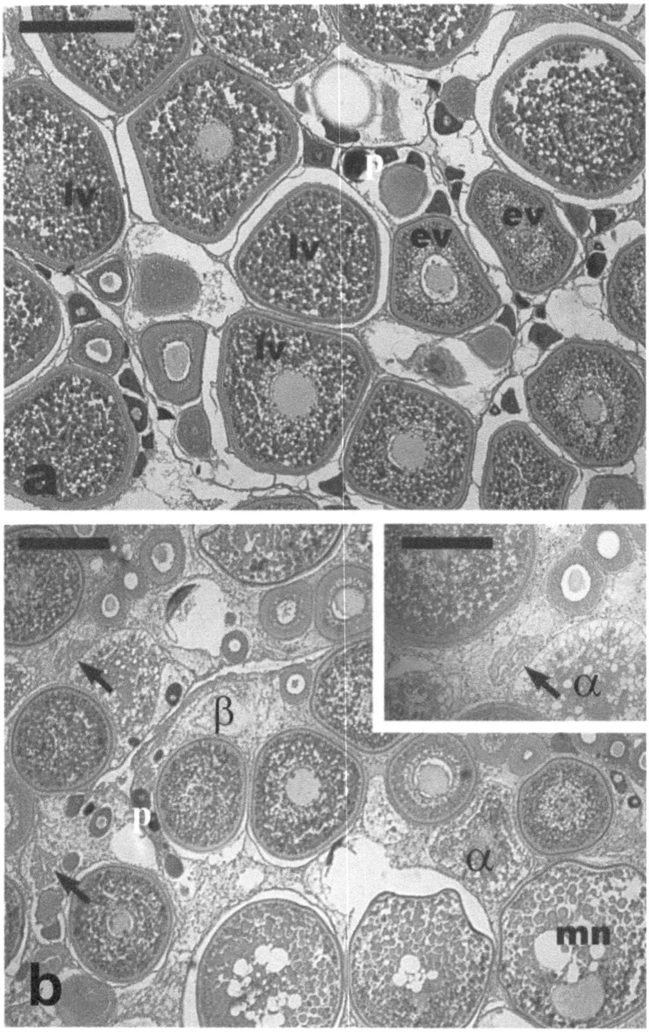

FIGURE 7.3 Micrographs of the gonads from two adult Atlantic bluefin tuna specimens. (A) Ovary from an active non-spawning individual showing both primary and vitellogenic oocytes. (B) Ovary from a spawning specimen showing primary oocytes, migratory nucleus stage oocytes, and postovulatory follicles. Haematoxylin-eosin staining. Magnification bars = 500 μm; bar in inset of (B) = 300 μm. Arrow, postovulatory follicle; α, vitellogenic follicle in early stage of atresia (α-atretic follicle); β, vitellogenic follicle in an advanced stage of atresia (β-atretic follicle); ev, early vitellogenic oocyte; lv, late vitellogenic oocyte; p, perinucleolar stage oocyte; mn, migratory nucleus stage oocyte. *Modified from Heinisch et al. (2008).*

ultrastructurally similar to oogonia but they are found in association with pre-follicle cells that send thin processes between adjacent oocytes and progressively surround them. Primary oocyte growth, also known as the pre-vitellogenesis stage (20–120 μm in diameter), includes two stages of oocyte development: the chromatin-nucleolus and peri-nucleolus stages. Chromatin-nucleolus oocytes (20–40 μm in diameter) are rarely observed in ovarian sections. This stage is characterized by a basophilic ooplasm (due to ribosome abundance) showing the appearance of Balbiani's vitellin body and a large nucleus with chromatin strands and a large nucleolus. Oocytes at the peri-nucleolus stage (40–110 μm in diameter) have a basophilic ooplasm and a large euchromatic nucleus containing numerous nucleoli adjoining the nuclear envelope. Mitochondria, Golgi complexes, and nuages are visible in close proximity to the nucleus. Oocytes at this stage are surrounded by a continuous layer of flattened granulosa cells. Membrane processes developed from the oocyte extend toward granulosa cells and vice versa, and the formation of the zona radiata (vitelline envelope) starts as a two-layered structure: a moderately electron-dense outer layer and a more electron-dense inner layer. The two thecal layers (theca externa and theca interna) are identifiable at this stage of follicle development. Lipid stage oocytes (110–220 μm in diameter) exhibit a weak ooplasm basophily and are characterized by an increasing amount of small lipid droplets. Ribosomes, mitochondria, and dictyosomes are found in the ooplasm. Finely granular chromatin is present in the nucleus and the nuclear envelope contains numerous pores. The zona radiata increases in thickness and becomes visible by light microscopy as Pas + material. Vitellogenic oocytes (up to 500 μm in diameter) are characterized by vitellogenin uptake and by vitellogenin-derived yolk protein deposited in the ooplasm in the form of yolk granules (platelets). The use of anti-vitellogenin antibodies in immunohistochemical staining (Susca et al., 2001) allows the detection of small amounts of yolk proteins in oocytes with a minimum diameter of 220 μm. Early vitellogenic oocytes are characterized by the presence of spherical acidophilic granules and Pas + cortical alveoli in the peripheral ooplasm. The zona radiata progressively increases in thickness (up to 12 μm in late vitellogenic oocytes) and the granulosa cells become cubic. In oocytes at advanced stage of vitellogenesis, yolk globules increase both in size and number and progressively fill the entire ooplasm. Oocyte microvilli pass through the zona radiata pore canals and extend into the intercellular spaces of granulosa cells. At this stage of vitellogenesis, transmission electron microscopy data has demonstrated an intense oocyte uptake of exogenous substances. The ooplasm is rich in clathrin-coated vesicles (indicative of receptor-mediated endocytosis) containing electron-dense material, which appears to bud from membrane invaginations. Endocytotic vesicles lose their clathrin coat and fuse with each other to form yolk platelets, which increase progressively in size. Large lipid droplets are interspersed among yolk platelets. After the completion of vitellogenesis, oocytes enter post-vitellogenesis or oocyte maturation, which is characterized by the resumption of meiosis, including migration of the nucleus (also referred to at this stage as the Germinal Vesicle, GV) toward the animal pole followed by the breakdown (BD) of the nuclear envelope. Lipid droplets fuse with each other (coalescence), eventually forming one large lipid globule that occupies the center of the oocyte. Vitellogenin-derived yolk proteins that are accumulated within yolk granules undergo a proteolytic cleavage (Pousis et al., 2011), which increases the concentration of free amino acids and small peptides in the oocyte, thereby providing an osmotic mechanism for water influx into the oocyte (oocyte hydration) and resulting in both the formation of a fluid yolk mass and the rapid increase of oocyte size (diameter of ABFT mature oocytes ≈ 1 mm). After ovulation, follicular cells constitute convoluted post-ovulatory follicles, characterized by hypertrophic granulosa cells that delimit an irregular lumen. Post-ovulatory follicles are short-lived structures and degenerate rapidly, becoming indistinguishable from atretic follicles and then from connective stroma. Atretic follicles are normally found in ovaries in vitellogenic and post-vitellogenic development. Follicles in the early stage of atresia (α-atretic stage) are easily distinguishable because of their irregular shape and fragmentation of the zona radiata. The nuclear envelope of an atretic oocyte disintegrates and yolk granules lose their structural integrity and are progressively phagocytized by granulosa cells, which invade the oocyte. After the complete degradation of the oocyte, atretic follicles (now called β-atretic follicles) consist of only disorganized granulosa and thecal cells showing pyknotic nuclei.

7.2.4 Gonadal Cycle and Sexual Maturity

The description of naturally occurring gametogenesis, gonadal activity, and reproductive cycles in ABFT, as described in this Section 7.2, is strictly dependent on the availability of wild fish, which is generally limited to a few months per year due to both fish accessibility and severe restrictions imposed by fishery regulations. ABFT has been managed by the ICCAT as two populations, the eastern and western populations, with known breeding areas in the Mediterranean Sea and the Gulf of Mexico, respectively. Most of the available information on the temporal pattern of ABFT gonadal maturation refers to the eastern ABFT stock and, in particular, comes from individuals captured in the Mediterranean Sea. In this basin, ABFT have been found to be reproductively inactive from August to March, when only unyolked oocytes are present in the ovaries, and mainly spermatogonia and meiotic cells have been found in the seminiferous epithelium. From mid-May to mid-June, ovaries are in full vitellogenesis and testes are in advanced spermatogenesis, showing all spermatogenetic stages. The reproductive period in the Mediterranean Sea, as evident by the presence of individuals with ovaries containing oocytes at final maturation stage and/or postovulatory follicles, extends from May through July, coinciding with a rise (>23 °C) in sea surface temperature. According to Heinisch et al. (2008), ABFT shows a characteristic spatial-temporal pattern of gonadal maturation across the Mediterranean Sea with an initial peak of spawning activity during mid-May at eastern locations (Levantine Sea), and during June and July at central (Malta) and western (Balearic Islands) locations, respectively. Data on gonad maturation of the western ABFT stock are limited and indicate that spawning occurs in April–May in the Gulf of Mexico (Baglin, 1982), although the presence of additional, unknown, spawning grounds in the western Atlantic has not been excluded (Lutcavage et al., 1999; Goldstein et al., 2007; Cermeno et al., 2015).

Based on the presence of yolked oocytes as a sign of activation of the reproductive axis and providing substantial confirmation of historical data which is based on macroscopic observations of the gonads, Corriero et al. (2005) reported that 50% of ABFT in the Mediterranean Sea reach sexual maturity at 104 cm Straight Fork Length (SFL), corresponding to an age between 3 and 4 years, and 100% at 130 cm SFL (5 years of age). In the western Atlantic, histological examination of ovaries from females points to delayed maturation schedules, and individuals have been reported to reach sexual maturity between age 7 and 8 (Baglin, 1982; Goldstein et al., 2007). However, by means of a new molecular approach that uses pituitary ratios of follicle stimulating hormone to luteinizing hormone as markers of sexual maturity, Heinisch et al. (2014) demonstrated that western ABFT mature at a considerably smaller size, similar to that of the eastern Atlantic population.

7.3 CAPTIVITY EFFECTS ON GAMETOGENESIS AND ENDOCRINOLOGY

7.3.1 Adult Fish

Studies focused on assessing the reproductive potential of adult ABFT, captured at the Balearic Islands and held in floating cages on the coast of Murcia, demonstrated the onset and progression of gametogenesis and the hormones that control it in captivity (De la Gandara et al., 2009, 2010). Endocrine parameters including the hypo-physiotropic gonadotropin-releasing hormone (GnRH1) and the maturational gonadotropin luteinizing hormone (LH) peak in June, during which the highest gonado-somatic index (GSI) values were recorded (Figure 7.4). Furthermore, the levels of both LH and GnRH1 in captive fish did not vary significantly ($P > 0.05$) from those measured in wild ABFT that were collected during June along the Maltese coastline (Figure 7.5B,C). Altogether, these findings indicate that the reproductive endocrine system in captive ABFT is functional and undergoing physiological adjustments in anticipation of the natural spawning season in this region.

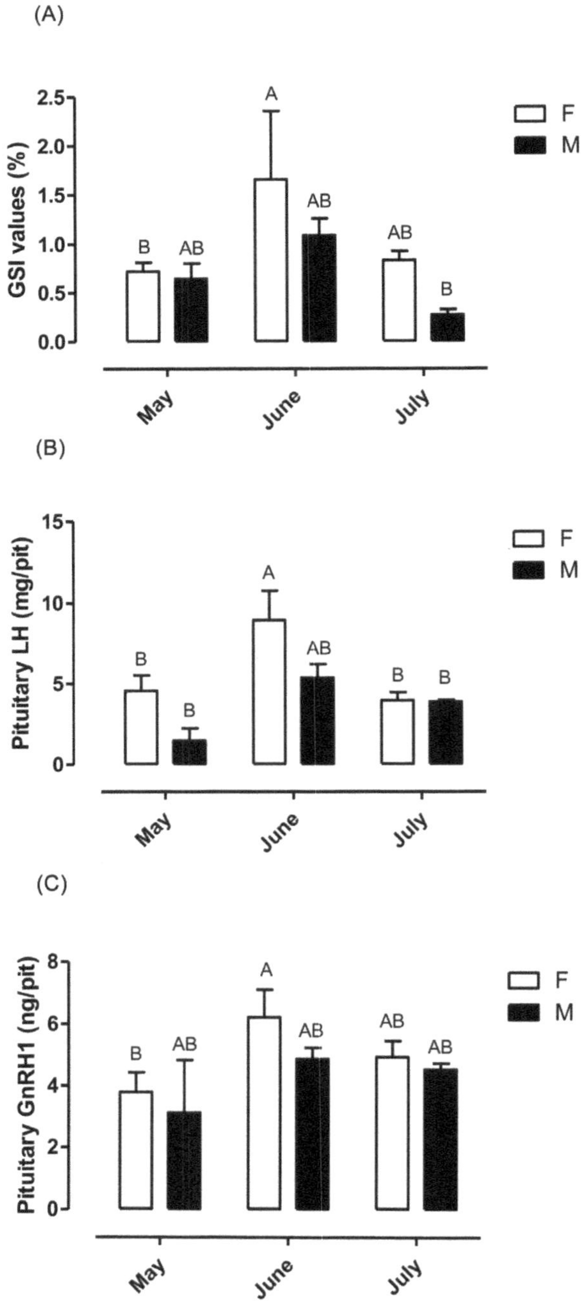

FIGURE 7.4 GSI values (A), pituitary LH content (B) and pituitary GnRH1 levels (C) in adult captive Atlantic bluefin tuna during the species' natural reproductive season in the Mediterranean Sea (May–July 2003). Levels (mean ± SEM) are expressed as percentage (%), total amount (mg) per pituitary, and ng per pituitary, respectively. Groups with different letters are significantly different ($P < 0.05$).

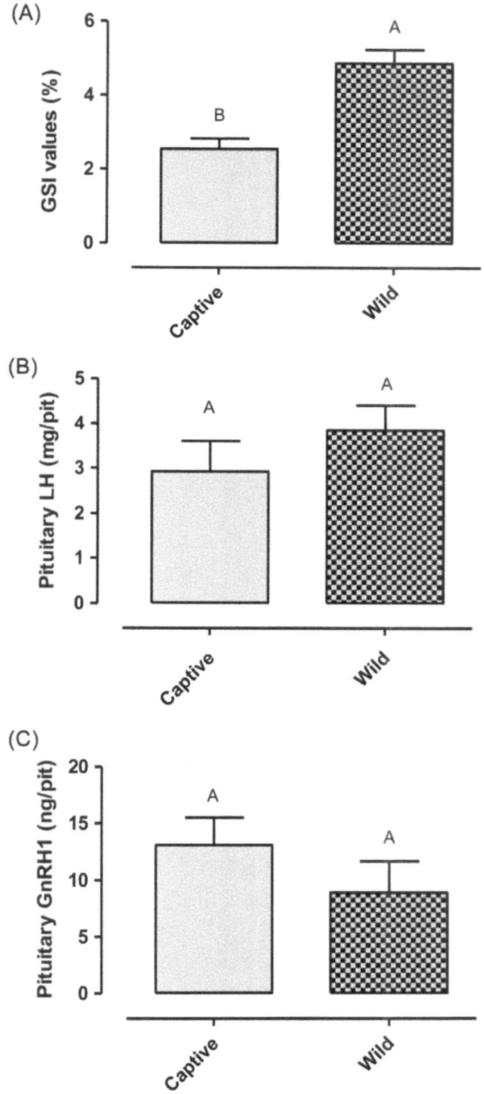

FIGURE 7.5 GSI values (A), pituitary LH content (B) and pituitary GnRH1 levels (C) in captive-reared versus wild-caught adult Atlantic bluefin tuna during the peak of the gametogenic cycle in the Mediterranean Sea (June 2004). Levels (mean ± SEM) are expressed as percentage (%), total amount (mg) per pituitary, and ng per pituitary, respectively. Groups with different letters are significantly different ($P < 0.05$).

Nonetheless, although gametogenesis did take place in captivity, the GSI values of the caged fish were relatively low compared to wild fish (Figures 7.5A and 7.6). A series of studies demonstrated that both spermatogenesis (Corriero et al., 2007, 2009; Zupa et al., 2013) and oogenesis

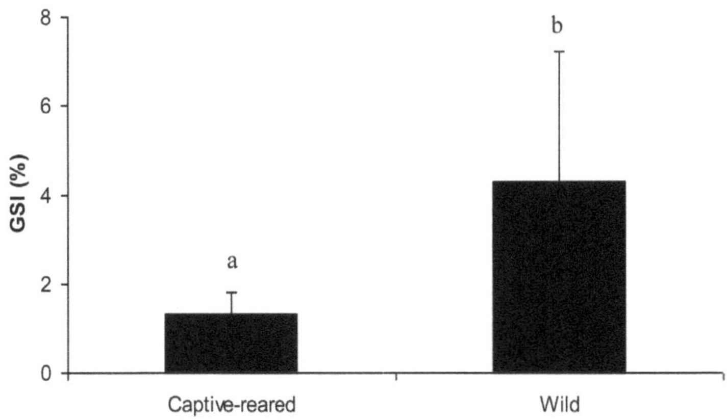

FIGURE 7.6 Mean (± s.e.m.) gonado-somatic index (GSI) of male ABFT broodfish. Comparison between individuals reared in captivity (pooled controls and GnRHa-treated) and wild spawners. Different letters represent statistically different means (ANOVA, $P \leq 0.05$). *Modified from Corriero et al. (2009).*

(Corriero et al., 2007, 2011; Pousis et al., 2011, 2012) are somewhat impaired in captive conditions. In a histological study carried out by Corriero et al. (2007) on 36 ABFT males reared for 1–3 years under commercial conditions in a tuna farm located in the western Mediterranean and sampled during the 2004 and 2005 reproductive seasons, the vast majority (>72%) of the fish were found to be at early spermatogenesis, while less than 5% were classified as spent. Yet, distinctively from captive fish, the histological appearance of the testis in mature wild males (sampled on the spawning grounds during the reproductive season) showed noticeably enhanced testicular development (large seminiferous lobules entirely filled with spermatozoa).

In a comparative study of male germ cell proliferation and apoptosis in wild and captive-reared ABFT, Zupa et al. (2013) reported that the spermatogonial proliferation rate of captive ABFT sampled in May was higher than in wild individuals, while the number of meiotic cysts was lower (Figure 7.7). In June, the density of proliferating spermatogonia and spermatocytes was higher in captive fish than in wild spawners. The authors concluded that rearing in captivity had resulted in a shift of the spermatogenetic process "with a high spermatogonial division in May and predominance of meiosis in June, when the wild spawners were undergoing their last phase of spermatogenesis (i.e. spermiogenesis), with high prevalence of spermatids and spermatozoa." In the same study, apoptotic spermatogonia and primary spermatocytes were detected in the testes of both wild and captive-reared individuals. The density of apoptotic germ cells was found to be significantly higher in captive-reared than in wild ABFT throughout both periods (May and June) examined during the reproductive cycle. The increased apoptotic rate in captive ABFT appears to be the result of reduced androgen levels, particularly those of 11-ketotestosterone (11-KT), the

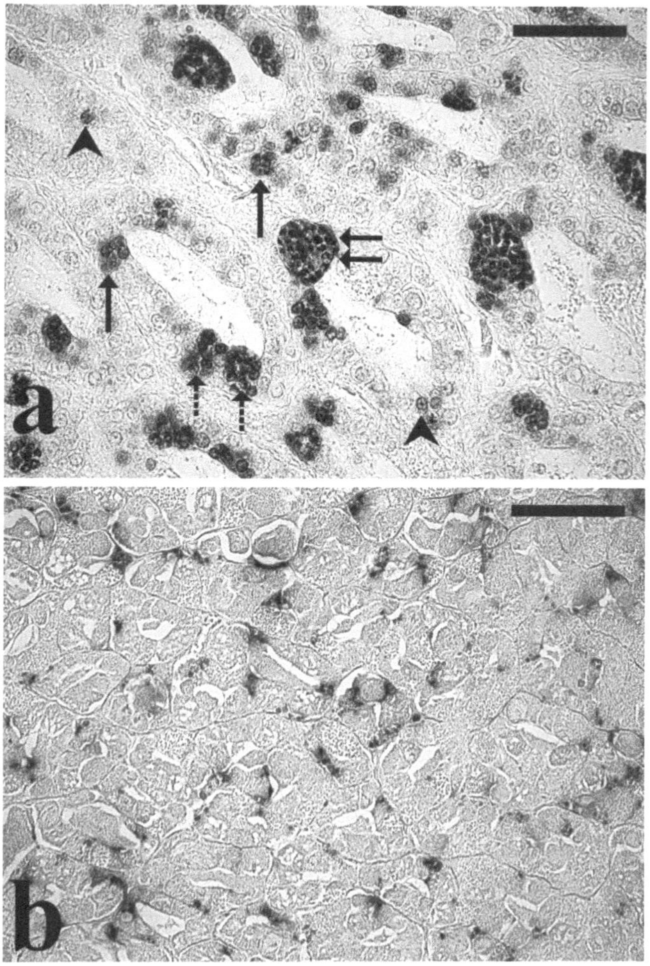

FIGURE 7.7 (A) Micrographs of testis sections from two adult Atlantic bluefin tuna specimens. (A) Immunolabeling with antibodies against Proliferating Cell Nuclear Antigen (PCNA) showing nuclei of proliferating cells stained in brown. Magnification bar = 50 μm. Arrowhead: undifferentiated spermatogonium; arrow: type A spermatogonial cyst; dashed arrow: type B spermatogonial cysts; double arrow: spermatocyte cysts. (B) Staining with the terminal deoxynucleotidyl transferase-mediated dUTP Nick End Labelling (TUNEL) method. Apoptotic cells and apoptotic bodies appear as dark blue dots. Magnification bar = 200 μm. *Reproduced from Zupa et al. (2013).*

principal androgen that stimulates spermatogenesis in fish. In fact, in a previous study Corriero et al. (2009) highlighted the close association between high circulating levels of 11-KT and reduced germ cell apoptosis in captive-reared ABFT treated with GnRHa, compared to untreated control fish. It is likely that the relatively low androgen levels typifying captive ABFT reflect a reduced secretion of pituitary gonadotropin caused by captive conditions, as was proposed for other species of fish (Zohar, 1989).

The effects of tuna captivity and farm rearing conditions on oogenesis are more pronounced than those observed on spermatogenesis. A thorough analysis of ovaries derived from ABFT reared in the western Mediterranean clearly demonstrated that captivity does not prevent vitellogenesis and oocyte growth, since yolk granule formation and oocyte diameter of fully vitellogenic oocytes were similar to wild fish (Corriero et al., 2007). This finding was confirmed further by Pousis et al. (2011, 2012), who demonstrated an absence of any significant differences in vitellogenin (Vg) and Vg receptor (VgR) gene expression, as well as in oocyte yolk accumulation between wild and captive-reared individuals. However, compared to wild fish (analyzed per kg body weight) farmed females consistently exhibit: (i) a relatively lower gonadal mass, (ii) lower numbers of vitellogenic oocytes, (iii) reduced ability to undergo oocyte maturation, and (iv) a higher proportion of atretic oocytes. In this regard, it is interesting to note that acute stress is an additional effector contributing to impaired vitellogenesis. Thus, farmed ABFT individuals have been shown to undergo a massive degeneration of vitellogenic oocytes (up to 100% α-atresia) and a dramatic loss of ovarian mass 24 h following an exposure to crowding-induced severe acute stress (Corriero et al., 2011).

7.3.2 Puberty

In light of increasing interest in the domestication of ABFT, first sexual maturity in captivity was studied by documenting its occurrence and by characterizing the gonadotropins (FSH and LH), the central regulators of gonadal development and gamete maturation (Berkovich et al., 2013). Histological sections of gonads obtained from 2- and 3-year-old (2Y and 3Y, respectively) farmed ABFT juveniles revealed that females of both age classes were sexually immature, showing the perinucleolar stage as the most advanced oocyte stage in the ovaries. The testes of all 2Y males showed a germinal epithelium composed of only spermatogonia. In contrast, the 3Y males were already sexually mature with testes containing all the spermatogenetic stages including spermatozoa in the lumen of seminiferous lobules. An earlier puberty in males, as compared to females, is a common feature of many fish species (Carrillo et al., 1995; Jakupsstovu and Haug, 1988; Holland et al., 2000; Saillant et al., 2003; Hurvitz et al., 2005; Cao et al., 2009), including the Pacific bluefin tuna (Sawada et al., 2007). Moreover, this phenomenon becomes even more pronounced under intensive culture conditions where ample food availability facilitates faster growth rates compared to those in wild populations (reviewed by Carrillo et al., 2009; Taranger et al., 2010). Thus, it is not yet clear whether the observed puberty in 3Y ABFT males is a general feature also occurring in wild populations or a phenomenon induced by the culture conditions.

The expression and accumulation patterns of LH in the pituitary showed a steady increase of this hormone, concomitant with fish age, reaching higher levels in adult females compared to males of the same age class (Berkovich et al., 2013). Conversely, the pituitary FSH levels were elevated only in 2Y and adult fish. The pituitary FSH to LH ratio was consistently higher (>1) in immature ABFT compared to maturing or pubertal fish, resembling the situation in other fish species and higher vertebrates. Nevertheless, the results suggest that a rise in the LH storage level above a minimum threshold may be an indicator of the onset of puberty in ABFT females. Higher pituitary LH levels in sexually mature females versus males may further support this notion. Future studies on the effects of captivity and hormonal treatments on precocious maturity may allow for improved handling of this species in a controlled environment, which would lead to more cost-effective broodstock management.

7.4 HORMONAL MANIPULATION OF REPRODUCTIVE FUNCTIONS

Like many other fishes reared in captivity, ABFT do not spawn consistently when maintained in cages or tanks, thus necessitating the use of exogenous hormones. Early studies examining the potential of ABFT to mature and complete its reproductive cycle in captivity in the Mediterranean indicated that wild fish caught from the fishery and maintained for at least one year in sea cages underwent complete gametogenesis during the May–July reproductive season (Corriero et al., 2007; Gordoa et al., 2009; Mylonas et al., 2007; Zupa et al., 2013). However, while some males were producing releasable sperm of good quality (Suquet et al., 2010), all females failed to undergo oocyte maturation, and hence ovulation and spawning. This is a typical reproductive dysfunction of fishes in captivity (Mylonas et al., 2011; Zohar, 1989; Zohar and Mylonas, 2001), and in ABFT it can be overcome with the exogenous administration of GnRHa (see Section 7.5.1). It is important, however, to point out that in some instances, after years of acclimatization to the captive conditions, ABFT broodstocks have also been shown to spontaneously complete their reproductive maturation and to produce large amounts of high-quality eggs without the employment of any exogenous hormones (Gordoa et al., 2009), as described later (see Section 7.5.2).

Hormonal manipulation of reproductive function in captive fishes ranges from the complete stimulation of gametogenesis practiced in freshwater eels (*Anguilla* spp.) (Miura et al., 2002) to simply the synchronization of ovulation to improve *in vitro* fertilization operations in salmonids (Zohar, 1989, 1996). Nevertheless, the most common application of exogenous hormones in aquaculture includes the enhancement of sperm production (Mylonas et al., 2016) and the induction of oocyte maturation, ovulation, and spawning (Mylonas and Zohar, 2001; Mylonas et al., 2010b). A controlled-release

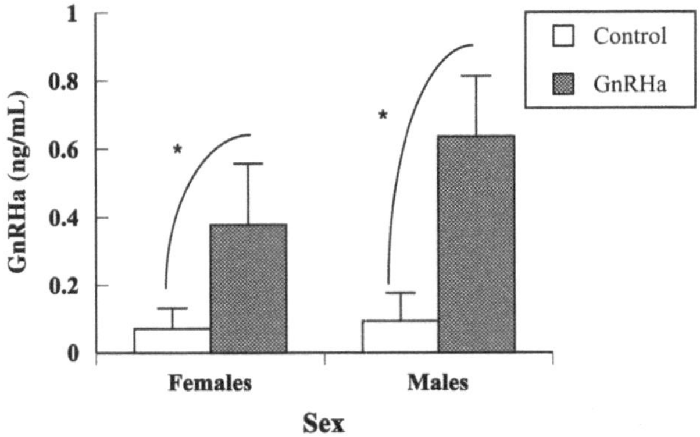

FIGURE 7.8 Mean (±SEM) plasma GnRHa levels of captive-reared Atlantic bluefin tuna males and females ($n = 7$ fish per group for the control females and males; $n = 10$ GnRHa-implanted females and 16 males), 2–8 days after treatment with GnRHa implants. In both sexes, there was a statistically significant elevation (two-way ANOVA, $P < 0.02$) in plasma GnRHa in the implanted fish. *Reproduced from Mylonas et al. (2007).*

delivery system (implant) for GnRHa (Mylonas and Zohar, 2001) was developed specifically for ABFT, releasing the hormone for a period of 2–3 weeks, and was used to induce maturation and spawning (Mylonas et al., 2007). In both males and females, administration of the GnRHa implants resulted in a significant elevation of GnRHa levels in the plasma (Figure 7.8), a decrease of pituitary LH content and a concomitant increase in plasma LH (Rosenfeld et al., 2012). These changes demonstrated that (i) during gametogenesis LH is being produced and stored in high amounts in the pituitary and (ii) the administered GnRHa in the provided dose and mode of release and the ensuing surge of plasma GnRHa were appropriate to induce the release of the stored endogenous LH. The plasma LH profiles paralleled the kinetics of the GnRHa released from the implants (Mylonas et al., 2007). In response to the GnRHa-induced increased plasma LH, there were significant changes in steroidogenesis including increases in plasma 17 β-estradiol (E_2) and 11-KT in females and males, respectively. Furthermore, a significant peak in the plasma of free 17, 20 β-dihydroxy-4-pregnen-3-one (17, 20β-P) levels occurred simultaneously in males and females at day 8 after treatment. Histological sections of the ovaries in the GnRHa-induced females contained oocytes at the final stages of maturation and germinal vesicle migration (Corriero et al., 2007), suggesting the role of 17, 20β-P as a maturation-inducing steroid in ABFT, similar to other perciform species (Nagahama, 1997). Interestingly, the stimulatory effects of sustained-release GnRHa implants on the pituitary–gonad axis elicit reproductive hormone dynamics during final oocyte maturation that display striking similarity between ABFT and gilthead sea bream (*Sparus aurata*), both of which are asynchronous spawners (seabream: Zohar and Gordin, 1979; Zohar et al.,

1995; ABFT: Schaefer, 2001; Medina et al., 2002; Corriero et al., 2003). The more noticeable similarities include the LH response to sustained release of GnRHa (Zohar, 1988, 1996; Zohar et al., 1990), and the association of the plasma 17, 20β-P levels and the maturation stage of the oocytes during a daily spawning cycle (Gothilf et al., 1997). Surprisingly, even under sustained GnRHa-induced conditions whereby long-term elevated plasma LH levels appear to override the natural daily fluctuation of plasma LH, in both gilthead sea bream (Zohar et al., 1995) and ABFT (De Metrio, 2010) the females still manage to retain their typical daily spawning cycles.

Due to the relatively late reproductive maturation of ABFT ($>$4 years of age), advancing the onset of puberty, via hormonal and/or environmental manipulations, will simplify broodstock management and spawning efforts and, in turn, seed production for the aquaculture of this species. Fueled by our expanded understanding of the regulation of puberty in teleost fish (Zohar et al., 2010), a series of *in vivo* and *in vitro* trials were carried out to evaluate the (i) dynamics of the maturational process of the brain–pituitary–gonadal (BPG) axis (Berkovich et al., 2013), and (ii) responsiveness of the BPG to exogenous hormones, in ABFT undergoing pubertal development (Berkovich et al., 2011). Different hormonal treatments, including GnRHa, kisspeptin decapeptide (Kiss), and E_2 administered through controlled-release implants, were tested as potential inducers of puberty in captive ABFT. In general, no clear effects of the different hormonal treatments were observed at the macroscopic (GSI) or microscopic (oocyte and seminiferous tubule diameter) levels. Nevertheless, further *in vitro* studies demonstrated the capacity of recombinant FSH (rFSH) and, to a lesser extent, rLH to stimulate cell proliferation in immature ovarian and testicular fragments of ABFT. Both rFSH and rLH have failed to stimulate steroidogenesis, yet the *in vivo* pretreatment with Kiss-containing implants appeared to potentiate rFSH-stimulated steroidogenesis in immature testes. Future research is required to study the effects of environmental and hormonal manipulations on the timing of puberty, with the goal of inducing precocious sexual maturation and enabling spawning of ABFT at relatively small body size. Working with smaller broodstock is expected to facilitate their husbandry and handling in confined environments, and to greatly improve the cost-effectiveness of ABFT broodstock operations, seed production, and in turn its farming profitability.

7.5 SPAWNING OF CAPTIVE-REARED BROODSTOCKS

7.5.1 Induced Spawning

As mentioned above, development of any industrial aquaculture operation requires absolute control of the reproductive cycle of the particular fish of interest. Since the spawning season in captive ABFT is very short (about 1–2 months) and not consistent, predictable or reliable, induced

FIGURE 7.9 Spawning induction in ABFT using polymer-based controlled release GnRHa implants. (A) GnRHa implant assemblies with a dart-head and colored Floy tag (dart-heads and tags both from Floy Tag & Mfg. Inc.). (B) An implant/tag assembly mounted on the arrow of the spear-gun. Note the broodstock cage with the curtains (background) surrounding the cage perimeter to retain spawned eggs for collection. (C) A diver with the spear-gun ready to administer the GnRHa implant to the broodstock. (D) An implanted broodstock tuna inside a SELFDOTT cage in Cartagena, Spain. *Photo credits: (A) Jorge Gomezjurado; (B) Kali Tuna; (C) Kali Tuna; (D) IOE.*

spawning methods are essential for the early expansion of the industry (Corriero et al., 2007, 2009; Mylonas et al., 2007). In the first efforts to induce spawning in ABFT, sexually mature migrating fish (5–12 years old) were captured in the Mediterranean Sea and reared in floating cages. Around the time of ABFT spawning in the wild (June–July), GnRHa was administered to broodstock fish (both males and females) via polymer-based controlled release delivery systems (implants), administered to the fish underwater, as they were swimming (Mylonas et al., 2007). Typically, 2–3 GnRHa implants (3–4 mm in diameter each) were mounted on the needle tip of a spear gun, along with a visible tag, and shot into the swimming fish by free-divers (Figure 7.9). The implants were designed to release the GnRHa into the fish for a period of 2–3 weeks, at a total dose of 50–75 micrograms of the hormone per kg body weight. For two consecutive years during the natural spawning period (June–July), GnRHa-implanted fish were

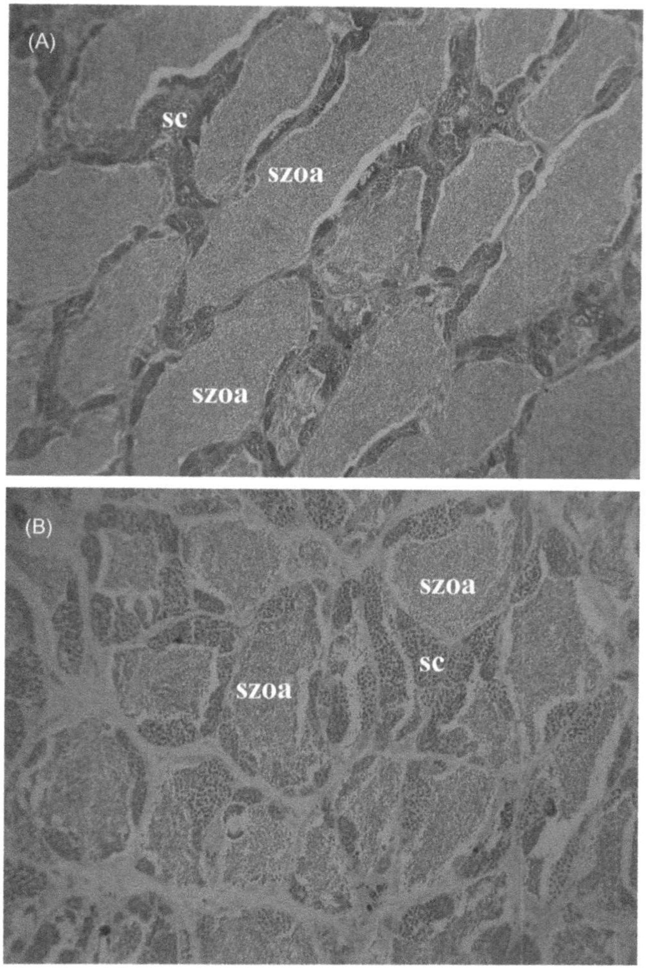

FIGURE 7.10 Photomicrographs of histological sections from testes of captive-reared Atlantic bluefin tuna in response to treatment with GnRHa implants at the peak of the reproductive season. (A) Non-implanted control; (B) GnRH-implanted male. Both males contained large numbers of spermatozoa (szoa) in their testes, as well as spermatocysts with gametes at an earlier stage of development (sc).

sacrificed at different times after treatment in order to monitor the effect of the hormonal implants on inducing final oocyte maturation and ovulation in females and spermiation in males (Mylonas et al., 2007). In males, the GnRHa implants did not result in any differences in the histological appearance of testes (Figure 7.10) and almost all fish (both GnRHa-treated and controls) contained intra-testicular spermatozoa (Corriero et al., 2007). There was an increase, however, in the proportion of spermiating males; in control males ($n = 17$) it was only 12% compared to 26% for the GnRHa-treated males ($n = 19$). A much stronger and definitive effect was observed in female ABFT, where GnRHa implantation induced final oocyte maturation (germinal vesicle migration and coalescence of lipid droplets and yolk globules) 2–8 days after treatment (Figure 7.11) in 63% of the fish, and 88% had post-ovulatory follicles in the ovaries, compared to 0% and 21%, respectively, for control females. In addition, two females were found to have ovulated eggs in the ovaries at the time they were sampled and, after insemination *in vitro* with sperm from spermiating males, viable embryos and larvae were produced for the first time in captivity in ABFT (Mylonas et al., 2007). In the same experiments, fertilized eggs were collected from the cages housing the treated fish (Mylonas et al., 2007), demonstrating that the developed GnRHa implants could induce oocyte maturation, ovulation, spermiation, and spawning in captive-reared ABFT maintained in sea cages. Later, another captive stock of ABFT in Italy was induced to spawn using the same GnRHa therapy, resulting in the production of 20 million eggs over four consecutive daily spawns with a fertilization rate of 80% for the first two spawns (De Metrio et al., 2010). These eggs drove the first larval rearing trials and resulted in the first-ever production of cultured ABFT juvenile up to the size of 8 cm in total length at 60 days post-hatching. In 2009, another captive-reared broodstock in Spain was induced to spawn using the same GnRHa implants, resulting in daily spawning for 17 days (reflecting the fact that ABFT is a batch spawner, see Section 7.2.3) and producing a total of 140 million fertilized eggs (Figure 7.12B; De la Gandara et al., 2010). These eggs were sent to various research hatcheries around the Mediterranean Sea, and larval rearing trials produced a small number of fingerlings, demonstrating the feasibility of the aquaculture production of ABFT. More recently, the GnRHa implants were used to extend the spawning season well beyond its natural duration in the Adriatic Sea (Zohar et al., unpublished results). After initiation of some sporadic spawning during the month of June, broodstock ABFT held in sea cages were treated with GnRHa implants several times, at intervals of a few weeks, from early July through the middle of August. These treatments resulted in multiple massive spawning events with eggs of excellent quality and in a spawning season that lasted until the end of August (a total duration of two and a half months). A few million of these eggs were successfully shipped to the United States (Maryland) and led to the first-ever aquaculture production of bluefin tuna juveniles in North America (Zohar et al., unpublished results).

One question relevant to both induced and spontaneous spawning (see Section 7.5.2) is how many individuals from a group of broodstock actually spawn and contribute to egg production. Using microsatellite-based genotyping of broodstock and eggs SELFDOTT investigators (Cilia Antoniou, HCMR, unpublished data) showed that out of 25 genotyped individual ABFT, 15 individuals contributed to the majority (87.8%) of the eggs produced, while the remaining 10 breeders contributed to only 13.2% of the

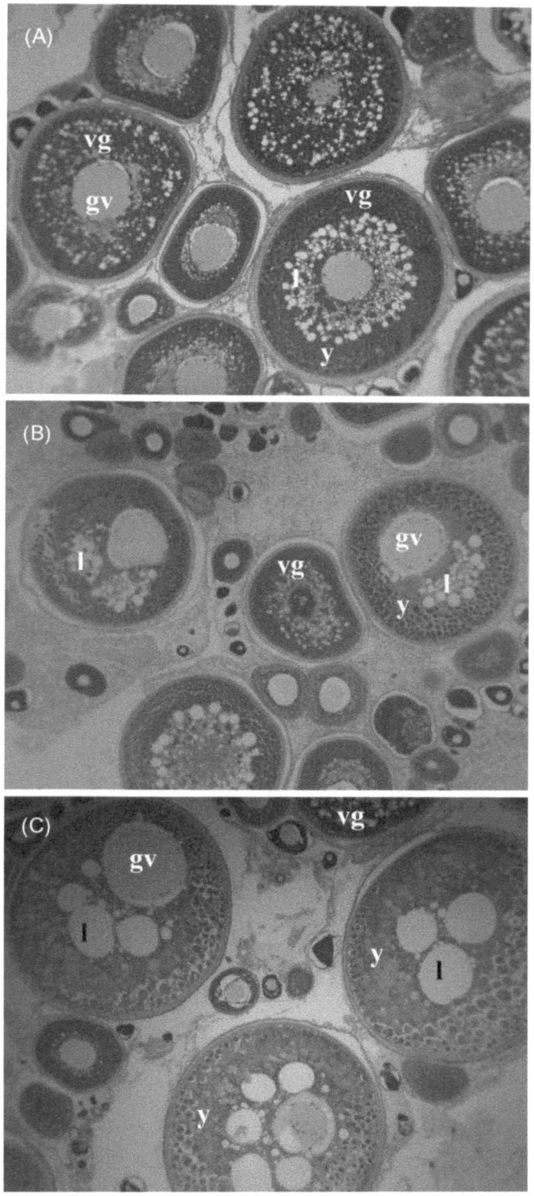

FIGURE 7.11 Photomicrographs of histological sections from ovaries of captive-reared Atlantic bluefin tuna in response to treatment with GnRHa implants at the peak of the reproductive season. (A) Non-implanted control female with oocytes at the stage of vitellogenesis (vg), with a centrally located nucleus (germinal vesicle, gv) and dispersed lipid droplets (l) and yolk globules (y). (B) GnRHa-implanted female with oocytes at the stage of early oocyte maturation, with the gv migrating to the periphery and the lipid droplets coalescing into larger droplets. (C) Another GnRHa-implanted female with oocytes at the stage of advanced oocyte maturation, with almost completely coalesced lipid (l) droplets and coalescing yolk (y) globules.

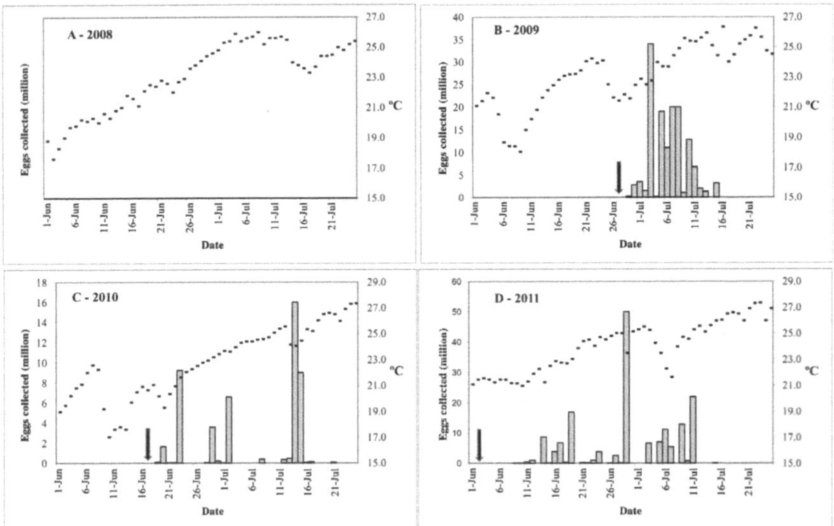

FIGURE 7.12 Daily egg collection (bars) and water temperature (dotted lines) throughout the four spawning seasons of the SELFDOTT project. The black arrows indicate the day when the egg collection system was deployed. Implantation with GnRHa was undertaken only in 2008 and 2009, since the fish started spawning spontaneously prior to the planned implantation time in 2010 and 2011.

eggs. Although it was not known which of the 25 studied individuals were induced to spawn (only 15 fish were implanted with GnRHa in this study), it is interesting to note that the majority of spawning was attributed to 15 individuals of the total population. While this study drew no conclusions about the relative contribution to the spawning of spontaneous or induced fish, similar research is required to support decisions on breeder selection, broodstock management and spawning induction, especially in situations where the size of the broodstock is limited, such as will be the case in future land-based broodstock operations. It is interesting to note that a study using the same genotyping approach in a southern bluefin tuna broodstock, held in a land-based tank, demonstrated that only a few females in the tank participated in spawning, and most of them in response to treatment with GnRHa implants (Knibb et al., in preparation).

Considered together, GnRHa implants were found to be an efficient tool to initiate, synchronize and prolong the duration of ABFT spawning, thus helping broodstock management and providing better predictability of egg supply. Since the early demonstration of their efficiency, GnRHa implants have been used by various commercial and research operations with the objective of inducing massive spawning of ABFT broodstock at the beginning or throughout the spawning season (especially when fish do not spawn spontaneously or spawn very small quantities of eggs), as well as to extend the length of the spawning season.

7.5.2 Spontaneous Spawning

In the framework of a European project (7FP SELFDOTT) coordinated by the Spanish Institute of Oceanography (Instituto Español de Oceanografia, IEO), 38 wild ABFT (~60 kg in bodyweight) were captured in June 2007 in the Balearic Sea using commercial purse-seiners. The fish were transferred to a 25 meter diameter, 20 meter-deep floating cage at El Gorguel Bay (Cartagena-Murcia) on the southeast coast of Spain. This broodstock population was fed to satiation once a day, 6 days per week, with raw fish consisting mainly of mackerel (*Scomber scombrus*) and Spanish mackerel (*S. japonicus*). The wild-caught broodstock adapted well to captivity, accepting the feed provided from their arrival at the culture facilities and with a low mortality rate (5%) over the 4 years of rearing. Spawning activity and the amount of eggs produced by these fish were monitored from 2008 to 2011 (Figure 7.12; De la Gandara et al., 2011). In 2008 and 2009, fish were treated with GnRHa implants, as described in Section 7.5.1. In 2008, no spawning events were detected and no eggs were collected (Figure 7.12A), which may reflect the fact that the fish were implanted and egg collectors deployed very late in the season (July 15), at a time when the spawning season at this site is typically over (see Figure 7.12, years 2009–2011; De la Gandara et al., 2009). In 2009, massive spawning started between June 29 and 30, roughly 48–72 h after the administration of GnRHa implants, and lasted 17 consecutive days, with a daily maximum egg collection of 34 million eggs and a total collection of 140 million. During this period, water temperature ranged between 22 and 28 °C on the surface and between 19 and 27 °C at the bottom of the cage (Figure 7.12B).

In the following year (17 June 2010), spontaneous spawning was observed prior to the time GnRHa implantation was planned. From this day forward, and more or less on a daily basis, viable eggs were collected for a period of 26 days, thus GnRHa treatment was not employed for this year. During the 2010 reproductive season, a total of 48 million eggs were collected from this stock. In 2011, eggs were again produced spontaneously without any GnRHa treatment on 9 June 2011, about one week after installing the egg collector. Spontaneous spawning lasted for 37 days and the total number of eggs collected was 162 million (Figure 7.12D). During the 2011 spawning period, the water temperature ranged between 21 and 26 °C. In this ABFT broodstock over the monitoring period, spawning typically occurred before dawn, usually from around 02:00 to 03:00 h, but often continued past dawn. Spawning followed intense courtship behavior that occurred at some depth, which therefore was not possible to fully observe and/or describe in detail.

The above ABFT broodstock was maintained past the end of the SELFDOTT project in November 2011, through funding by various research and development projects to IEO and the Ricardo Fuentes e Hijos Group, in order to establish hatchery/nursery production of ABFT. In the spawning seasons of 2012, 2013, and 2014, spawning was spontaneous and large numbers of ABFT eggs were produced, as monitored by the husbandry personnel. However, daily egg fecundity was not documented as it was done during the SELFDOTT project, since eggs were collected only when they were required by the hatchery for larval rearing experiments.

Gordoa et al. (2009), Gordoa (2010), and Gordoa and Carreras (2014) also reported spontaneous spawning events in another wild-caught ABFT broodstock, just after their capture from the wild using a purse seine, and also in an ABFT stock after one year in captivity. These stocks spawned in transport cages as they were being towed from the capture area around the Balearic Islands (Spain, western Mediterranean) to the company fattening facility off the coast of eastern Spain, and again as they were towed back from the fattening site to the capture area during the next year's spawning season (to allow spontaneously spawned eggs to survive in the wild). Also, spontaneous spawning has been reported in a commercial ABFT fattening operation in Malta (Deguara, 2011), again using wild-caught individuals after some years of acclimation, and in Turkey (Kilic Aquaculture) as well as in cages with broodstock that were initially captured in the wild as ~10 kg fish in the Adriatic Sea (Kali Tuna).

As a result of the spontaneous spawning of these wild-caught ABFT stocks maintained by commercial fattening operations during the last 5 years, larval rearing/nursery trials have finally succeeded in the production of high-quality juveniles for sea cage rearing. Consequently, in November 2014 the first-ever hatchery-produced ABFT were sent to the market after ~3 years of rearing at a size of ~20 kg (De la Gandara, personal communication). Therefore, the feasibility of aquaculture production of ABFT, from captive spawning and hatchery to the market, has been demonstrated, as has been achieved earlier with the Pacific bluefin tuna in Japan (Sawada et al., 2005).

Thus, it is clear that captive ABFT broodstock held at low densities in floating cages can spontaneously spawn large amounts of high-quality eggs. However, in most cases spontaneous spawnings are short in duration (3–4 weeks) and in some situations (i) do not occur consistently, (ii) cease for several days or weeks, and/or (iii) consist of small number of eggs as only a small number of females may be spawning out of the total broodstock. In these situations and until a better environmental control of ABFT broodstocks is achieved, GnRHa implants can be used successfully to initiate massive spawning, extend the duration of the spawning season and improve the consistency of egg supply.

7.6 REPRODUCTIVE BIOLOGY AND SPAWNING IN ABFT—THE FUTURE

As in other commercially important marine fish, understanding the endocrine and reproductive cycles of captive ABFT, and comparing them to those of wild fish, enabled the development of hormonally induced spawning and opened the egg production bottleneck that had previously inhibited the farming of this fish. Captive breeding is expected to lead, in the very near future, to completely closing the life cycle (eggs to broodstock to eggs) of ABFT in aquaculture. However, it is also clear from the successful spontaneous spawning obtained over the past few years in various captive-reared broodstocks (Spain, Malta, Turkey, Croatia) that ABFT collected from the capture fishery can complete their reproductive cycle naturally in confinement and produce large numbers of high quality eggs if maintained optimally. Successful spawning, however, is subject to the combined effect of proper husbandry, nutrition, and the existence of the appropriate environmental conditions (mainly water temperature), because captive broodstocks are unable to migrate and select the desired temperature that is required for maturation and spawning. Furthermore, successful egg collection also requires the existence of very calm weather, weak currents and/or the deployment of effective egg collection systems, which may be a limiting factor in some locations. This is why spawning of ABFT is very seasonal and lasts for only 4–8 weeks in any given broodstock operation, during which time availability of eggs and their numbers are still quite inconsistent. This makes it difficult to plan larval rearing and/or juvenile production, which is a major hurdle to the development of predictable and dependable hatchery, nursery, and grow-out operations. It is clear that an efficient and cost-effective ABFT aquaculture industry cannot develop without on-demand year-round production of seeds (fertilized eggs and viable offspring). For that to happen, broodstock operations will have to be placed on-land, allowing full control of all environmental parameters to which fish are exposed. Broodstock management protocols in such facilities will enable predictable and consistent egg production all year round. Three to four land-based tanks, exposed to phase-shifted environmental conditions, will result in fish with fully developed gonads throughout most of the year. Simulating the photoperiod, temperature, salinity, and water quality required for spawning will most probably result in some fish spawning spontaneously. Such a land-based facility has been operating for the southern bluefin tuna in Australia since 2008 (Chapter 10 of this book and Knibb et al., in preparation) and will start to operate during late 2015 for ABFT in Europe (IEO, Spain).

However, as is the case in other fish species held in captivity and from indications obtained thus far from ABFT, not all the fish in a given captive broodstock are expected to spawn, which can be confirmed through genotyping the eggs and the broodstock. To increase the efficiency and maximize productivity of land-based operations, spawning induction therapies will continue to be used to ensure that most, if not all, fish held in the tanks are actively spawning and contributing to egg production. Actively spawning fish (as indicated through DNA fingerprinting) should be tagged so they are not hormonally treated unnecessarily or repeatedly. To be efficient, spawning induction therapies should be applied only during the very short window of time when the broodstock are reproductively competent and responsive to exogenous hormonal stimulation, that is, when fish reach the final stages of spermatogenesis or vitellogenesis and before the resumption of gonadal apoptosis and atresia. Predicting the optimal timing of treatment, as well as the female to male ratio, will necessitate more research toward determining hormonal and gonadal status and competence, through analysis of molecular and/or endocrine indicators in fish mucosal samples obtained through gentle and noninvasive mucosal swabs. New approaches discussed in this chapter to accelerate the onset of puberty will enable the production of tuna eggs from smaller fish, thus reducing the cost and increasing the versatility and efficiency of land-based broodstock operations.

The very significant progress, reviewed in this chapter, in understanding the reproductive biology of ABFT and obtaining large numbers of fertilized eggs, together with future development of advanced land-based broodstock operations, will drive year-round availability of fertilized ABFT eggs to enable and support the establishment of an efficient, cost-effective, and environmentally sustainable aquaculture industry for this species.

REFERENCES

Abascal, F.J., Medina, A., 2005. Ultrastructure of oogenesis in the bluefin tuna, *Thunnus thynnus*. J. Morphol. 264 (2), 149–160.

Abascal, F.J., Megina, C., Medina, A., 2004. Testicular development in migrant and spawning bluefin tuna (*Thunnus thynnus* L.) from the eastern Atlantic and Mediterranean. Fish. Bull. 102, 407–417.

Aristotle, 1994. The History of Animals. Kaktos, Athens, Greece (in Greek).

Baglin Jr., R.E., 1982. Reproductive biology of western Atlantic bluefin tuna. Fish. Bull. 80, 121–134.

Berkovich, N., Corriero, A., Santamaria, N., Mylonas, C.C., Bridges C.R., Vassallo-Aguis, R., et al., 2011. The neuroendocrine control of pubertal development in Atlantic bluefin tuna (*Thunnus thynnus*). Ninth International Symposium on Reproductive Physiology of Fish, Cochin, India.

Berkovich, N., Corriero, A., Santamaria, N., Mylonas, C.C., Vassallo-Aguis, R., de la Gándara, F., et al., 2013. GnRHa-mediated stimulation of the reproductive endocrine axis in captive Atlantic bluefin tuna, *Thunnus thynnus*. Gen. Comp. Endocrinol. 175, 55–64.

Billard, R., 1986. Spermatogenesis and spermatology of some teleost fish species. Reprod. Nutr. Dev. 26 (4), 877–920.

Block, B.A., Teo, S.L.O., Walli, A., Boustany, A., Stokesbury, M.J.W., Farwell, C.J., et al., 2005. Electronic tagging and population structure of Atlantic bluefin tuna. Nature 434, 1121–1127.

Cao, H., Zhou, L., Zhang, Y.-Z., Wei, Q.-W., Chen, X.-H., Gui, J.-F., 2009. Molecular characterization of the Chinese sturgeon gonadotropins and cellular distribution in pituitaries of mature and immature individuals. Mol. Cell. Endocrinol. 303, 34–42.

Carrillo, M., Zanuy, S., Prat, F., Cerda, J., Ramos, J., Mananos, L., et al., 1995. Sea bass (*Dicentrachus labrax*). In: Bromage, N.R., Roberts, R.J. (Eds.), Broodstock Management and Egg and Larval Quality. Blackwell Science Publishers, Oxford, pp. 138–168.

Carrillo, M., Zanuy, S., Felip, A., Bayarri, M.J., Moles, G., Gomez, A., 2009. Hormonal and environmental control of puberty in perciform fish: the case of sea bass. Ann. NY Acad. Sci. 1163, 49–59.

Cermeno, P., Quilez-Badia, G., Ospina-Alvarez, A., Sainz-Trapaga, S., Boustany, A.M., Seitz, A.C., et al., 2015. Electronic tagging of Atlantic Bluefin tuna (*Thunnus thynnus*, L.) reveals habitat use and behaviors in the Mediterranean Sea. PLOS One 10 (2), e0116638.

Corriero, A., Desantis, S., Deflorio, M., Acone, F., Bridges, C.R., de la Serna, J.M., et al., 2003. Histological investigations on the ovarian cycle of the bluefin tuna in the western and central Mediterranean. J. Fish Biol. 63, 108–119.

Corriero, A., Karakulak, S., Santamaria, C.A., Deflorio, M., Spedicato, D., Addis, P., et al., 2005. Size and age at sexual maturity of female bluefin tuna (*Thunnus thynnus* L. 1758) from the Mediterranean Sea. J. Appl. Ichthyol. 21, 483–486.

Corriero, A., Medina, A., Mylonas, C.C., Abascal, F.J., Deflorio, M., Aragón, L., et al., 2007. Histological study of the effects of treatment with gonadotropin-releasing hormone agonist (GnRHa) on the reproductive maturation of captive-reared Atlantic bluefin tuna (*Thunnus thynnus* L.). Aquaculture 272, 675–686.

Corriero, A., Medina, A., Mylonas, C.C., Bridges, C.R., Santamaria, N., Deflorio, M., et al., 2009. Proliferation and apoptosis of male germ cells in captive Atlantic bluefin tuna (*Thunnus thynnus* L.) treated with gonadotroping releasing hormone agonist (GnRHa). Anim. Reprod. Sci. 116, 346–357.

Corriero, A., Zupa, R., Bello, G., Mylonas, C.C., Deflorio, M., Genovese, S., et al., 2011. Evidence that severe acute stress and starvation induce rapid atresia of ovarian vitellogenic follicles in Atlantic bluefin tuna, *Thunnus thynnus* (L.) (Osteichthyes: Scombridae). J. Fish Dis. 34 (11), 853–860.

Cort, J.L., Liorzou, B., 1991. Migration—eastern Atlantic and Mediterranean. In: Deriso, R.B., Bayliff, W.H. (Eds.), World Meeting on Stock Assessment of Bluefin Tunas: Strengths and Weaknesses. Inter-America Tropical Tuna Commision, pp. 130–132.

Deguara, S., 2011. Natural spawning of captive Atlantic bluefin tuna, *Thunnus thynnus* L., in farm cages in Malta. SCRS SCRS/2011/, 1–4.

De la Gandara, F., Mylonas, C.C., Covès, D., Bridges, C.R., (Eds.), 2009. SELFDOTT Report 2008. 139 pp. http://hdl.handle.net/10508/355.

De la Gandara, F., Mylonas, C.C., Covès, D., Bridges, C.R., (Eds.), 2010. SELFDOTT Report 2009. 279 pp. <http://hdl.handle.net/10508/356>.

De la Gandara, F., Ortega, A., Belmonte, A., Mylonas, C.C., 2011 Spontaneous spawning of Atlantic bluefin tuna *Thunnus thynnus* kept in captivity. In: Proceedings of the EAS2011, Rhodes (Greece), pp. 249–250.

De Metrio, G., Bridges, C.R., Mylonas, C.C., Caggiano, M., Deflorio, M., Santamaria, N., et al., 2010. Spawning induction and large-scale collection of fertilized eggs in captive Atlantic bluefin tuna (*Thunnus thynnus* L.) and the first larval rearing efforts. J. Appl. Ichthyol. 26, 596–599.

Goldstein, J., Heppell, S., Cooper, A., Brault, S., Lutcavage, M., 2007. Reproductive status and body condition of Atlantic bluefin tuna in the Gulf of Maine, 2000–2002. Mar. Biol. 151, 2063–2075.

Gordoa, A., 2010. The Atlantic bluefin tuna: study of the temporal pattern of spawning in the western Mediterranean region and reproductive capacity in captivity. Collect. Vol. Sci. Pap. ICCAT 65 (3), 837–847.

Gordoa, A., Carrerras, G., 2014. Determination of temporal spawning patterns and hatching time in response to temperature of Atlantic bluefin tuna (*Thunnus thynnus*) in the Western Mediterranean. PLoS ONE 9 (3), e90691.

Gordoa, A., Olivar, M.P., Arevalo, R., Viñas, J., Moli, B., Illas, X., 2009. Determination of Atlantic bluefin tuna (*Thunnus thynnus*) spawning time within a transport cage in the western Mediterranean. ICES J. Mar. Sci. 66, 2205–2210.

Gothilf, Y., Meiri, I., Elizur, A., Zohar, Y., 1997. Preovulatory changes in the levels of three gonadotropin-releasing hormone-encoding messenger ribonucleic acids (mRNAs), gonadotropin beta subunit mRNAs, plasma gonadotropin, and steroids in the female gilthead seabream, *Sparus aurata*. Biol. Repro. 57, 1145–1154.

Grier, H.J., Linton, J.R., Leatherland, J.F., de Vlaming, V.L., 1980. Structural evidence for two different testicular types in teleost fishes. Am. J. Anat. 159, 331–345.

Heinisch, G., Corriero, A., Medina, A., Abascal, F.J., De La Serna, J.M., Vassallo-Agius, R., et al., 2008. Spatial-temporal pattern of bluefin tuna (*Thunnus thynnus* L. 1758) gonad maturation across the Mediterranean Sea. Mar. Biol. 154, 623–630.

Heinisch, G., Rosenfeld, H., Knapp, J.M., Gordin, H., Lutcavage, M., 2014. Sexual maturity in western Atlantic bluefin tuna. Sci. Rep. 47205.

Holland, M.C., Hassin, S., Zohar, Y., 2000. Gonadal development and plasma steroid levels during pubertal development in captive-reared striped bass, *Morone saxatilis*. J. Exp. Zool. 286, 49–63.

Hurvitz, A., Degani, G., Goldberg, D., Din, S.Y., Jackson, K., Levavi-Sivan, B., 2005. Cloning of FSHb, LHb, and glycoprotein a subunits from the Russian Sturgeon (*Acipenser gueldenstaedtii*), b-subunit mRNA expression, gonad development, and steroid levels in immature fish. Gen. Comp. Endocrinol. 140, 61–73.

ICCAT, 2005. Report for the biennial period, 2004–2005. Part I (2004), vol. 2. ICCAT http://www.iccat.es, Madrid, Spain.

ICCAT, 2008. Report of the 2008 Atlantic Bluefin Tuna Stock Assessment Session (SCI-003/2008). International Committee for the Conservation of Atlantic Tuna, Madrid, Spain.

Jakupsstovu, S.H., Haug, T., 1988. Growth, sexual maturation, and spawning season of Atlantic halibut, *Hippoglossus hippoglossus*, in Faroese waters. Fish. Res. 6, 201–215.

Knibb, W., Miller, A., Deichmann, M., Lamont, R., Nguyen, N.H., Foster, C., et al. Statistical analysis of potential factors associated with spawning in captive Southern bluefin tuna, *Thunnus maccoyii* (in preparation).

Lutcavage, M.E., Brill, R.W., Skomal, G.B., Chase, B.C., Howey, P.W., 1999. Results of pop-up satellite tagging of spawning size class fish in the Gulf of Maine: do North Atlantic bluefin tuna spawn in the mid-Atlantic?. Can. J. Fish. Aquat. Sci. 56, 173–177.

Medina, A., Abascal, F.J., Megina, C., Garcia, A., 2002. Stereological assessment of the reproductive status of female Atlantic northern bluefin tuna during migration to Mediterranean spawning grounds through the Strait of Gibraltar. J. Fish Biol. 60, 203–217.

Miura, T., Ando, N., Miura, C., Yamauchi, K., 2002. Comparative studies between *in vivo* and *in vitro* spermatogenesis of Japanese eel (*Anguilla japonica*). Zoolog. Sci. 19, 321–329.

Mylonas, C.C., Zohar, Y., 2001. Use of GnRHa-delivery systems for the control of reproduction in fish. Rev. Fish Biol. Fish. 10, 463–491.

Mylonas, C.C., Bridges, C., Gordin, H., Rios, A.B., Garcia, A., de la Gandara, F., et al., 2007. Preparation and administration of gonadotropin-releasing hormone agonist (GnRHa) implants for the artificial control of reproductive maturation in captive-reared Atlantic bluefin tuna (*Thunnus thynnus thynnus*). Rev. Fish. Sci. 15, 183–210.

Mylonas, C.C., De La Gandara, F., Corriero, A., Rios, A.B., 2010a. Atlantic bluefin tuna (*Thunnus thynnus*) farming and fattening in the Mediterranean Sea. Rev. Fish. Sci. 18, 266–280.

Mylonas, C.C., Fostier, A., Zanuy, S., 2010b. Broodstock management and hormonal manipulations of fish reproduction. Gen. Comp. Endocrinol. 165, 516–534.

Mylonas, C.C., Zohar, Y., Pankhurst, N.W., Kagawa, H., 2011. Reproduction and broodstock management. In: Pavlidis, M., Mylonas, C.C. (Eds.), Sparidae: Biology and Aquaculture of Gilthead Seabream and Related Species. Blackwell Science Publishers, London, pp. 95–131.

Mylonas, C.C., Duncan, N.J., Asturiano, J.F., 2016. Hormonal manipulations for the enhancement of spermiation in cultured fish and evaluation of sperm quality. Aquaculture (under review).

Nagahama, Y., 1997. 17a,20b-Dihydroxy-4-pregnen-3-one, a maturation-inducing hormone in fish oocytes: mechanisms of synthesis and action. Steroids 62, 190–196.

Ottolenghi, F., 2008. Capture-based aquaculture of bluefin tuna. In: Lovatelli, A., Holthus, P.F. (Eds.), Capture-Based Aquaculture. Food and Agriculture Organization of the United Nations, Rome, pp. 169–182.

Pousis, C., De Giorgi, C., Mylonas, C.C., Bridges, C.R., Zupa, R., Vassallo-Agius, R., et al., 2011. Comparative study of liver vitellogenin gene expression and oocyte yolk accumulation in wild and captive Atlantic bluefin tuna (*Thunnus thynnus* L.). Anim. Reprod. Sci. 123, 98–105.

Pousis, C., Santamaria, N., Vassallo-Agius, R., Zupa, R., Bello, G., Bridges, C.R., et al., 2012. Expression of vitellogenin receptor gene in the ovary of wild and captive Atlantic bluefin tuna (*Thunnus thynnus*). Anim. Reprod. Sci. 132 (1-2), 101–110.

Rooker, J.R., Bremer, J.R.A., Block, B.A., Dewar, H., De Metrio, G., Corriero, A., et al., 2007. Life history and stock structure of Atlantic bluefin tuna (*Thunnus thynnus*). Rev. Fish. Sci. 15, 265–310.

Rosenfeld, H., Mylonas, C.C., Bridges, C.R., Heinisch, G., Corriero, A., Vassallo-Agius, R., et al., 2012. GnRHa-mediated stimulation of the reproductive endocrine axis in captive Atlantic bluefin tuna, *Thunnus thynnus*. Gen. Comp. Endocrinol. 175, 55–64.

Safina, C., 1995. The worlds imperiled fisheries. Sci. Am. November 1995, 49–53.

Saillant, E., Chatain, B., Menu, B., Fauvel, C., Vidal, M.O., Fostier, A., 2003. Sexual differentiation and juvenile intersexuality in the European sea bass (*Dicentrachus labrax*). J. Zool. 260, 53–63.

Sarasquete, C., Cardenas, S., De Canales, M.L.G., Pascual, E., 2002. Oogenesis in the bluefin tuna, *Thunnus thynnus* L., A histological and histochemical study. Histol. Histopathol. 17, 775–788.

Sawada, Y., Okada, T., Miyashita, S., Murata, O., Kumai, H., 2005. Completion of the Pacific bluefin tuna *Thunnus orientalis* (Temminck et Schlegel) life cycle. Aquaculture Res. 36, 413–421.

Sawada, Y., Seoka, M., Kato, K., Tamura, T., Nakatini, M., Hayashi, S., et al., 2007. Testes maturation of reared Pacific bluefin tuna *Thunnus orientalis* at two-plus years old. Fish. Sci. 73, 1070–1077.

Schaefer, K.M., 2001. Reproductive biology of tunas. In: Block, B.A., Stevens, E.D. (Eds.), Tuna: Physiology, Ecology, and Evolution. Academic Press, San Diego, pp. 225–270.

Suquet, M., Cosson, J., de la Gándara, F., Mylonas, C.C., Papadaki, M., Lallemant, S., et al., 2010. Sperm features of captive Atlantic bluefin tuna (*Thunnus thynnus*). J. Appl. Ichthyol. 26, 775–778.

Susca, V., Corriero, A., Bridges, C.R., De Metrio, G., 2001. Study of the sexual maturity of female bluefin tuna: purification and partial characterization of vitellogenin and its use in an enzyme-linked immunosorbent assay. J. Fish Biol. 58, 815–831.

Taranger, G.L., Carrillo, M., Schulz, R.W., Fontaine, P., Zanuy, S., Felip, A., et al., 2010. Control of puberty in farmed fish. Gen. Comp. Endocrinol. 165, 483–515.

Zohar, Y., 1988. Gonadotropin releasing hormone in spawning induction in teleosts: basic and applied considerations. In: Zohar, Y., Breton, B. (Eds.), Reproduction in Fish - Basic and Applied Aspects in Endocrinology and Genetics. INRA Press, Paris, pp. 47–62.

Zohar, Y., 1989. Fish reproduction: its physiology and artificial manipulation. In: Shilo, M., Sarig, S. (Eds.), Fish Culture in Warm Water Systems: Problems and Trends. CRC Press, Florida, pp. 65–119.

Zohar, Y., 1996. New approaches for the manipulation of ovulation and spawning in farmed fish. Bull. Natl. Res. Inst. Aquaculture(Suppl. 2), , 43–48.

Zohar, Y., Gordin, H., 1979. Spawning kinetics in the gilthead sea bream, *Spaurs aurata* (L.) after low doses of human chorionic gonadotrpoin. J. Fish. Biol. 15, 665–670.

Zohar, Y., Mylonas, C.C., 2001. Endocrine manipulations of spawning in cultured fish: from hormones to genes. Aquaculture 197, 99–136.

Zohar, Y., Pagelson, G., Gothilf, Y., Dickhoff, W.W., Swanson, P., Duguay, S., et al., 1990. Controlled release of gonadotropin releasing hormones for the manipulation of spawning in farmed fish. Control. Rel. Bioact. Mater. 17, 51–52.

Zohar, Y., Harel, M., Hassin, S., Tandler, A., 1995. Broodstock management and manipulation of spawning in the gilthead seabream, *Sparus aurata*. In: Bromage, N., Roberts, R.J. (Eds.), Broodstock Management and Egg and Larval Quality. Blackwell Science Press, London, pp. 94–117.

Zohar, Y., Muñoz-Cueto, J.A., Lareyre, J.J., Elizur, A., Kah, O., 2010. Neuroendocrinology of reproduction in teleost fish. Gen. Comp. Endocrinol. 165, 438–455.

Zupa, P., Fauvel, C., Mylonas, C.C., Santamaria, N., Valentini, L., Pousis, C., et al., 2013. Comparative analysis of male germ cell proliferation and apoptosis in wild and captive Atlantic bluefin tuna (*Thunnus thynnus*). J. Appl. Ichthyol. 29, 71–81.

Chapter 8

Tuna Farming in Japan and Mexico

Alejandro Buentello[1], Manabu Seoka[2], Keitaro Kato[3] and Gavin J. Partridge[4]
[1]Ichthus Unlimited, LLC, West Des Moines, IA, USA, [2]Skretting ARC/Japan, Fukuoka, Japan, [3]Fisheries Laboratory of Kinki University, Kinki University, Wakayama, Japan, [4]Australian Centre for Applied Aquaculture Research, Challenger Institute of Technology, Fremantle, WA, Australia

8.1 INTRODUCTION

Tuna is one of the most important seafood commodities in the world, with annual global production being ~3.5 million metric tons (MT), and accounting for ~5% of total fishery products destined for human consumption (FAO, 2010). One-third of the landed tuna is sold as fresh, chilled, or frozen fish and is exported to the major tuna markets of Japan, the United States, and the European Union (EU; Paquotte, 2003). Two nations, Japan and Taiwan, dominate Pacific bluefin tuna (PBFT, *Thunnus orientalis*) fishing, with the largest volume caught. Other countries with important tuna fleets (fishing for several scombrid species) are Indonesia, the Philippines, Spain, Republic of Korea, Papua New Guinea, France, Ecuador, Mexico, Maldives, Islamic Republic of Iran, the United States, Seychelles, Venezuela, Sri Lanka, Colombia, China, Vanuatu, Panama, and Ghana (FAO, 2008).

That this chapter encompasses tuna farming activities in both Japan and Mexico is fitting because, at least for culture activities with PBFT, both nations share the same foundation with a common Pacific-wide stock of which Japan and Mexico presently harvest about 75% and 15% of its total volume, respectively (Figure 8.1), with other nations such as South Korea, Taiwan, and the United States harvesting only 5%, 4%, and 1%, respectively (ISC, 2007).

D.D. Benetti, G.J. Partridge, A. Buentello (Eds): Advances in Tuna Aquaculture.
DOI: http://dx.doi.org/10.1016/B978-0-12-411459-3.00007-2

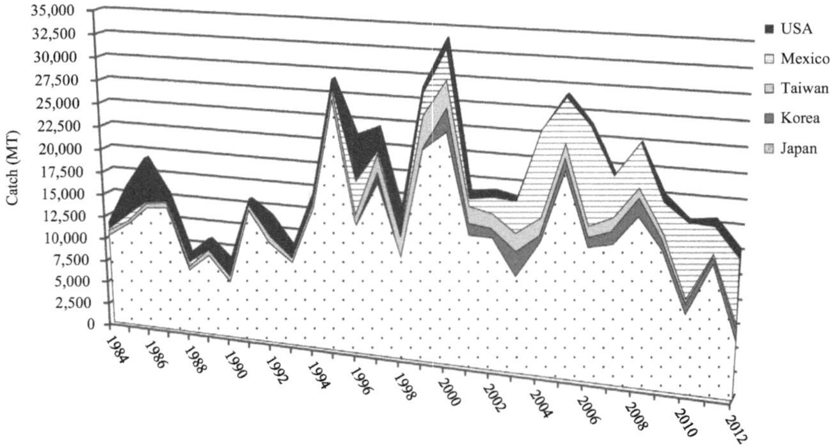

FIGURE 8.1 Annual landings of PBFT from 1984–2012. EPO (black tones) and WPO (grey tones) in the North Pacific Ocean. *Modified from IATTC, 2014.*

8.1.1 Global Tuna Sushi and Sashimi Consumption

Sushi is a traditional Japanese food consisting of a steam-cooked rice ball combined with raw fish slices (sashimi) and tuna sashimi is an essential item in Japanese restaurants. According to a survey of the Japanese Ministry of Agriculture, Forestry and Fisheries (JMAFF, 2013), it has been estimated that globally there are about 55,000 Japanese-style restaurants, including those located in Japan. A report from Japan's External Trade Organization (JETRO, 2010) indicates that the number of Japanese-style restaurants in the United States grew from ~3000 in 1992 to ~14,000 in 2010, emphasizing the great expansion and popularity of Japanese-style food. Another JMAFF survey reported that over 50% of Japanese-style restaurants in the United States and the United Kingdom are sushi-style restaurants or restaurants serving sushi and/or sashimi as one of the main menu items (JMAFF, 2006).

The Organization for the Promotion of Responsible Tuna Fisheries (OPRT) identifies Japan as the largest buyer of tuna sashimi, consuming over 300,000 MT/y, followed by the United States (90,000 MT/y), South Korea (20,000 MT/y), China (10,000 MT/y), and both Taiwan and the EU (8000 MT/y each; OPRT, 2015). The sizeable Japanese consumption is partly due to Japanese people enjoying tuna sashimi not only in restaurants but also at home. Among tuna sashimi, prime grade PBFT, ABFT (*Thunnus thynnus*), and SBFT (*Thunnus maccoyii*) sashimi are considered most favorably in Japan because of these species' unique flavor. The next class is big eye tuna (*Thunnus obesus*) followed by YFT (*Thunnus albacares*), which are mainly sold in sushi bars and supermarkets.

The topics addressed in this chapter describe efforts to resolve some of the most critical issues currently limiting the success and permanence of the tuna aquaculture industry. Our intent is to provide information and insights that will support tuna aquaculture in a sustainable manner.

8.2 FARMING IN JAPAN

The first effort to farm PBFT was carried out in Japan during the 1970s (Miyashita, 2002). This project was a collaboration between the National Research Institute for Far Seas Fisheries, jointly with Tokai University, the Fisheries Research Institutes of Shizuoka Mie Nagasaki, Kochi, and Kagoshima Prefectures and Kinki University's Oshima Experimental Station and Fisheries Laboratory (Wakayama Prefecture). This 5-year project aimed at developing technologies for culturing PBFT and evaluating the possible industrialization of these technologies. Although it was primarily aimed at raising captured fish for use as future brood fish, the lessons learned were directly applicable to farming.

In the case of Kinki University, young PBFT were initially captured by small fixed nets installed around Kushimoto, Oshima, and Shionomisaki, in the Wakayama Prefecture; however, this method resulted in only small catches and very low survival rates due to the stress of confining these fragile animals in the small fixed nets. Towline fishing was then tested on the basis it could catch a relatively high number of fish. Nevertheless, the survival rate using this method was less than 30% due to the handling stress associated with detaching the young animals from the fishing hooks. Young tuna are very sensitive to handling and their skin is easily damaged by contact with human hands.

This first project was completed in 1975, at which time all participating organizations, with the exception of the Fisheries Research Institutes of Kochi Prefecture and Kinki University, terminated their efforts to culture tuna. The effort of the Fisheries Research Institutes of Kochi Prefecture terminated in 1981, but Kinki University continued to develop a variety of farming technologies independently (Sawada, 2005). For example, fishing methods were improved by removing the barbs of the hooks to facilitate the fish's smooth release then placing them into small buckets (18–20 L). Once the fish had calmed down, they were then transferred into transport tanks on the fishing boat. Through these efforts, survival rate improved to about 80% by 1974 and the young tuna captured in that year were grown successfully to adult stages in a net-cage (Kumai, 2012).

In a more modern industrial manner, tuna farming in Japan began at a similar time to the rest of the world, with the first commercial harvests of PBFT reported in 1993 (Tada, 2010). Production between this time and 2010 was reported by this author to be between 2000 and 7000 MT/yr. It was not until 2010 that JMAFF introduced mandatory reporting for tuna farmers and the first reports were released in 2011. Key production figures from these reports from 2011 to 2014 are summarized in Table 8.1. Tuna farming in Japan is now a ~¥30 billion/year industry (~US$250 million; JFA, 2015).

TABLE 8.1 PBFT Domestic Production 2011–2014

	2011	2012	2013	2014
Companies (#)	94	94	92	95
Farms (#)	137	140	147	150
Cages (#)	949	1191	1362	1375
Wild juveniles stocked (#)	534,000	206,000	347,000	221,000
Hatchery-produced juveniles stocked (#)	141,000	268,000	264,000	298,000
Total production (tons)	9043	9592	10,396	14,713
Production via hatchery-produced juveniles (tons)	NR	244	276	387

2014 data only a forecast. Early juveniles <8 cm TL.

As previously described in this book, the practice of tuna farming differs in Japan to the rest of the world. The main difference is that the farming cycle begins with young wild fish weighing 100 g to 2 kg. These fish are caught in near shore waters associated with the Kuroshio and Tsushima currents in areas surrounding the Kii and Satuma Peninsulas, in the Shikoku and Bungosuido areas and around the Goto, Tsushima, and Iki Islands (JFA, 2015). Fish are caught either by purse-seine net or by trolling barbless hooks then either held in local farms for acclimation to captive conditions for a few months before being transferred to the farming sites or directly transferred to these sites via well boats. The wells within the transfer boats often have stripes painted on the walls aimed at minimizing collision deaths that are described elsewhere throughout this book. Masuma et al. (2008) reported that survival between capture and stocking into farming pens is approximately 60%. As is shown in Table 8.1, between 200,000 and 500,000 juvenile fish are caught from the wild each year. The variation is due primarily to weather and poor catches occurring in 2005, 2009, 2010, and 2012 (JFA, 2015).

Farming takes place in 14 of Japan's 47 prefectures. All farms are located in southern Japan where suitable water temperatures occur. The minimum water temperature is ideally no less than 13°C, although some farming sites do experience temperatures as low at 10°C (JFA, 2015). Sites are selected that face the open ocean, with good flushing to supply highly oxygenated water and are typically located in water of 30 to 50 m depth. Sites near major rivers should be avoided to prevent low salinity and other problems associated with run off. Cages are typically 10 to 15 m deep and are either round (20 to 30 m in diameter) or square (40 to 80 m per side). Fish are fed manually or using mechanical feeding devises with locally caught bait fish being dispensed

TABLE 8.2 Annual Production of PBFT in Japan (2013) by Prefecture

Prefecture	Production (MT)	% of total
Kagoshima	3222	31.0
Nagasaki	3070	29.5
Kochi	1163	11.2
Mie	696	6.7
Ehime	682	6.6
Oita	492	4.7
Wakayama	203	2.0
Others	868	8.3
Total	10,396	100

(either fresh or frozen). The farming period in Japan is typically 2 to 3 years and fish are grown to a market size of 30 to 70 kg (JFA, 2015).

As shown in Table 8.1, the Japanese tuna farming industry currently comprises over 90 enterprises which operate about 150 tuna farms with a total of over 1300 cages. Major fisheries enterprises, such as Nissui Group, Maruha Nichiro Group, and Kyokuyo Group, and major trading companies, such as Mitsubishi Corporation (Mitsushima Suisan) and Sojitz Corporation (Sojitz Tuna Farm Takashima), are engaged in large-scale production (Minato Shinbun, 2012).

The total annual aquaculture production of PBFT by prefecture in 2013 is shown in Table 8.2 and totaled 10,396 MT, of which only 276 MT originated from hatchery produced juveniles (JMAFF, 2013). The Kagoshima and Nagasaki prefectures together produce approximately 60% of the country's total production. Minor prefectures producing ~8.3% of the tonage in 2013 include Kyoto, Yamaguchi, Kumamoto, and Okinawa.

The Nagasaki Prefecture has the highest number of farms (62) with only 24 farms located in the Kagoshima Prefecture. In 2013, these farms were owned by tuna farming companies of which 40 are ascribed to the Nagasaki Prefecture and 8 belong to Kagoshima (JMAFF, 2013). The concentration of tuna farms in this area follows favorable environmental conditions for PBFT production with the majority of farms in the Kagoshima Prefecture being located in a subtropical climate which supports an increase of over 50% in growth rates as compared to that possible in other PBFT culturing regions in Japan.

TABLE 8.3 Sample PBFT Operation Costs in Japan (US$ in 2006)

Income	1,202,737.0
Sales amount (kg)	41,343.0
Number of shipments	1029.0
Unit price/kg	29.0
Expenses	1,295,592.0
Feed costs	564,675.0
Employment costs	174,600.0
Young fish	114,708.0
Young fish transportation	29,408.0
Fuel cost	7933.0
Repair costs	25,917.0
Depreciation cost	36,975.0
Fish box and ice for shipping	19,642.0
Transport cost for shipping	82,700.0
Rental expense	82,600.0
Selling costs and administrative charge	70,692.0
Initial inventory	825,025.0
Final inventory	1,421,058.0
Profit	503,283.0
Profitability index (%)	41.8

8.2.1 Production Inputs and Costs

Ono (2013) analyzed several management strategies for PBFT aquaculture companies in Japan and summarized the yearly operational costs in Table 8.3. For 2006, an undisclosed farm reported shipping 1029 PBFT (41,343 kg) to Japanese markets at a unit price of US$29 per kg. Therefore, total annual income for this farm was US$1.2 million. Expenses including feed, payroll, young fish, etc., added up to US$1.3 million. Initial and final inventory of the year was 0.8 and 1.4 million, respectively. Thus, the profit margin of the year was 0.5 million which accounts for approximately 42% of the total income.

8.3 CLOSED-CYCLE CULTIVATION OF PBFT IN JAPAN

The term "closed-cycle cultivation" refers to cultivation in which fish of all life stages are kept in artificial rearing conditions, and most commercially cultured seawater fish, such as seabream (*Pagrus major* and *Sparus aurata*), European seabass *Dicentrarchus labrax* and Japanese flounder *Paralichthys olivaceus*, are produced using the closed-cycle cultivation technologies. Aquaculture production using wild fish depletes wild fish stocks, causing an unstable supply of fish for both fisheries and aquaculture. Spawning of captive and domesticated PBFT has been made possible in Japan utilizing closed-cycle cultivation technology which allows for more sustainable production of PBFT juveniles. In addition, the genetic improvement of cultured marine fish including red sea bream *Pagrus major* (Murata et al., 1996), Japanese flounder (Araki et al., 2011), and yellowtail *Seriola quinqueradiata* (Yoshida et al., 2011) has been achieved in Japan by selection of desirable traits for aquaculture in targeted species. The entire genome of PBFT has recently been sequenced (Nakamura et al., 2013) and now that the life cycle has been closed, attempts are underway for genetic selection to improve production characteristics of farmed PBFT (Agawa et al., 2009).

The captive propagation of tuna species using closed-cycle technologies is essential for fisheries management considerations and also for the improvement of production efficiencies with cultured marine fish. Here, we focus on the approaches used to complete the closed-cycle cultivation of PBFT at the Fisheries Laboratory of Kinki University and two federal government (Fisheries Research Agency) facilities; the Amami Laboratory (Research Center for Tuna Aquaculture) and the Seikai National Fisheries Research Institute.

8.3.1 Fisheries Laboratory of Kinki University

The Fisheries Laboratory of Kinki University was established in 1948 for cultivating the seas and increasing seafood production. By 1954, techniques for the ranching culture of yellowtail in net cages had been developed by Kinki University, and the technology transferred to commercial aquaculture operations. This technology has been applied in several regions of Japan for the culture of many species of marine fish. In 1965, the world's first successful juvenile production of Japanese flounder was achieved in the laboratory and the techniques developed for this species were expanded to 17 other commercially important species of fish, including yellowtail, greater amberjack *Seriola dumerili*, striped jack *Pseudocaranx dentex*, kelp grouper *Epinephelus bruneus*, and PBFT.

As previously described, Kinki University has been at the forefront of developing culture techniques for PBFT since 1970 and was the only group to continue PBFT farming efforts following the first project that was completed in 1975. In 1979, the PBFT reared from juveniles sourced in 1974 spawned spontaneously in a net-cage at the Oshima station and fertilized eggs were collected. This was the world's first reported spawning and egg collection of PBFT in captivity. Eggs collected from spawns in 1980 and

1982 (more than 10 batches) were utilized for larval rearing efforts. However, survival rates were extremely low and the longest survival period attained was only 57 days, during which time fish grew to 98 mm in total length. This outcome was not surprising given the status of marine fish larval rearing at that time. Spawning of PBFT at Kinki University ceased for 11 years thereafter. The reason(s) for the interruption in spawning remain unclear, but the geographic location and resulting environmental parameters at the Oshima station appear to have been sub-optimal for PBFT spawning during that period. Kinki University therefore established a second station in 2001 at Amami Island in the Kagoshima Prefecture (tropical climate) to optimize PBFT spawning (Murata and Miyashita, 2012). Further details regarding the consistency and reliability of spawning of PBFT in Japan are given in sections below. In 1994, the 12-year hiatus in spawning at the Oshima station came to an end, and 7-year-old PBFT brood fish started spawning again. Spawning activity by the brood fish was observed during four of five spawning seasons until 1998.

8.3.2 Larval and Juvenile Rearing of PBFT

Larval and juvenile rearing was attempted with many batches of eggs collected from the Oshima station between 1994 and 1998. Survival rates in these early attempts were also very low (<0.4%; Sawada et al., 2005) and important mortality events were identified in four distinct critical periods. The first critical mortality event is "the surface death of larvae," documented to occur within 10 days after hatching (dah) (Sakamoto et al., 2012). Surface mortalities were observed from 1 to 4 dah as the larvae became trapped by surface tension between the air and seawater interface when swimming to the water surface to first fill their swim bladders. The correlation between physical surface tension and biological swimming thrust was not then understood. This source of mortality tended to be more prevalent for larvae obtained from low-quality eggs. Research demonstrated that adding a film of oil on the water surface prevents surface entrapment; however, it also interrupts or delays the inflation of the swim bladder.

The second source of high mortality is "sinking death," caused in part by the physical contact of the PBFT larvae with the bottom of the rearing tank, resulting in abrasive damage to the fragile skin. This particular mortality appears at 4–10 dah, and is more problematic in terms of the mortality it causes than surface death (Masuma et al., 2011). It has been traced to the larval tuna's inability to maintain their position in the water column by swimming against the gravitational force and is more prevalent at nighttime. The sinking speed of larvae depends on their body density which in turn is controlled by the swim bladder's volume, deflating during the day, and inflating at night. A larger air volume in the swim bladder translates into a lesser body density. Because the body density of PBFT larvae is higher than seawater, even with an inflated swim bladder, and because foraging activity ceases during nighttime, tuna larvae stop swimming and the gravitational pull becomes larger than the swimming force; thus, the larvae tend to sink to the bottom of the rearing tank (Takashi et al., 2006). This mortality can be reduced by controlling water circulation via aeration (Nakagawa et al., 2011) and water upwelling to prevent sinking of larvae during the night. Water upwelling is described in further detail in a section below. Failure of the larvae to inflate their swim bladder further exacerbates sinking death (Ishibashi, 2012).

A third source of mortality is caused by the cannibalistic behavior of larvae and juveniles, beginning from ~10 dah (Ishibashi, 2012). The digestive system of larval PBFT develops earlier than in other cultivated marine fish species such as sea bream, sea bass, and flounder. The gastric gland and pyloric caeca of larval PBFT differentiate at 10 and 15 dah, respectively (Miyashita et al., 1998). Moreover, aggressive behavior toward peers is noted in larval PBFT when they reach a total length of about 7 mm (~10 dah). PBFT larvae therefore initiate cannibalistic aggression at a very early stage compared to other fish. Aggressive behavior and cannibalism is observed not only in larval stages, but also in the early juvenile stages. To minimize the effects of cannibalism, it is important to switch the live prey offered, from *Artemia* spp., to newly hatched fish larvae such as Japanese parrot fish *Oplegnathus fasciatus* or red sea bream (Sawada et al., 2005). Feeding on fish larvae aligns with the natural piscivorous behavior of tuna larvae but also has the benefit of overcoming the "growth failure" associated with feeding *Artemia*, as outlined in further detail below. Although feeding yolk sac is effective, it has many disadvantages including the costs associated with maintaining other broodstock as well as acting as a vector for the transmission of viral diseases (Masuma et al., 2011). Size grading is another effective measure in reducing cannibalism because larger fish tend to attack smaller fish. However, as described by Partridge (2013), size grading juvenile tuna is difficult as they do not respond well to handling due to their sensitive skin.

The fourth type of death is due to collision with tank and/or walls, beginning from ~30 dah, as detailed in the section below.

8.3.3 Collision Death and Survival of Juveniles after Transport and Stocking into Oceanic Net Pens

Juvenile fish aged 30 to 80 dah, frequently collide with the tank walls or net-cage and the number of live fish therefore dwindles over time (Miyashita, 2002). One plausible reason for this occurrence is that juvenile PBFT of this age undergo rapid development of the caudal fin, which provides the forward thrust for swimming, but lack sufficiently developed pectoral, pelvic, dorsal, and anal fins to control their tridimensional swimming direction (Miyashita, 2002). Detailed documentation of contact or collision deaths after burst swimming has revealed that the majority of dead fish exhibit skull fractures, vertebral dislocations, abrasions, and other injuries on the head and sides of their body. Collision mortality of juveniles is not a particular problem in the culture of red sea bream, sea bass, yellowtail, amberjack, or flounder.

Improvement of PBFT juvenile survival after transport from land-based tanks into oceanic net pens is one of the most challenging issues in the production of hatchery-reared PBFT (Miyashita, 2002). This transfer is required to decrease the opportunity for contact or collision of the fish against walls and to increase the swimming space; however, early attempts to move hatchery-produced PBFT to net pens resulted in 100% mortality within a few days after the transportation event. Initially, this mortality was attributed to stress and/or traumatic injury incurred during the hauling process. However, a number of studies by researchers at Kinki University have demonstrated that the major underlying factor contributing to these mortalities is the poor scotopic vision (vision under low light conditions) of PBFT juveniles which results in tank, net and/or congener collisions at night. Survival measured 23 days post transport was improved to 73% simply by providing 200–3000 lux of light intensity at the water surface via overhead lamps; whereas, unlit controls averaged 12% survival (Ishibashi et al., 2009). The scotopic visual threshold of juvenile PBFT is 11.1 log photons/cm^2/s, which is 40 times inferior to the scotopic threshold of other marine teleosts such as juvenile red sea bream, tiger puffer *Takifugu rubripes*, greater amberjack, and convict band grouper *Epinephelus septemfasciatus* (Ishibashi et al., 2009). Honryo et al. (2013, 2014), recommended an optimized nighttime lighting period lasting for 8–12 days after transport, which also results in an improved growth performance and lower stress levels, as estimated using whole-body cortisol and glucose readings. There seemed to be no further improvements in growth or stress indicators by prolonging nighttime illumination beyond day 12 after transport.

Although improvements have been made over the years, even with nocturnal illumination, survival after transport remains low at Kinki University facilities, averaging 69 ± 13% (44–88%) after 10 days and only 46 ± 11% (27–63%) 30 days after transfer (Okada et al., 2014). These low values suggest the possibility that other factors may contribute to PBFT juvenile mortality. As a case in point, a 2012 transport exercise conducted with 9441 PBFT juveniles used nocturnal illumination starting just before dusk and terminating shortly after dawn (~1700 to 0800 h; Okada et al., 2014). A total of 4449 fish were dead by day 28 after the transfer event (48% mortality) with over 50% of these losses occurring during the first week after stocking. This experience corroborates previous observations at Kinki University where documented mortalities range beyond the normal collision (structural injuries) or dermal abrasion. These include observations of poor growth, recorded mostly during the first week after stocking and are perhaps related to the difficulty in monitoring feeding activity at the oceanic net pen and/or in the transport vessel, which is significantly more complex than in land-based facilities. Observable symptoms include emaciated fish with no tissue deposition and weight gain. In addition, the ingestion of foreign organic/inorganic particles including tiny pieces of tree branches, wood, Styrofoam, and rust from the cage's steel frame results from the animals' inability to discriminate between genuine food items and particles that trigger their natural feeding responses and only become available to them during transport or stocking events. Strategies to provide suitable nourishment to juvenile PBFT during transit time should result in further improvements in survival at this critical period.

Despite remaining challenges, the cumulative effects of the strategies described above have resulted in incremental and substantial increases in the survival of larval and juvenile PBFT, year after year. From the two PBFT laboratory-produced groups in 1995 and 1996, 17 and 35 fish survived until 1997, respectively. In 2002, they reached 6 and 7 years old and started spawning. The world's first closed-cycle cultivation of PBFT was achieved at this moment (Sawada et al., 2005). It took 32 years to attain this goal from the beginning of research on PBFT farming. Kinki University continues to develop larval and juvenile rearing techniques to improve survival. The third generation of PBFT were produced in 2007, and 1500 young PBFT were shipped to a commercial PBFT farm. Kinki University has progressively increased its PBFT juvenile ($\sim$30 cm TL) production, with shipments to tuna farms of 5000, 30,000, and 80,000 juvenile fish in 2008, 2009, and 2012, respectively; and, the number of shipped PBFT is now more than 10% of the total number of cultured PBFT per year in Japan. Beginning in 1986, a number of private companies also started commercial hatchery operations in Japan including Maruha Nichiro, Nippon Formula Feed Manufacturing Company, Takuyo Ltd., the Nissui Group, Mitsushima Suisan, and Kaneko Sangyo. However, production data for private efforts is not yet readily available to the public.

The hatchery methods used to produce PBFT at Kinki University are described by Sawada et al. (2005) and were reviewed by Partridge (2013). Larval rearing tanks are typically quite large in size (20 to 30 m^3). They are supplied with UV filtered water to exclude viral nervous necrosis (VNN). Tanks are initially static, but flow increases as the larvae grow up to an exchange rate of 500%/day. Water temperature is typically maintained between 23°C and 26°C. Newly hatched larvae measure approximately 3 mm in total length and begin feeding on 3 dah. Detailed descriptions of the morphological development of PBFT can be found in Kaji et al. (1996) and Kawamura et al. (2003). PBFT larvae are reared at lower densities than most marine fish, typically less than 10 larvae per liter (Kaji et al., 1996). The first feed offered is enriched rotifers (until 19 to 25 dah) and in the presence of green water (Nakagawa et al., 2007). Feeding enriched *Artemia* typically commences around 10 dah and whilst historically *Artemia* were fed until around 25 dah, this feeding period has now been reduced in favor of the early introduction of other fish larvae as described above, as extended feeding on *Artemia* results in growth failure. This failure has been attributed mainly to a lower concentration of key phospholipid fractions in *Artemia* compared to yolk sac larvae (Seoka et al., 2007, 2008). Larvae are typically weaned off live fish and onto a diet of minced fish; however, recent advances in weaning diets specifically for PBFT are showing great promise.

8.3.4 The Fisheries Research Agency

Known formerly as the Japan Sea Farming Association, the Fisheries Research Agency (FRA) is the national research and development agency for wild fisheries and aquaculture.

The FRA established in 1985 the Yaeyama Station at Ishigaki Island in the Okinawa Prefecture and from 1985 to 1997 focused on developing techniques for PBFT reproduction, with rearing of juvenile and young in captivity (Masuma, 2008). Young PBFT were transported to Ishigaki Island over a long distance (1300–1500 km) and time (74–113 h) by boat. The growth of PBFT reared in the Yaeyama Station was faster than that at other farming sites in Japan, suggesting greater climatic suitability of subtropical regions such as this for PBFT farming. However, broodstock raised in the Yaeyama Station never spawned, raising doubts about suitability of conditions at the Yaeyama Station for the spawning of PBFT where the sea surface temperatures ranged from 20°C to 31°C.

In 1994, the FRA established the Amami Station at Amami Island in the Kagoshima Prefecture (Masuma, 2008). A unique characteristic of the Amami Station is that it has not only normal floating net-cages for broodstock but also a net wall at the mouth of a natural 14 hectare fjord (Figure 8.2) to

FIGURE 8.2 Amami Station: (1) natural fjord, (2) land-based facility, (3) metal cable and netting for the enclosure of PBFT brood stock, (4) circular cages with surface tarp-ring, (5) and (6) feeding and feed items—horse mackerel with vitamin-carrageenan premix, (7) egg collector, (8) concrete tanks for the production of fish larvae as prey for PBFT larvae. *Photographs by A. Buentello with permission of S. Shiozawa.*

create a more natural environment for the rearing of PBFT broodstock. Although no remarkable differences in PBFT growth have been observed between the net-cage animals versus those kept in the natural bay, survival of PBFT in the bay has been considerably higher than that in the net-cages. Motility of sperm obtained from the brood fish was also higher from those animals maintained in the bay than those maintained in the net-cages (Masuma, 2008).

Spawning of captive PBFT at the Amami Station started in 1997, and has occurred every year since. Research insights regarding spawning ecology, behavior, and other aspects of PBFT in captivity have been obtained. Further, the optimized sequence of feed items, management of water quality in rearing tanks, and countermeasures against VNN (a betanodavirus) infections in seedling production of PBFT have all been characterized and published by scientific personnel working at this facility.

Masuma et al. (2008, 2011) described measurements of GSI and DNA analyses of spawned eggs and concluded from these measurements that the rate of maturity in PBFT broodstock held in cages throughout Japan is generally quite low and that only a limited number of fish contribute to egg production. This was identified by these authors as a major constraint to expansion. These authors also noted that of the nine sites holding PBFT broodstock in sea cages, those in Amami Island have the most stable and consistent spawning which typically occur during June and July, but sometimes as early as May and until as late as November. Seawater temperature at Amami ranges from 20°C to 28°C throughout the year and spawning typically commences when water temperature reaches approximately 23°C. It was noted that if water temperatures rise too slowly or too late in the season that spawning may be delayed or may not occur at all. These data suggest that the water temperatures at Amami are the most appropriate for inducing final oocyte maturation and spawning in PBFT and also point to the critical importance of locating broodstock in sites with the most appropriate temperature regimes. Although males are observed to mature as young as 2 years in captivity and females as young as 3, most specimens mature at an age of at least 5 years.

To complement the technical developments achieved for spawning PBFT in sea cages and to improve the predictability of egg collection from PBFT broodstock, the FRA launched the construction of a new land-based broodstock facility for tuna aquaculture at the Seikai National Fisheries Research Institute in the Nagasaki Prefecture (FRA, 2011). The facility, with a floor space of 4500 m^2, began construction in 2011 at a cost of ~¥2.1 billion, (~US$21.2 million). The construction project included two spawning tanks each 1880 m^3 in volume (20 m in diameter and 6 m depth) and was successfully completed in 2012. The facility uses state of the art recirculating systems that can control water temperature between 13°C and 30°C and LED lighting to replicate daylight and moonlight conditions. In May and June of

2013, two hundred 2-year PBFT were transported from the Amami Station via sea, with 127 fish arriving alive. A further 40 died between May and November, leaving 89 fish (41 and 48 in the two tanks). These brood fish are fed commercial pellets, frozen chub mackerel and Japanese common squid 3 days per week. The first spawning of 200,000 eggs was observed at this new facility on May 16, 2014, in one of the tanks when water temperature was at 20.2°C. By controlling the spawning season within these tanks, stable egg supply is expected to be achieved at this facility in the near future. The plan is to ship fertilized eggs obtained from this facility back to the hatchery facilities at Amami Station for larval rearing. The objective for this new facility is to produce sufficient eggs to grow 100,000 3-month-old juveniles per year.

Larval rearing protocols at FRA hatcheries are similar to those described above for Kinki University. A new indoor larval rearing system was commissioned at the Amami Station in 2008 at a cost of ¥600 million (~ US$6 million; Shiozawa 2013, personal communication). It comprises $7 \times 50\ m^3$ octagonal larval rearing tanks and $6 \times 20\ m^3$ larval rearing tanks. Tanks are dark green in color and have integrated titanium heaters to maintain a stable rearing temperature of ~26°C. In an effort to reduce sinking death, Masuma et al. (2011), reported using a 1 m^3/h water pump at the base of the 50 m^3 tanks to create upward and anticlockwise current. Survival rates as high as 65% during the first 10 days were achieved by using these water-driven upwelling; however, these authors noted that these survival rates were not consistently achieved and that more refinement and research is still required.

8.3.5 Conclusions and Future Outlook

Japan has a long history of tuna farming and significant achievements have been made throughout this 40-year history. Particular progress has been made in recent years with significant numbers of hatchery-produced fish now being put to sea for commercial farming by both Kinki University, government agencies, and several private companies. Despite such great progress, challenges still remain at all stages.

The number of PBFT eggs collected has increased greatly in recent years to reach nearly 1 billion/year at Kinki University. This has allowed many more juvenile fish to be produced; however, broodstock maturation and spawning remain inconsistent and further research is required to optimize spawning and to have more fish within each broodstock cage contributing to overall spawning. The new land-based brood system at Nagasaki aims to improve spawning reliability by providing a controlled environment and aims at producing 100,000 3-month-old PBFT juveniles per year which will significantly contribute to close-cycle PBFT farming production.

Larval rearing remains challenging and further improvements to survival are required to reduce the cost of juvenile production. Nutrition during all larval and juvenile stages must be improved, including widespread implementation of effective weaning diets. Providing yolk sac larvae is very effective food for PBFT larvae, but it is very expensive and labor-intensive and has the disadvantage of introducing viral diseases such as VNN to the PBFT larvae. While the number of hatchery produced fish stocked into farming cages is increasing, approximately 50% of these still die within weeks of being stocked. Overcoming this mortality will immediately result in a doubling of the production of artificially produced fish.

In recent years, there has been greater collaboration between Japanese organizations working on tuna aquaculture and this will result in further gains that will reduce the pressure on wild fisheries and therefore make tuna farming a more sustainable industry.

8.4 TUNA FARMING IN MEXICO

Inspired by the Australian experience, the first operation on the American continent was established in 1997 at Cedros Island, Baja California (BC), Mexico (Del Moral Simanek and Vaca Rodriguez, 2009). At the time, Atunera Nair was Mexico's largest tuna fishing operator but only managed to produce 64 MT of ranched tuna over a 3-year period. Today, a dozen PBFT operations exist in Ensenada, BC; one yellowfin tuna YFT venture near Isla Isabela offshore of San Blas, Nayarit, terminated operations in 2004 (Hendrix 2015, personal communication); and, two more YFT ranches recently ceased activities in La Paz, Baja California Sur (BCS), due to the increasing difficulties in securing enough wild fish of appropriate size to stock their net pens (Buentello et al., 2009). Although PBFT, YFT, skipjack tuna (*Katsuwonus pelamis*), albacore tuna (*T. alalunga*), big eye tuna *T. obesus*, black skipjack tuna *Euthynnus lineatus*, and striped bonito *Sarda orientalis* all occur naturally in the Eastern Pacific Ocean (east of 150°W; EPO) and waters of Mexico's Economic Exclusive Zone, farming of tuna in Mexico thus far has involved only two species, namely, PBFT and YFT (Figure 8.3).

Markets for tuna meat produced in Mexico are well established in Japan and rapidly expanding in the United States, EU and Asia. In Mexico, tuna farming is generally conducted by capturing juvenile and sub-adult wild PBFT or YFT (wild sourcing), and holding them up to 8 months in specially designed oceanic cages (Buentello, 2006). During this time, the tuna are fed mackerel, sardine/squid rations, which promote rapid growth both in length and weight, and change significantly the proximate composition of their flesh. Farmed tuna are then harvested and shipped fresh to sushi/sashimi markets throughout the world typically reaching international destinations less than 72 h from harvest time.

FIGURE 8.3 YFT in oceanic cages in La Paz Bay, BCS, Mexico.

Although this industry is certainly lucrative, it is still striving to become well established and several constraints presently limit its full potential. Closed-cycle mariculture—stocking cages with laboratory-produced juveniles (F1 and beyond)—is a promising alternative to deliver high-quality tuna meat in a sustainable way. Although some Japanese tuna farmers have started to use hatchery-produced tuna juveniles from Kinki University for farming operations (Masuma et al., 2008), the global standard among tuna farming countries—including Mexico—is still based on wild-caught fish (Mylonas et al., 2010). For YFT, artificial propagation has been attempted in Panama and Indonesia (Wexler et al., 2003; Margulies et al., 2007; Morris, 2011) with a modest level of success, but YFT fattening operations are now problematic in Mexico due to the lack of sufficient amounts of stocking-size fish, obtained from the wild, to place in their cages.

Coupled with normal "learning curve" issues and complications in handling a new aquaculture species, the success and permanence of the companies currently farming tuna are at risk due to limited number of fish available for tuna ranching. Irrespective of location, each of these companies, on average, provides 25 permanent jobs and over 50 temporary jobs as well as providing business to several spin-off industries. Therefore, the successful establishment of such companies is a priority for the Mexican government at the federal level as well as state governments in which these operations are established.

8.5 PRODUCTION

The 2013 official statistical almanac of aquaculture and fisheries published by the Mexican government (Anuario Estadistico de Pesca y Acuacultura,

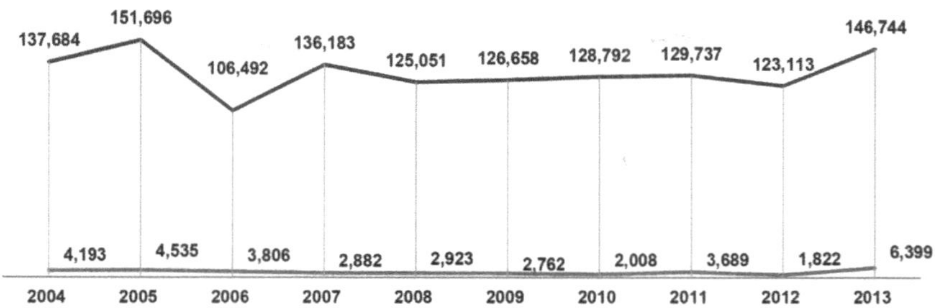

FIGURE 8.4 Annual production of scombrid species by fisheries ▬ and aquaculture ▬ (AEPA, 2013).

AEPA, 2013) reports that between 1800 to 6400 MT of scombrids (mostly PBFT) per year are produced by ranching in Mexico (Figure 8.4).

This represents only 1.5–4.4% of the total captured volume (tuna fisheries) in the country. The vast majority (>75%) of the scombrid capture by the Mexican fleet is YFT.

8.6 PRODUCTION BY GEOGRAPHIC AREA IN MEXICO

Four states, namely Sinaloa, Colima, Chiapas, and Baja California (Figure 8.5), are responsible for 94.4% of Mexican tuna fisheries production (all species), mainly because tuna fleets—including vessels servicing both tuna-ranching operations (to which are sold live fish from high-seas "sets" on tuna schools) and the cannery industry—are located in the ports of Mazatlan, Manzanillo, Puerto Madero, and Ensenada, respectively. Tuna captures in the Pacific by the Mexican fleet represent 99.2% of the country's total (128,818 MT in 2013; AEPA, 2013) (Figure 8.5).

Starting in 1998 both the Mexican regulatory agency (SAGARPA) and the monitoring agency for the distribution and abundance of scombrids in the EPO, the Inter-American Tropical Tuna Commission (IATTC), agreed on specific resolutions calling for fishery restrictions in this area, and typically spanning from November to January each year (IATTC, 1999). Therefore, it is worrisome that captures still occurred in January and December 2013 (AEPA, 2013; Figure 8.4).

For PBFT, this is a complex issue because the stock is harvested by several countries in the Pacific Ocean, with 85% of the total capture occurring in the Western Pacific Ocean (WPO). Matsukawa (2006) explains that an evolutionary response takes place when the population exceeds the carrying capacity in the WPO, causing trans-Pacific migration (~15%, but quite variable) of advanced PBFT juveniles. Furthermore, migration of PBFT into the EPO increases in years when the abundance of sardines in EPO may be declining. Therefore, there is a complex and unpredictable migration of mostly "small specimens." As explained in the previous sections, Japan utilizes significant amounts of wild-caught PBFT juveniles to stock their tuna-ranching operations adding complexity to the overall equation.

Mexican tuna ranchers argue that once a tuna is captured from the wild, its destiny (whether cannery or net pen) is irrelevant, from the perspective of fisheries management and sustainability. This set of stakeholders proposes that regulatory actions be implemented on captures occurring everywhere for that species (e.g., fish introduced into the global tuna market system by the Japanese, South Korean, and Taiwanese fleets also) and oppose regulations enforced only for those PBFT that embark on the trans-Pacific migration to California and Baja California (<15% of the total; Matsukawa, 2006) and end up available to Mexican ranches. In their view, a stringent quota system—at one point the Mexican regulatory agency CONAPESCA limited to 400 MT the amount of fish a single tuna ranch could stock in its cage system—does not result in sound fisheries management, and the unilateral enforcement of these regulations, if the WPO does not abide by the same rules, only hurts the Mexican rancher but does little for the overall preservation of PBFT (Del Moral Simanek and Vaca Rodriguez, 2009). Ideally, every nation harvesting PBFT—whether the 85% remaining in WPO waters or the 15% migrating to the EPO—should do so responsibly and with a global vision, because, regardless of the fishing fleet flag, overfishing is occurring and the PBFT stock is already overexploited, as judged on reference point ratios (IATTC, 2014).

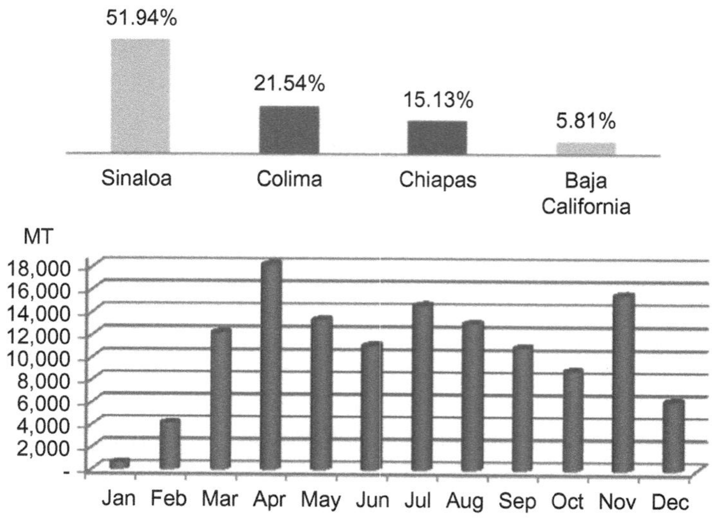

FIGURE 8.5 Main producers of wild-captured scombrids and seasonality of capture (MT) in EPO (AEPA, 2013).

8.6.1 Tuna Farming Scale and Structure

Official statistics (AEPA, 2013) indicate that in 2013 there were only nine active tuna-ranching concessions in Mexico, with ~576 hectares allocated for net-pen mooring (each concession ~ 64 hectares or 158 acres). However, since 2005 over 35 applications have been submitted to the Federal Fisheries Bureau (CONAPESCA – SAGARPA) and approved, solely for the Pacific coast. These concessions ranged in latitude from Colima (18° 47′ 00.03″ N; 104° 00′ 42.19″ W) to just short of the border with the United States, near Tijuana (32° 33′ 25.69″ N; 117° 16′ 51.31″ W) (Figure 8.6). For Mexico this activity represented around 1100 direct permanent jobs and several times that in temporary jobs and spin-off industry revenue.

As is the case in other countries where tuna ranching occurs, the primary market for Mexican tuna is Japan, because many Japanese consider sashimi as the most exquisite delicacy of the native cuisine (Kazuko and Fukoka, 2003). Of the 35,000 MT of tuna imported by Japan in 2005, about 5000 MT originated in Mexico (Del Moral Simanek and Vaca Rodriguez, 2009). However, both Mexico and the United States are beginning to experience an emergence of the US market for sashimi products. While the activity takes place in Mexico, the investment capital is in fact provided by Japan, Iceland, Australia, the United States, the United Kingdom, Canada, with only a minor portion originating in Mexico itself.

Mexico is particularly inviting for tuna ranching because of its temperate-subtropical climate; abundant supply of locally caught sardine, mackerel, and squid; proximity to major international airports in the United States; lack of stringent regulations; and low labor costs. According to the Mexican

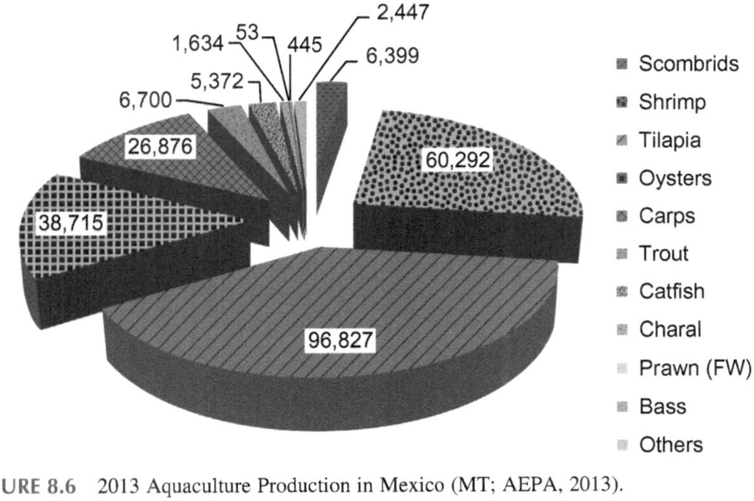

FIGURE 8.6 2013 Aquaculture Production in Mexico (MT; AEPA, 2013).

government, in 2013, tuna ranches in the country produced 6399 MT (Figure 8.5) of sashimi-grade tuna with 82% of that being produced in Baja California, 16% in Baja California Sur, and 2% in Jalisco. This level of production is just under that of rainbow trout (6700 MT), but with tilapia, shrimp, oysters, and carps being much more important, volume wise (Figure 8.5).

8.6.2 Key Players in Mexican Farmed Tuna

Mexican tuna ranching operations are concentrated in the northern part of the state of Baja California, with ranches located off/near Coronado Island, Salsipuedes Bay, Todos Santos Bay, Puerto Escondido, Santo Tomas, Soledad Bay, and Cedros Island. For PBFT, ranching occurs in the proximity of the city of Ensenada, with most of the processing activities conducted at Puerto el Sauzal, located less than 2 h from the US border, making it quite practical to select fish from the available stock, slaughter, process, pack, and ship fresh-chilled PBFT to Los Angeles, where air cargo flights are readily available to Japanese and American markets.

After the pioneering efforts of Atunera Nair in 1997, Maricultura del Norte initiated operations in 1998 and, for a short period of time, held three separate concessions in two different states, each with over 15,100-MT cages. By 2003, there were already five companies in northern Baja California; this increased to 10 in 2007, including Baja Aquafarms, Acuacultura de Baja California, Rancho Marino Guadalupe, Bajamachi, and Operadora Maritima de Oriente (Figure 8.7).

In later years, some of these companies ceased operations or changed their activities to focus on other marine species for which laboratory produced fish are more readily available.

In 2004, a Japanese company called Dohsui, formerly Hokkaido Suisan Seizo, invested heavily in tuna farms in La Paz, BCS. Mitsubishi, one of world's most diversified transnationals, is also the world's largest tuna trader and owns a tuna ranch in La Paz Bay called "Aquacultura de Baja California Sur". In the same state, a YFT venture (Oceanic Aquaculture) just recently secured a concession site outside Magdalena Bay in a location known as Santa Maria Bay. "*Thunnus* Acuicola de Nayarit" was established in 2000 by an investment group from San Antonio, Texas, as a fully permitted farm for YFT sited off Isla Isabela on the Pacific Coast of Mexico (21° 50′ 59.89″ N; 105° 53′ 08.02″ W) ~40 nautical miles NW from San Blas, Nayarit (Hendrix 2015, personal communication). Because the island is a national park and designated bird sanctuary, it could not be used as base of operations; so, a steel barge was converted to provide such base, with quarters for a crew of 22, full kitchen, complete dive locker, office with phone and internet service, two generators, two −20°C freezers with holding capacity for 80 MT of frozen sardines or squid. This ranch targeted small YFT (10–15 kg) for stocking, with intended harvest of >36-kg animals. The

FIGURE 8.7 Approximate locations of PBFT and YFT ranches in Northwest Mexico.

ranch operated for 2.5 years when a category 5 hurricane ("Kenna," winds of 252 km/h or higher) severely damaged the barge and two net pens. These were rebuilt and sold to form the first tuna project in the state of Jalisco (Mexican investment) now located in Punta Tavito in the municipality of Cabo Corriente, which resumed operations under the name "Maricultura de la Bahia." This new YFT ranch started operations with three 100-MT capacity net pens and later increased stocking capacity to 20 cages.

That not all approved concessions are operating today can be explained in part by the case study of La Paz Bay, BCS, which exemplifies some of the greatest challenges confronting those brave investors wanting to pursue tuna ranching in Mexico. From 2003 to 2005, CONAPESCA – SAGARPA granted at least seven concessions for YFT ranches within La Paz Bay (Figure 8.8). Using conservative estimates based on capture data (FAO, 2005) of the EPO, investing companies structured comprehensive business plans, complied with all existing regulations and started operations in 2004.

An analysis of data extracted from logbooks of the Mexican tuna purse-seine fleet from 2004 to 2006 reveals a downward slope in total catches of YFT, the target species in La Paz, indicating either poor monitoring by IATTC or periods of exceptionally low abundance of YFT. At the same time, a smaller number of purse-seine sets was made in the EPO, and the average capture size of fish taken in those sets significantly decreased (to ~9 kg in average), as compared to the average size of captured YFT obtained in previous years (~20 kg in 1991, IATTC, 2000).

In conclusion, a chief limitation today for the Mexican tuna industry lurks in its inability to secure enough stocking-size fish to fill the existing net pens. In this limitation also lies a unique opportunity for those able to develop refined hatchery technologies for the production of juvenile PBFT and YFT.

8.7 MARKET PRICES

According to AEPA (2013), shrimp and tilapia represent the biggest source of aquaculture revenue for Mexico (52% and 23%, respectively; Figure 8.9).

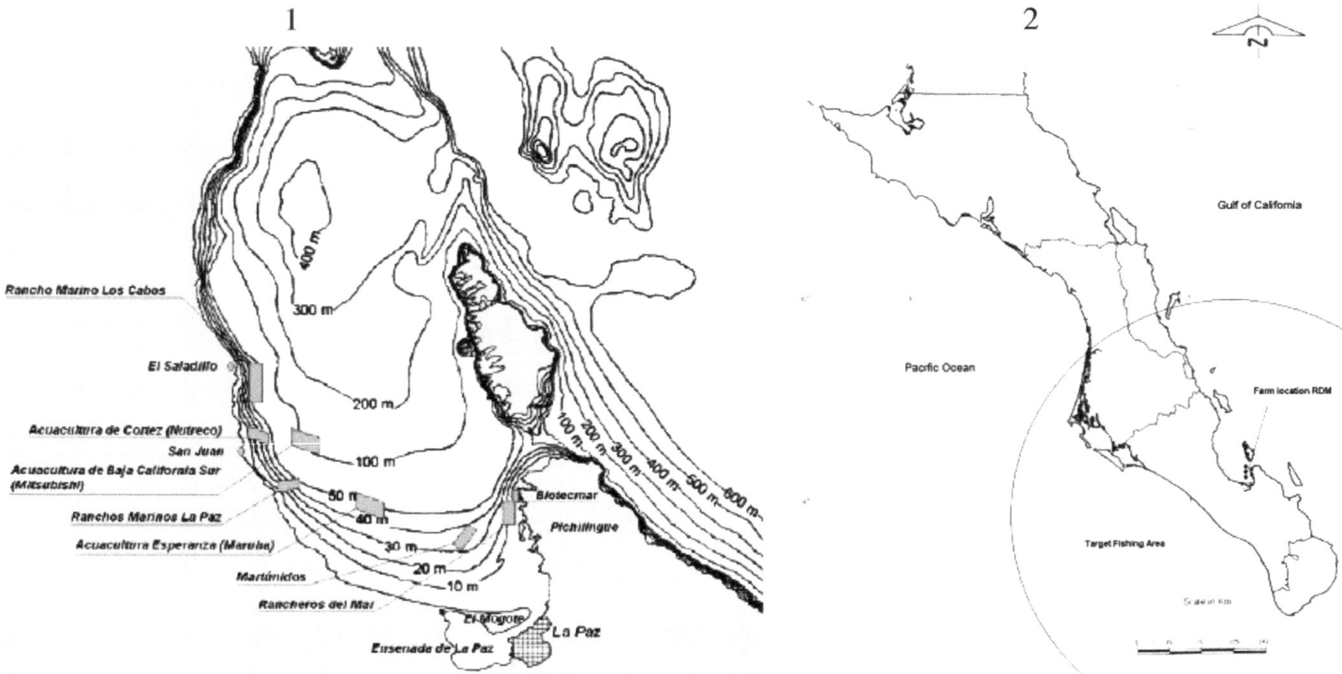

FIGURE 8.8 Location of (1) approved concessions for YFT farms within La Paz Bay and (2) location in the Baja California Peninsula.

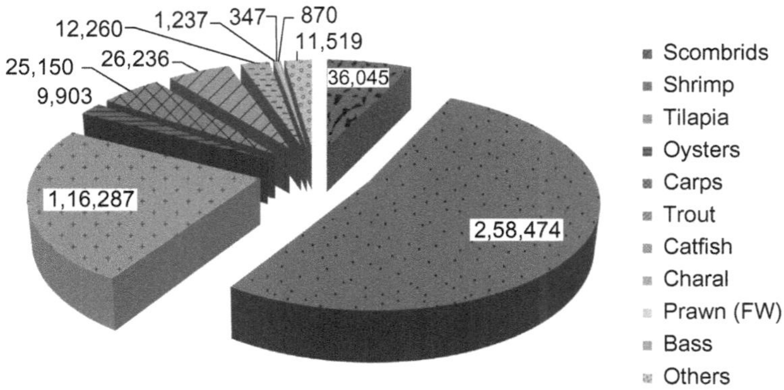

FIGURE 8.9 2013 Aquaculture Production in Mexico (US$ × 1000; AEPA, 2013).

However, tuna production in dollars per kilogram is significantly more attractive than any other aquaculture product in the country.

Assembling accurate, consistent data on sale volume and price of Mexican ranched tuna is not a straightforward task, nor is a steady horizon of increasingly high production guaranteed, since many factors influence the output. For example, the wholesale price (received at the Mexican border before trans-shipment to Japan) for a lot of 94 fresh tuna from the ranches averaged US$21.19 per kg and ranged from $18.18 to $31.40 per kg for the period February 23–March 8, 2007 (Sonu, 2007).

Tuna fed in net pens and slaughtered using the Ikejime method (very rapid destruction of the brain and severing of spinal cord, immediate bleeding, evisceration and gill removal, and placement in an ice slurry, all ≤ 60 s from harvest) typically fetches prices 15- to 30-fold those obtained at tuna canneries (US$3–5 per kg), and auction prices at Japanese markets such as those in Tsukiji, Nagoya, and Osaka are substantially higher. Although record prices are often paid for PBFT and/or ABFT (e.g., ~ US$1.76 million was paid in January 2013 at the Tokyo auction for a single 222 kg PBFT; Revkin, 2013) these annual singularities are often publicity stunts designed to attract the public's attention but even top-quality fish do not typically fetch those towering prices.

Mexico faces unique marketing challenges in that typical sizes of PBFT captured in the vicinity of Ensenada (<500 mile radius) are small—advanced juveniles migrating from the WPO—and even after the fattening period PBFT ranched in Mexico simply cannot compete with the larger tunas ranched in other countries. Retail prices also do not coincide with those derived from AEPA (2013) data in which 6,399,000 kg fetched US$36 million (Figures 8.6 and 8.9, respectively), which equates to about US$5.6 per kg. Another report that brings into question the accuracy of the official data comes from Del Moral Simanek and Vaca Rodriguez (2009), who stated that in 2006 Mexican tuna farmers generated revenue for over US$74 million for their annual production of sashimi grade PBFT which was 4350 MT (US$17 per kg)—this is in contrast with the official volume reported by AEPA (2013) which for the same year was 3806 MT (Figure 8.4). The difference (−544 MT) is worrisome because the combined aquaculture production of both YFT and PBFT should be more than the volume for individual species, not less.

8.8 FUTURE PROSPECTS

Further expansion of PBFT capture-based aquaculture in either Japan or Mexico is small, but considered viable in the short term if improvement to fisheries management strategies—including significant reductions in Illegal, Unreported, and Unregulated fishing—are implemented. However, in the long term, sustainability certainly will depend on the economically viable mastering of the full life cycle (i.e., reproduction and larval rearing), survival of juveniles for net pen stocking and improvements in the artificial feed formulation to reduce baitfish consumption and improve feed conversion ratios, thus mitigating the environmental footprint of tuna ranching. International scientific collaboration to improve culture technologies should resolve present and future bottle neck issues. In addition, market diversification and expansion beyond the Japanese market should result in increased sale volumes perhaps at somewhat reduced prices. Furthermore, farmers need to follow best procedures to ensure traceability of traded tuna.

REFERENCES

AEPA. Anuario Estadistico de Aquaculture y Pesca. 2013. Secretaria de Agricultura, Ganaderia, Desarrollo Rural, Pesca y Alimentacion – Comision Nacional de Acuacultura y Pesca. Gobierno de Mexico, 299 pp.

Agawa, Y., Honryo, T., Kurata, M., Okada, T., Sawada, Y. 2009. Introduction of selective breeding of Pacific bluefin tuna *Thunnus orientalis* and its molecular biological approaches. In: Allan, G., Booth, M., Mair, G., Clarke, S., Biswas, A. (Eds.), Proceedings of the 2nd Global COE Program Symposium of Kinki University, Sustainable Aquaculture of the Bluefin and Yellowfin Tuna – Closing the Life Cycle for Commercial Production. Kinki University, SARDI, Adelaide South Australia, pp. 8–11.

Araki, K., Ozaki, T., Okauchi, M., Okamoto, H., Hara, M., Yamada, T., et al., 2011. Current flounder breeding in Japan. Fish Genet. Breed. Sci. 40, 67–69 (in Japanese).

Buentello, J.A., 2006. The blue revolution in southern Baja California. Yellowfin tuna, *Thunnus albacares* aquaculture. Panorama Acuicola 11 (3), 60–67.

Buentello, J.A., Neill, W.H., Gatlin III, D.M., 2009. Tuna aquaculture faces challenges in continued growth. Glob. Aquacult. Advocateissue.

Del Moral Simanek, R., Vaca Rodriguez, J.G., 2009. Administration of bluefin tuna fishery in Baja California. A global vision. Frontera Norte 21 (41), 151–175 (in Spanish).

FAO, Food and Agriculture Organization of the United Nations, 2008. Tuna Fisheries and Utilization. Fisheries and Aquaculture Department, Rome, Text by Jacek Majkowski.

FAO, Food and Agriculture Organization of the United Nations. 2010. The state of world fisheries and aquaculture 2010. Rome, 218 pp.

FRA. Fisheries Research Agency. 2011. Homepage. http://www.fra.affrc.go.jp/english/eindex.html.

Honryo, T., Kurata, M., Okada, T., Ishibashi, Y., 2013. Effects of night-time light intensity on the survival rate and stress responses in juvenile Pacific bluefin tuna *Thunnus orientalis* (Temminck and Schlegel). Aquacult. Res. 44, 1058–1065.

Honryo, T., Okada, T., Kurata, M., Tamura, T., Ishibashi, Y., 2014. Optimal periods of night-time lighting in the sea cage culture of Pacific bluefin tuna, *Thunnus orientalis*, juvenile (Temminck and Schlegel). Aquacult. Res. 45, 1109–1115.

IATTC. 1999. Inter-American Tropical Tuna Commission. Annual report. La Jolla, CA. 184 pp.

IATTC. 2000. Inter-American Tropical Tuna Commission. Quarterly Reports (Oct–Dec 2000) of the Inter-American Tropical Tuna Commission. La Jolla, California, 32 pp.

IATTC. 2014. Inter-American Tropical Tuna Commission. Fishery Status Report No. 12. Tunas and billfishes in the eastern Pacific Ocean in 2013. La Jolla, California, 177 pp.

ISC. International Scientific Committee for Tuna and Tuna-like Species in the North Pacific Ocean. 2007. Report of the Pacific bluefin tuna working group workshop. Shimizu, Japan, 47 pp.

Ishibashi, Y., 2012. Fingerling production-II. Flexion larvae to juveniles. In: Kumai, H., Sakamoto, W., Ono, S. (Eds.), Full Life Cycled Aquaculture of the Pacific Bluefin Tuna. Agriculture and Forestry Statistics Publishing, Inc., Tokyo, Japan, pp. 39–59.

Ishibashi, Y., Honryo, T., Saida, K., Hagiwara, A., Miyashita, S., Sawada, Y., et al., 2009. Artificial lighting prevents high night-time mortality of juvenile Pacific bluefin tuna, *Thunnus orientalis*, caused by poor scotopic vision. Aquaculture 293, 157–163.

JETRO. Japan External Trade Organization. 2010. http://www.jetro.go.jp/jfile/report/07000203/japanese_restaurant.pdf.

JFA. Japanese Fisheries Agency. 2015. Marino-Forum 21 General Incorporated Association. http://www.yousyokugyojyou.net/index4.htm.

JMAFF. Japanese Ministry of Agriculture, Forestry and Fisheries. 2006. http://www.maff.go.jp/j/shokusan/sanki/easia/e_sesaku/japanese_food/kaigi/01/pdf/data3.pdf.

JMAFF. Japanese Ministry of Agriculture, Forestry and Fisheries. 2013. http://www.maff.go.jp and http://www.jfa.maff.go.jp/j/press/saibai/140331.html.

Kaji, T., Tanaka, M., Takahashi, Y., Oka, M., Ishibashi, N., 1996. Preliminary observations on development of Pacific bluefin tuna *Thunnus thynnus* (Scombridae) larvae reared in the laboratory, with special reference to the digestive system. Mar. Freshw. Res. 47, 261–269.

Kasuko, E., Fukuoka, Y., 2003. Japanese Cooking. Hermes House, Japan.

Kawamura, G., Masuma, S., Tezuka, N., Koiso, M., Jinbo, T., Namba, K., 2003. Morphogenesis of sense organs in the bluefin tuna *Thunnus orientalis*. In: Browman, H.I., Skiftesvik, A.B. (Eds.), The Big Fish Bang. Proceedings of the 26th Annual Larval Fish Conference. Institute of Marine Research, Bergen, Norway.

Kumai, H., 2012. History, current status and perspective of bluefin tuna aquaculture. In: Kumai, H., Miyashita, S., Sakamoto, S., Ono, S. (Eds.), Full Life-Cycle Aquaculture of the Pacific Bluefin Tuna. Agriculture and Forestry Statistics Publishing Inc., Tokyo, pp. 1–12.

Margulies, D., Suter, J.M., Hunt, S.S., Olson, R.J., Scholey, V.P., Wexler, J.B., et al., 2007. Spawning and early development of captive yellowfin tuna (*Thunnus albacares*). Fish. Bull. 105, 249–265.

Masuma, S., 2008. Development of techniques of stock enhancement for Pacific bluefin tuna *Thunnus orientalis* by the Fisheries Research Agency (formerly, Japan Sea Farming Association). J. Fish. Technol. 1, 21–36.

Masuma, S., Miyashita, S., Yamamoto, H., Kumai, H., 2008. Status of bluefin tuna farming, broodstock management, breeding and fingerling production in Japan. Rev. Fish. Sci. 16, 385–390.

Masuma, S., Takebe, T., Sakakura, Y., 2011. A review of the broodstock management and larviculture of the Pacific northern bluefin tuna in Japan. Aquaculture 315, 2–8.

Matsukawa, Y., 2006. Carrying capacity and survival strategy for the Pacific bluefin tuna, *Thunnus orientalis*, in the Western Pacific. Fish. Oceanogr. 15 (2), 104–112.

Minato Shinbun, 2012. Homepage. http://www.minato-yamaguchi.co.jp/minato/index.html.

Miyashita, S., 2002. Studies on the seedling production of the Pacific bluefin tuna *Thunnus thynnus orientalis*. Bull. Fish. Lab. Kinki Univ. 8, 1–171.

Miyashita, S., Kato, K., Sawada, Y., Murata, O., Ishitani, Y., Shimizu, K., et al., 1998. Development of digestive system and digestive enzyme activities of larval and juvenile bluefin tuna, *Thunnus thynnus*, reared in the laboratory. Suisanzoshoku 46, 111–120.

Morris, T., 2011. Marine fish culture in Mexico: amberjack, yellowtail farming developing. Glob. Aquacult. Advocate, 76–77.

Murata, U., Miyashita, S., 2012. Maturation, spawning and hatching. In: Kumai, H., Miyashita, S., Sakamoto, S., Ono, S. (Eds.), Full Life-Cycle Aquaculture of the Pacific Bluefin Tuna. Agriculture and Forestry Statistics Publishing Inc., Tokyo, pp. 13–24.

Murata, O., Harada, T., Miyashita, S., Izumi, K., Maeda, S., Kato, K., et al., 1996. Selective breeding for growth in red sea bream. Fish. Sci. 62, 845–849.

Mylonas, C.C., De la Gándara, F., Corriero, A., Belmonte Ríos, A., 2010. Atlantic bluefin tuna (*Thunnus thynnus*) farming and fattening in the Mediterranean Sea. Rev. Fish. Sci. 3 (18), 266–280.

Nakagawa, Y., Eguchi, M., Miyashita, S., 2007. Pacific bluefin tuna, *Thunnus orientalis*, larvae utilize energy and nutrients of microbial loop. Aquaculture 267, 83–93.

Nakagawa, Y., Kurata, M., Sawada, Y., Sakamoto, W., Miyashita, S., 2011. Enhancement of survival rate of Pacific bluefin tuna (*Thunnus orientalis*) larvae by aeration control in rearing tank. Aquat. Living Resour. 24, 403–410.

Nakamura, Y., Mori, K., Saitoh, K., Oshima, K., Mekuchi, M., Sugaya, T., et al., 2013. Evolutionary changes of multiple visual pigment genes in the complete genome of Pacific bluefin tuna. Proc. Natl. Acad. Sci. 110, 11061–11066.

Okada, T., Honryo, T., Sawada, Y., Agawa, Y., Miyashita, S., Ishibashi, Y., 2014. The cause of death of juvenile Pacific bluefin tuna (*Thunnus orientalis*) reared in sea net cages. Aquacult. Eng. 59, 23–25.

Ono, S., 2013. Economic Analysis of Fish Aquaculture. Agriculture and Forestry Statistics Publishing Inc., Tokyo (in Japanese).

OPRT. Organization for the Promotion of Responsible Tuna Fisheries. 2015. http://oprt.or.jp/eng/data/global-tuna-supply-sashimi-tuna/ and http://www.oprt.or.jp/pdf/KOBEnihongo.pdf.

Paquotte, P. 2003. Tuna in the international market for seafood. In: Bridges, C.R., Gordin, H., Garcia, A. (Eds.) International Symposium on Domestication of Bluefin Tuna. Cartagena, Spain. pp. 12–18.

Partridge, G.J., 2013. Closed-cycle hatchery production of tuna. In: Allan, G., Burnell, G. (Eds.), Advances in Aquaculture Hatchery Technology. Woodhead Publishing Limited, Cambridge, England, p. 626.

Revkin, A.C. 2013. Bluefin tuna sold for record $1.76 million in Tokyo "auction." New York Times, January 4, 2013.

Sakamoto, W., Kurata, M., Takaoka, O., 2012. In: Kumai, H., Miyashita, S., Sakamoto, S., Ono, S. (Eds.), Full-Life Cycle Aquaculture of the Pacific Bluefin Tuna. Agriculture and Forestry Statistics Publishing Inc., Tokyo, pp. 31–38.

Sawada, Y., 2005. In: Kumai, H. (Ed.), Bluefin Tuna in "Aquaculture System – Marine Fish.". Kouseisha-kouseikaku, Tokyo, pp. 173–203.

Sawada, Y., Okada, T., Miyashita, S., Murata, O., Kumai, H., 2005. Completion of the Pacific bluefin tuna *Thunnus orientalis* (Temminck et Schlegel) life cycle. Aquacult. Res. 36, 413–421.

Seoka, M., Kurata, M., Kumai, H., 2007. Effect of docosahexaenoic acid enrichment in Artemia on growth of Pacific bluefin tuna *Thunnus orientalis* larvae. Aquaculture 270, 193–199.

Seoka, M., Kurata, M., Tamagawa, R., Biswas, A.K., Biswas, B.K., Seok Kian Yong, A., et al., 2008. Dietary supplementation of salmon roe phospholipid enhances the growth and survival of Pacific bluefin tuna *Thunnus orientalis* larvae and juveniles. Aquaculture 275, 225–234.

Sonu, S. 2007. Wholesale prices at Tokyo central wholesale market. Southwest regional office, National Marine Fisheries Service. http://swr.nmfs.noaa.gov/fmd/sunee/twprice/twp23807.htm>.

Tada, M., 2010. Challenges and opportunities for the full cycle farmed tuna in japan. In: Miyashita, S., Takii, K., Sakamoto, W., Biswas, A. (Eds.), 'Joint International Symposium of Kinki University and Setouchi Town on The 40th Anniversary of Pacific Bluefin Tuna Aquaculture. Kinki University Press, Setouchi Town, Japan, pp. 40–44.

Takashi, Y., Kohno, H., Sakamoto, W., Miyashita, S., Murata, O., Sawada, Y., 2006. Diel and ontogenetic body density change in Pacific bluefin tuna, *Thunnus orientalis* (Temminck and Schlegel), larvae. Aquacult. Res. 37, 1172–1179.

Wexler, J.B., Scholey, V.P., Olson, R.J., Margulies, D., Nakazawa, A., Suter,, J.M., 2003. Tank culture of yellowfin tuna, *Thunnus albacares*: developing a spawning population for research purposes. Aquaculture 220, 327–353.

Yoshida, K., Nakagawa, M., Hotta, T., Tsuzaki, T., Nagakura, Y., Hattori, K., et al., 2011. Breeding of yellowtail in Goto Center for Stock Enhancement "Cooperation of Tokyo University of Marine Science and Technology, National Research Institute of Aquaculture.". Fish Genet. Breed. Sci. 40, 71–73 (in Japanese).

Chapter 9

Ranching of Southern Bluefin Tuna in Australia

David Ellis and Ilse Kiessling
David Ellis and Associates, Port Lincoln, SA, Australia

9.1 INTRODUCTION

Southern bluefin tuna (*Thunnus maccoyii*) (SBFT) belong to the family Scombridae, which includes all mackerels and tunas (Lagler et al., 1977). The specific name, maccoyii, was bestowed by de Castelnau (1872) noting, "the flesh of this fish is not eaten, or at least is not esteemed as food." The species is one of the largest bony fishes, living up to 40 years, growing to a length of 2.25 m, and weighing over 200 kg (Patterson et al., 2009, 2010). One specimen that washed up on a beach at Glenelg, South Australia in 1890 was reported to have weighed over 350 kg (Serventy, 1956).

SBFT are a single, highly migratory stock (Patterson et al., 2009, 2010) found primarily between the latitudes of 30° and 50° South (Collette and Nauen, 1983). The only known spawning ground for SBFT is located in the warm oceanic waters south of Java, Indonesia in the north-east Indian Ocean between 10°–20° South and 105°–120° East. The spawning season spans from September to April. Following spawning, developing juveniles move with the Leeuwin Current along Australia's north-western coast to the south-west tip of Australia and east into the Great Australian Bight or west toward South Africa (Campbell, 2001; Patterson et al., 2009) (Figure 9.1). It is uncertain when SBFT reach spawning age, but it is generally understood that females between the ages of 8 and 12 produce several million eggs each spawning event (Hayes, 1997).

Tuna are distinguished by a number of anatomical and physiological adaptations that minimize anterior resistance and maximize caudal thrust (Bushnell and Jones, 1994; Dewar and Graham, 1994; Brill, 1996; Fitzgibbon et al., 2008). These adaptations include a streamlined shape designed for speed, maneuverability, drag reduction, and efficiencies in

D.D. Benetti, G.J. Partridge, A. Buentello (Eds): Advances in Tuna Aquaculture.
DOI: http://dx.doi.org/10.1016/B978-0-12-411459-3.00010-2

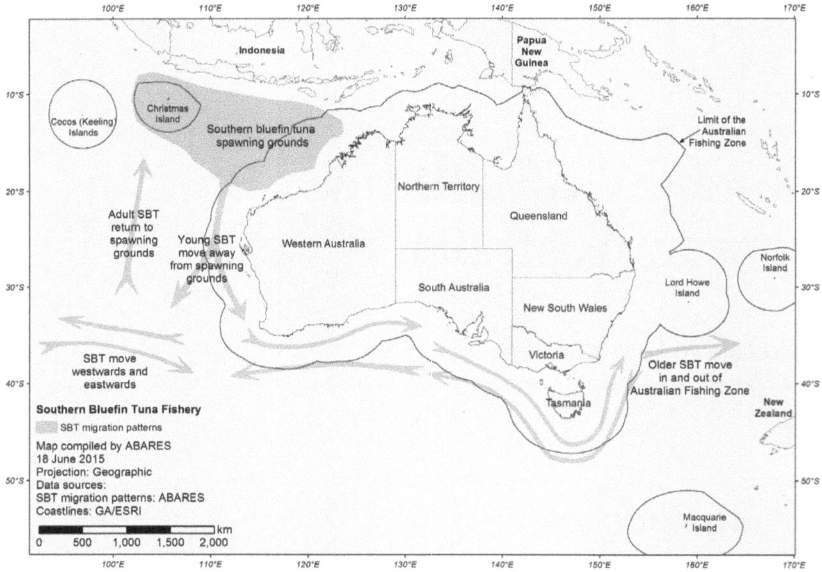

FIGURE 9.1 SBFT migration patterns. *Map provided by ABARES (Australian Bureau of Agricultural and Resource Economics and Sciences).*

locomotion (Magnuson, 1978; Dewar and Graham, 1994). The dorsal, pelvic, and pectoral fins of SBFT provide guidance but serve no role in propulsion. When moving at high speed (approximately 70 km/h, Wardle et al., 1989), SBFT use long propulsive beats of their tail and retract their fins within defined body grooves to minimize drag. Caudal keels along the top and bottom edges of the body act as spoilers to prevent turbulence (Attenborough, 1979; Altringham and Shadwick, 2001). These adaptations enable SBFT to feed rapaciously in the epipelagic layers of oceans, opportunistically targeting fish, crustaceans, cephalopods, salps, and other marine animals (Dickson, 1996; Young et al., 1996, 1997; Itoh et al., 2011).

9.2 HISTORY

The history of the SBFT industry in Australia began in 1936 when Mr. Stanley Fowler from the Commonwealth Scientific Industrial Research Organisation (CSIRO) initiated a survey using a military plane and fishing vessels to assess tuna stocks. The survey was undertaken to stimulate economic development as, until that time, SBFT were largely unexploited (Serventy, 1956). The survey was interrupted during World War II when the survey vessel, *F.R.V. Warreen*, was commandeered by the Royal Australian Navy for use in the war effort. At the end of the war, the *Warreen* was recommissioned as a survey vessel until 1951 when the Royal Australian

Navy acquired the vessel again. The SBFT fishery survey was continued using the *F.R.V. Stanley Fowler*, and the first commercial SBFT trolling fishery is believed to have started in 1949 (Serventy, 1956).

During the 1950s the South Australian state government provided financial support for the building of the purse-seine vessel, *F.V. Tacoma*, in Port Fairy, Victoria (Plevin, 2000) from where it made its way west to Port Lincoln, South Australia. The *F.V. Tacoma* caught its first catch of 10 tonnes of SBFT in February 1953, which was sent to the local Boston Bay cannery near Port Lincoln. In 1956 the South Australian Government involved experts from the United States using the *F.V. Tacoma* in a survey to determine the extent of the tuna fish stock (Plevin, 2000). The results of the survey initiated an expansion of the tuna fishing industry and fishermen began to use poling, long-line, and purse-seine gear to catch SBFT.

The 1960s and 1970s witnessed a rapid and unregulated increase in SBFT fishing with record numbers of fish caught. Spotting planes were introduced to locate and direct fishing vessels to schools of SBFT. Fishing vessels evolved from wood to metal, and fishing effort further increased through cooperation between purse-seine and pole-fishing vessels resulting in significantly more SBFT landings (Campbell, 2001). Between the 1950s and 1970s the SBFT fishery started to become very important to both the Port Lincoln economy and the South Australian state economy with large numbers of local residents employed in the Port Lincoln tuna cannery.

In 1979, fishery biologists began to warn that the SBFT stock was fully exploited, that parental biomass was being reduced at an alarming rate (standing at only 30% of its pre-exploitation size at that time), and that poor recruitment of juvenile SBFT would result (Geen and Nayer, 1989). Despite these warnings, fishing effort was maintained and the Australian catch of SBFT reached a peak of 21,000 tonnes in 1982 (Geen and Nayer, 1989). An Australian federal government inquiry in 1983 subsequently found the fishery was biologically overexploited and heavily over-capitalized (Geen and Nayer, 1989). In 1984 the Australian federal government imposed Individual Transferable Quotas (ITQs) to control the fishery's expansion.

To reduce pressures on declining stocks internationally, the two other countries targeting SBFT at the time, Japan and New Zealand, also began to apply quotas in 1985. Continuing concerns about tuna stocks and excessive quotas set between 1984 and 1988 led the Australian, Japanese, and New Zealand governments to further reduce catch limits and introduce annual reviews (Geen and Nayer, 1989), and the three countries established a joint voluntary management arrangement in 1989 limiting the combined total yearly quota for all three countries to 11,750 tonnes (Geen and Nayer, 1989).

Management of SBFT take by Australia, Japan, and New Zealand was formalized in 1994 with the entering into force of the Convention for the Conservation of SBFT. The Convention formed the Commission for the Conservation of SBFT (CCSBT). Today there are six members of the

CCSBT: Australia, Japan, New Zealand, Republic of Korea, Indonesia, and Taiwan. The Philippines, South Africa, and the European Community are cooperating nonmembers (CCSBT, 2015a).

With Australia's introduction of SBFT catch limits, Port Lincoln fishermen quickly realized that to remain in business, the value of their SBFT catch would need to increase. With the lucrative Japanese sashimi market in mind, the Port Lincoln fishing industry began to investigate options for increasing SBFT yield through farming of wild tuna (Ellis, 2013). A study was undertaken by the Tuna Boat Owners Association of Australia, the Federation of Japan Tuna Fisheries Co-operative Association, and the Japanese Overseas Fishery Cooperation Foundation, with the support of the South Australian state government and the Australian federal government (Government of South Australia, 1993; Jeffriess, 1999). The study showed that Australian SBFT tended to be larger and more adaptable than tuna species found in the northern hemisphere, and it described handling techniques and farming technology for SBFT that could be easily applied by local fishermen.

On the basis of the study, the industry changed from poling individual SBFT into vessel tanks to the purse-seine capture of schools of SBFT for on-growing in static ranching pontoons. This change in catching approach was developed and established by the Lukin and Sons fishing company and facilitated the rapid expansion and development of the Port Lincoln tuna ranching industry. As Australian ranching of SBFT developed, international market prices for SBFT grew steadily to reach a peak in 2002, and an increased value of Australian SBFT in premium Japanese sashimi markets was realized.

9.3 SBFT CAPTURE METHODS

The current Australian harvest of SBFT generally targets 2- to 3-year-old fish using purse-seine gear within the Great Australian Bight during the Australian summer. Spotting aircraft direct chum vessels to schools of SBFT observed from the air. To lead SBFT schools close to vessels towing specialized pontoons, deckhands on the chum vessel cast thawed and live local baitfish (*Sardinops sagax*) into the water. Once the school of SBFT is in close proximity to the towing pontoon, a commercial purse-seine vessel encircles the chum vessel and the SBFT school with a purse-seine net measuring 1000 m long by 130 m deep. The purse-seine vessel then retrieves the purse cable that runs along the bottom of the net containing the chum vessel and SBFT school in a net "bowl". Before the purse-seine net is hauled in, the chum vessel leaves the net enclosure. Smaller vessels powered by outboard engines help to keep the purse-seine net open while it is hauled and at the same time the towing pontoon is moved to meet the purse-seine net containing the SBFT (Figure 9.2).

FIGURE 9.2 Purse-seine vessel F.V. Independence with majority of net on board with purse-seine net connected to tow pontoon while small vessels hold the net open.

Once the towing pontoon and net (150 mm Badinotti 450 ply net, 45 m diameter by 12 m deep) and the purse-seine net are alongside each other, commercially trained divers connect the enclosures via a specially designed underwater gate. The purse-seine net continues to be hauled on board the purse-seine vessel thereby reducing the swimming area for the SBFT to a point where they move through the underwater gate from the purse-seine net into the towing pontoon. The pontoon containing the fish is then slowly towed back to Spencer Gulf on a journey that can take up to 3 weeks. Stocking densities in the towing pontoon are generally between 5 and 10 kg/m^3 and the SBFT are fed baitfish during their transport as weather and sea conditions permit.

9.4 SBFT RANCHING

Ranching of SBFT occurs near Port Lincoln, South Australia within the clean waters of Spencer Gulf. The ranching area is contained in a designated Lincoln Aquaculture Zone covering an area of 27,385 ha comprising an inner and outer sector. Leased area within the inner sector is limited to 1825 ha and a maximum biomass of 10,500 t. The outer sector is limited to 5000 ha and a maximum biomass of 14,000 t. Water depth in the inner sector is between 20 and 24 m and wind driven swells are generally 1–2 m. Water depths in the outer sector are between 40 and 50 m and oceanic swells tend to be between 2 and 2.5 m. Water temperatures across the zone reach a high of 23°C in summer and fall to 13°C in winter.

Leases and licenses for SBFT aquaculture are allocated by the state government agency, Primary Industries and Regions South Australia (PIRSA) through an Aquaculture Tenure Allocation Board. An individual or company must own a minimum of 60 t of tuna quota to be eligible for a lease. Lease sites are generally allocated at 1 ha per 3 t of tuna quota. A series of environmental performance measures must be met by operators annually to maintain lease access (PIRSA, 2015a,b,c).

Towing pontoons containing schools of SBFT netted in offshore waters are transported to the allocated ranching sites, where samples of 100 fish are taken from each tow pontoon by baited hook and handline. The length and weight of each sampled fish are recorded and an average weight for the 100 fish sample is determined. Following sampling, each towing pontoon is paired with a static ranching pontoon and the SBFT are encouraged through a gate (measuring 3 m × 3 m) by the hydraulic hauling of strategically placed ropes to dry the towing pontoon (Figure 9.3). Static ranching pontoons are typically 45 m in diameter and made up of a 450 mm diameter polyethylene buoyancy ring supporting a 415 ply knotless net (150 mm stretched mesh) with 10 m deep walls creating an enclosure of 15,906 m^3.

As the SBFT swim through the transfer gate from the towing pontoon into the static ranching pontoon they are observed and manually counted via video. Once the transfer is complete, the number of SBFT in each ranching pontoon is multiplied by the average sampled weight and total tonnage is recorded against the company's Statutory Fishing Rights represented as quota for the season managed by the Australian Fisheries Management Authority.

FIGURE 9.3 Tow pontoon connected to static ranching pontoon and SBFT being herded through transfer gate by the hauling of strategically placed ropes.

When the SBFT first arrive in the ranching pontoon, their appetite is voracious. To optimize growth, commercially ranched SBFT are generally fed two times per day, 6 days per week for up to 8 months as weather conditions allow (Gunn et al., 2002; Ellis, 2013). As water temperatures cool with the changing seasons, the appetite of the fish decreases (Glencross et al., 2002).

SBFT are harvested for processing as chilled/fresh or super frozen product. Harvesting involves the fish being crowded into a small purse-seine net within the static ranching pontoon where they are manually caught by divers and passed onto a fish travelator conveying them to the deck of a harvest vessel (Figure 9.4).

At the harvest vessel, the fish are slaughtered by a spike through the brain, followed by coring of the brain and insertion of a fiberglass rod into the exposed spinal column to destroy the central nervous system. SBFT are then bled by severing lateral blood vessels located close to the skin in the pectoral fin recess, and the gill basket and viscera are removed. The slaughtering process usually lasts less than 30 s. Harvested SBFT are placed in ice slurry and transported either to Japanese tuna longline/factory vessels for super freezing at −65°C or to onshore-processing facilities for chilled harvest or super freezing.

Ranching wild caught SBFT enables significant improvement to the value of harvested product through regular feeding over a relatively short time frame. The practice effectively doubles the pre-ranching weight of fish (Carter et al., 2010) and promotes fish conditioning (protein and fat content) attractive to the premium sashimi market. Japan consumes most of Australia's ranched

FIGURE 9.4 Diver catching a SBFT during harvest and placing it into a fish travelator.

FIGURE 9.5 Feeder cage moored in the middle of a static ranching pontoon.

SBFT , but in recent years markets for Australian ranched SBFT have grown domestically as well as in Korea, China, Singapore, and the United States.

9.5 FEEDING METHODS

When SBFT ranching in Australia first began, ranched fish were fed fresh and thawed baitfish delivered manually by shovel. As the ranching effort expanded and greater numbers of hungry SBFT required feeding, alternative methods of delivering food were developed. Feeding methods now include: fish pumps that transfer thawed baitfish to the center of a pontoon via a buoyed pipe; siphons that suck thawed or fresh baitfish from fish bins through an extended pipe to the center of the ranching pontoon; and the two methods currently most commonly employed in Australia's SBFT ranching industry, hydraulic rotating disc feeders and feeder cages.

The rotating disc feeder developed by Sarin Marine Farm spins thawed or freshly caught baitfish into the ranching pontoon. This method allows large volumes of fresh caught sardines to be delivered over a large area, thereby reducing feeding competition and scavenging from seabirds.

Feeder cages are small, lidded enclosures made of weld mesh or nylon netting moored in the center of a ranching pontoon (Figure 9.5). Frozen baitfish placed in the cage fall through the bottom of the cage as they thaw. This feeding technique has been significantly improved by the Blaslov Tuna Farming Company in partnership with AQ1 Systems, enabling feed to be monitored and regulated according to the appetite of the ranched SBFT using underwater cameras, and a door that can be opened and closed via a video link.

9.6 NUTRITION AND FEED MANAGEMENT

Ranched SBFT in Australia are generally fed baitfish. More than 20 species of domestic and internationally caught baitfish are used in the ranching industry (Ellis and Rough, 2005; van Barneveld and Ellis, 2007), although the local sardine (*Sardinops sagax*) accounts for over 50% of ranching feed requirements.

The size of individual baitfish used for feed is variable and proximate nutritional content ranges from <1 to 20% lipid and 13.3 to 20.1% protein (Ellis and Rough, 2005). Evidence shows that, depending on size, an individual SBFT can consume over 3 kg (or approximately 10% body weight) of baitfish per day during peak feeding periods (Ellis, 2013). An individual SBFT fed manufactured feeds (rather than baitfish) during the same period will only consume a maximum of 1.8 kg of feed per day (van Barneveld and Ellis, 2007; Ellis, 2013). It is hypothesized that the binding of manufactured feed, differences in feeding regimes, and relative absorption of nutrients, and the fat coating on manufactured feed pellets may influence the quantity of manufactured feed consumed by ranched SBFT (van Barneveld and Ellis, 2007).

The nutritional requirements of SBFT are not well determined. A small number of studies have investigated requirements based on the nutritional content of baitfish (Gunn et al., 2002; Ellis and Rough, 2005; van Barneveld and Ellis, 2007; Ellis, 2013), but most information is derived largely from analysis of manufactured feed (Clarke et al., 1997; Carter et al., 1998; Glencross et al., 1999, 2002; van Barneveld et al., 1999, 2009; Gordon et al., 2006a,b; van Barneveld and Ellis, 2007).

Species used in baitfish feed tend to be mixed to improve fish growth and product quality, the composition of the species mix often being determined by the preference, or "gut feel" of the farmer. Some farmers prefer to feed high lipid diets early in the ranching season (when feeding intake is high) to "pack the weight on" and condition SBFT for early marketing opportunities. Some farmers prefer to feed SBFT with low lipid/high protein baitfish species blends to allow the length of the SBFT to increase before switching to high lipid diets later in the ranching season to condition fish in the last phase of the grow-out cycle (Ellis, 2013). Other farmers prefer consistent use of a medium lipid/medium protein baitfish species blend throughout the ranching season, and some farmers associated with companies that own sardine fishing vessels only use local sardines for feed. The software program, "Formubait," has been developed to assist farmers determine the composition of baitfish feed to optimize production based on feed cost and fish growth (van Barneveld and Ellis, 2007).

The quality of SBFT product is assessed on its fat content, color, texture, chemical residue, and shelf life. To realize the full value of market-ready SBFT , Australia's tuna industry has invested heavily in research into feed that promotes quality but leaves little residue (Padula et al., 2007). Vitamins, in particular vitamins E, C, and selenium, and naturally occurring antioxidants have been shown to improve SBFT flesh color while increasing product shelf life (Thomas, 2007). These vitamins and antioxidants are provided by fresh local baitfish (Thomas et al., 2009), or can be delivered as supplements injected into baitfish feed using an Accujector 450 (D'Antignana et al., 2008), or included in manufactured diets.

There have been attempts over a number of years to fully replace baitfish feed with manufactured diets for ranched SBFT in Australia (Clarke et al., 1997; Carter et al., 1998; Glencross et al., 1999, 2002; van Barneveld et al., 1999, 2009; Gordon et al., 2006a,b; van Barneveld and Ellis, 2007). Most attempts have been unsuccessful due to problems with the weaning of SBFT to an "artificial" food source, the physical form and ingredients of manufactured diets, and manufacturing costs (van Barneveld and Ellis, 2007).

9.7 RANCHED SBFT HEALTH

Fish health has been a focus since the beginning of tuna ranching in Australia. The first comprehensive study of ranched SBFT identified four parasites of particular concern to SBFT in ranching environments: a scuticociliate (*Uronema nigricans*), copepods (*Caligus sp.*), blood fluke (*Cardicola forsteri*), and gill fluke (*Hexostoma thynni*) (Nowak et al., 2003).

Uronema nigricans, more recently identified as *Miamiensis avidus* (Nowak et al., 2013), is a scuticociliate parasite that causes fish mortality through "swimmers disease," a disease causing SBFT to swim erratically with sharp head movements at the water's surface. SBFT tend to be infected with this parasite when water temperatures are low. The reservoir(s) for this parasite is unknown.

Sea lice (*Caligus chiastos)* is associated with eye damage in SBFT . It is thought that sea lice either graze directly on the cornea of the SBFT , or that SBFT self-injure as they rub against pontoon nets in an attempt to remove the lice (Hayward et al., 2008, 2009, 2010). The reservoir of sea lice infecting ranched SBFT is a fish species (Degen's leatherjackets, *Thamnaconus degeni*) that scavenges on waste tuna feed (Hayward et al., 2011). To manage this parasite, farmers closely manage feed quantities and maximize the distance between the bottom of the net enclosure and sea floor (Kirchhoff et al., 2011a,b,c).

Blood flukes (*Cardicola forsteri)* infect ranched SBFT within 6–8 weeks of their confinement in the ranching pontoon (Aiken et al., 2006) and have been responsible for a significant number of mortalities in ranched SBFT (Nowak et al., 2013). Blood fluke is a sanguinicolid requiring two hosts to complete its life cycle: the terrebellid polychaete (*Longicarpus modestus)* and SBFT (Cribb et al., 2011). To manage blood fluke infections, praziquantel is dissolved in a solution of propylene glycol and water, injected into freshly caught sardines (*Sardinops sagax*) using an Accujector 450, and fed to ranched SBFT under veterinary supervision. This treatment does not leave residue, it has shown to significantly reduce the prevalence of adult blood flukes in SBFT hearts and minimize blood fluke infections, and SBFT readily accept the treated feed (Hardy-Smith et al., 2012). Blood fluke is also managed by locating SBFT ranching pontoons in deeper waters (between 40 and 50 m) to increase the distance between the ranched SBFT and the seafloor (Kirchhoff et al., 2011c) where the intermediate host lives in broken shell deposits. Together these treatments have been responsible for reducing mortalities of ranched SBFT from approximately 15% in 2010 to less than 1% in 2014.

The gill fluke, *Hexostoma thynni*, and gill copepods, *Pseudocycnus appendiculatus* and *Euryphorus brachyterus*, have also been identified in ranched SBFT but no relationship between the presence of these parasites and SBFT health has been observed (Nowak et al., 2010).

9.8 THE FUTURE OF SBFT RANCHING IN AUSTRALIA

Since 1990, Australian production of SBFT through ranching has grown steadily (Figure 9.6). SBFT ranching is now the single-most valuable sector of South Australia's aquaculture industry (PIRSA, 2012). Up to 9,000 tonnes of gilled and gutted SBFT are now produced annually with an estimated annual value of between AUD$150–$300 million (PIRSA, 2012). The SBFT ranching industry is a significant employer: more than 1,000 full-time

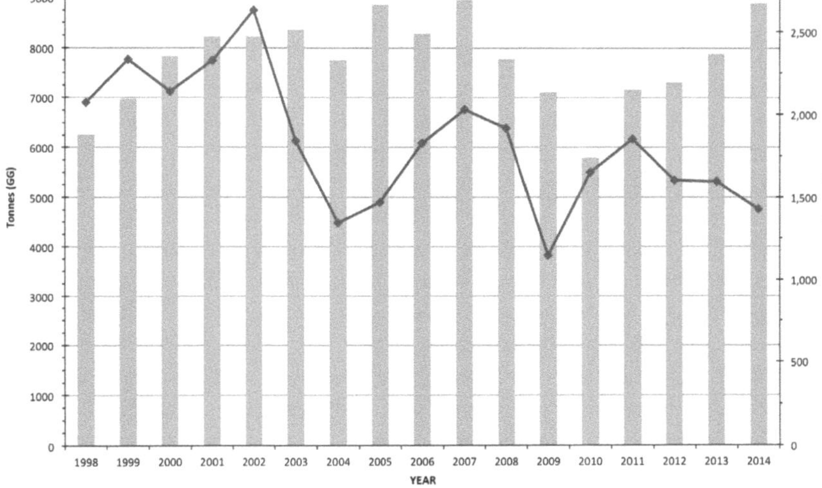

FIGURE 9.6 SBFT production volume and value. *Data provided by the Australian Southern Bluefin Tuna Industry Association.*

equivalent jobs were directly and indirectly generated by the industry in 2009/2010 (Econsearch, 2011).

To become the economic success that it is today the development of SBFT ranching in Australia has had to overcome a number of challenges. In 1996, in the early days of the industry, a mass mortality occurred coinciding with the remnants of a tropical cyclone that swept across southern Australia (Munday and Hallegraeff, 1998).

Rapid growth in the industry between 1996 and 2002 was possible with refinement of husbandry techniques, high demand from the Japanese sashimi market, limited competition, and high product values (peaking at an average of ¥2650 per kg) (Figure 9.6).

However in 2002, as commercial quantities of ranched bluefin tuna began to be produced by Japan, Mexico, and countries surrounding the Mediterranean Sea, an over-supply of premium quality bluefin product began to flood the market. Compounded by the global financial crisis in 2009, international market prices for ranched SBFT steeply declined (Figure 9.6). Since 2009, the price of ranched SBFT in the Japanese market has stabilized around ¥1500 per kg (Figure 9.6).

As the sustainability of global stocks of SBFT comes under increasing scrutiny, Australia's SBFT ranching industry will need to continue to evolve to maintain viability and success. Fundamental to the sustainability of the industry is the growing understanding of wild SBFT stocks and SBFT ecology (CCSBT, 2015b). On this foundation, future challenges include the need to diversify markets both domestically and internationally, improvements in product quality that will enable extended shelf life, and development of better links between production and market preferences. Fish health will remain a critical determinant in productivity, and greater understanding of the nutritional requirements of SBFT will help to minimize production costs through optimized feeding strategies (Montague et al., 2008).

Key to the future of SBFT ranching in Australia, however, is the replacement of baitfish feed with manufactured diets. The single biggest cost to SBFT ranching operations is feed. There are significant biosecurity risks and strict quarantine import requirements that must be met for imported baitfish. The nutritional content of baitfish feed is variable (Ellis and Rough, 2005), quantities of catch and timing of delivery of baitfish feed are often unpredictable, and freight costs are high. The use of manufactured feed in the SBFT ranching industry will allow a reliable, consistent, and more cost-effective feed source free from biosecurity risks. It will also better enable the industry to tailor SBFT products to market preferences while maintaining optimal health of fish through improved nutrition and husbandry practices. Current Australian research into manufactured feed shows promising results and may shape the development of the Australian SBFT ranching industry in the years ahead (Plentex, 2015).

9.9 CONCLUSION

Australia's SBFT ranching industry developed out of concerns about the impacts of wild harvest on SBFT populations and the need for technical innovation to enhance product value. Ranching of SBFT caught in the clean waters of the Great Australian Bight off southern Australia adds significant value to harvested products through regular feeding, effectively doubling the pre-ranching weight of fish over a relatively short time frame, and promoting product quality attractive to the premium sashimi market. The industry now earns hundreds of millions of dollars for local, regional, and national economies. While significant advances have been made in developing knowledge and technologies to support improvements in fish health, product quality, nutrition, and feed management, the future of the SBFT ranching industry in Australia depends on furthering understanding of SBFT populations, health, and nutritional requirements, in addition to continuing innovation in product development and manufactured feeds.

REFERENCES

Aiken, H.M., Hayward, C.J., Nowak, B.F., 2006. An epizootic and its declines of a blood fluke, *Cardicola forsteri*, in farmed southern bluefin tuna, *Thunnus maccoyii*. Aquaculture 254, 40–45.

Altringham, J.D., Shadwick, R.E., 2001. Swimming and muscle function. In: Block, B.A., Stevens, E.D. (Eds.), Tuna Physiology, Ecology, and Evolution. Academic Press, San Diego, pp. 313–344.

Attenborough, D., 1979. Life on Earth. Little, Brown and Company, Boston.

Brill, R.W., 1996. Selective advantages conferred by the high performance physiology of tunas, billfish, and dolphin fish. Comp. Biochem. Physiol. 113A, 3–15.

Bushnell, P.G., Jones, D.R., 1994. Cardiovascular and respiratory physiology of tuna: adaptations for support of exceptionally high metabolic rates. Environ. Biol. Fish. 40, 303–318.

Campbell, D., 2001. Change in fleet capacity and ownership of harvesting rights in the Australian southern bluefin tuna fishery. In: Shotten, R., (Ed.), Case Studies on the Effects of Transferable Fishing Rights on Fleet Capacity and Concentration of Quota Ownership. FAO Fisheries Technical Paper 412, pp 134–150.

Carter, C.G., Seeto, G.S., Smart, A., Clarke, S., van Barneveld, R.J., 1998. Correlates of growth in farmed juvenile southern bluefin tuna Thunnus maccoyii (Castelnau). Aquaculture 161, 107–119.

Carter, C., Nowak, B., Clarke, S., 2010. The tunas (Family: Scombridae). In: Le Francois, N.R., Jobling, M., Carter, C., Blier, P. (Eds.), Finfish Aquaculture Diversification. CAB International, Wallingford, pp. 432–449.

de Castelnau, F., 1872. Contribution to the ichthyology of Australia. Proc. Zool. Acclim. Soc. Vict. 1, 29–248.

CCSBT, 2015a. Origins of the convention Commission for the Conservation of Southern Bluefin Tuna. <http://www.ccsbt.org/site/origins_of_the_convention.php> (viewed May 2015).

CCSBT, 2015b. Origins of the convention Commission for the Conservation of Southern Bluefin Tuna. <http://www.ccsbt.org/site/research.php> (viewed May 2015).

Clarke, S., Smart, A., van Barneveld, R., Carter, C., 1997. The development and optimisation of manufactured feeds for farmed southern bluefin tuna. Austasia Aquaculture 11 (3), 59–62.

Collette, B.B., Nauen, C.E., 1983. FAO Species Catalogue. Vol. 2. Scombrids of the World. An Annotated and Illustrated Catalogue of Tunas, Mackerels, Bonitos and Related Species Known to Date. Food and Agriculture Organization of the United Nations, Rome.

Cribb, T.H., Adlard, R.D., Hayward, C.J., Bott, N.J., Ellis, D., Evans, D., et al., 2011. The life cycle of *Cardicola forsteri* (Trematoda: Aporocotylidae), a pathogen of ranched southern bluefin tuna, Thunnus maccoyii. Int. J. Parasitol. 41 (8), 861–870.

D'Antignana, T., Thomas, M., Bowyer, J., Chick, R., Thomas, P., 2008. Provide a Baitfish Supplement Diet for the Effective Delivery of Specific Nutrient Requirements to Juvenile Farmed SBT (*Thunnus maccoyii*) Aquafin CRC Project 2.2(2) (FRDC Project No. 2004/209).

Dewar, H., Graham, J.B., 1994. Studies of tropical tuna swimming performance in a large water tunnel—Energetics. J. Exp. Biol. 192, 13–31.

Dickson, K.A., 1996. Locomotor muscle of high-performance fishes: what do comparisons of tunas with ectothermic sister taxa reveal?. Comp. Biochem. Physiol. 113A (1), 39–49.

Econsearch. 2011. The Economic Impact of Aquaculture on the South Australian State and Regional Economies, 2009/10. <http://pir.sa.gov.au/__data/assets/pdf_file/0005/237227/AquaImpacts10_Final_110617.pdf>.

Ellis, D., Rough, K., 2005. Quality and Nutritional Evaluation of Baitfish used for SBT Farming (Including Baitfish Profiles). Aquafin CRC, Adelaide.

Ellis, D., 2013. The measurement of visceral temperature patterns and implications for feeding practices in ranched southern bluefin tuna *Thunnus maccoyii*. Unpublished MSc thesis.

Fitzgibbon, Q.P., Frappell, P., Clark, T., Buchanan, J., Ellis, D., Carter, C., et al., 2008. Activity Metabolism in Live-Held Southern Bluefin Tuna (*Thunnus maccoyii*). Aquafin CRC Project 1 A.7 (2), (FRDC Project No. 2005/200). SARDI, South Australia.

Geen, G., Nayer, M., 1989. Individual Transferable Quotas and the Southern Bluefin Tuna Fishery, Economic Paper. Occasional Paper 105, Project 62336. Australian Publishing and Printing Co, Fyshwick, A.C.T.

Glencross, B.D., van Barneveld, R.J., Carter, C.G., Clarke, S.M., 1999. On the path to a manufactured feed for farmed southern bluefin tuna. World Aquaculture Mag. 30 (3), 42–47.

Glencross, B.D., Clarke, S., Buchanan, J., Carter, C.G., van Barneveld, R.J., 2002. Temporal growth patterns of farmed juvenile southern bluefin tuna, *Thunnus maccoyii* (Castelnau) fed moist pellets. J. World Aquaculture Soc. 33, 138–145.

Gordon, R., Bayly, G., Foster, C., 2006a. Commercialisation Trials for a Manufactured Feed. Aquafin CRC Project 1 A.3, (FRDC Project No. 2001/201). Printed by SARDI, South Australia.

Gordon, R., Bayly, G., Foster, C., 2006b. Industry report on commercial pellet trial on southern bluefin tuna (*Thunnus maccoyii*) Skretting Australia, Cambridge, Tasmania.

Government of South Australia, 1993. Report on the Joint Feasibility Study Project for Southern Bluefin Tuna Farming in Boston Bay, Port Lincoln, South Australia. Government of South Australia, Tuna Boat Owners Association of Australia, Overseas Fishery Cooperation Foundation.

Gunn, J., Hartog, J., Rough, K., 2002. Experimental Analyses of the Effects of Ration and Feeding Frequency on the Thermodynamics, Energetics, Growth and Condition of Farmed Southern Bluefin Tuna. Fisheries Research and Development Corporation Final Report, Project 97/363.

Hardy-Smith, P., Ellis, D., Humphrey, J., Evans, M., Evans, D., Rough, K., et al., 2012. In vitro and in vivo efficacy of anthelmintic compounds against blood fluke (*Cardicola forsteri*). Aquaculture 337, 39–44.

Hayes, E.A., 1997. A Review of the Southern Bluefin Tuna Fishery: Implications for Ecologically Sustainable Management. A TRAFFIC Oceania Report.

Hayward, C.J., Aiken, H.M., Nowak, B.F., 2008. An epizootic of Caligus chiastos on farmed southern bluefin tuna *Thunnus maccoyii* off South Australia. Dis. Aquat. Org. 79, 57–63.

Hayward, C.J., Bott, N.J., Nowak, B.F., 2009. Seasonal epizootics of sea live, Caligus spp., on southern bluefin tuna, *Thunnus maccoyii* (Castelnau), in a long-term farming trial. J. Fish Dis. 32, 101–106.

Hayward, C.J., Ellis, D., Foote, D., Wilkinson, R.J., Crosbie, P.B.B., 2010. Concurrent epizootic hyperinfections of sea lice (predominantly *Caligus chiastos*) and blood flukes (*Cardicola forsteri*) in ranched southern bluefin tuna. Vet. Parasitol. 173 (1–2), 107–115.

Hayward, C.J., Svane, I., Lachimpadi, S.K., Itoh, N., Bott, N.J., Nowak, B.F., 2011. Sea lice infections of wild fishes near ranched southern bluefin tuna (*Thunnus maccoyii*) in South Australia. Aquaculture 320, 178–182.

Itoh, T., Kemps, H., Totterdell, J., 2011. Diet of young southern bluefin tuna (*Thunnus maccoyii*) in the southwestern coastal waters of Australia in summer. Fish Sci. 77, 337–344.

Jeffriess, B., 1999. Growout of Southern Bluefin Tuna. Fisheries Research and Development Corporation, Project No. 91/56.

Kirchhoff, N.T., D'Antignana, T., Leef, M., Hayward, C.J., Wilkinson, R.J., Nowak, B.F., 2011a. Effects of immunostimulants on ranched southern bluefin tuna *Thunnus maccoyii*: immune response, health and performance. J. Fish Biol. 79, 331–355.

Kirchhoff, N.T., Leef, M.J., Ellis, D., Purser, J., Nowak, B.F., 2011b. Effects of the first two months of ranching on the health of southern bluefin tuna *Thunnus maccoyii*. Aquaculture 315, 207–212.

Kirchhoff, N.T., Rough, K.M., Nowak, B.F., 2011c. Moving cages further offshore: effects on southern bluefin tuna, *T. maccoyii*, parasites, health, and performance. PLoS ONE 6 (8), e23705.

Lagler, K.F., Bardach, J.E., Miller, R.R., May Passino, D.R., 1977. Ichthyology, second ed. John Wiley and Sons, Brisbane.

Magnuson, J.J., 1978. Locomotion by scombrid fishes: hydromechanics, morphology, and behaviour. In: Hoar, W., Randall, D.J. (Eds.), Fish Physiology, vol. 7. Academic Press, New York, pp. 239–313.

Montague, P., Downes, E., Ham, J., 2008. Achievements and Impacts (2001–2008), the Cooperative Research Centre for the Sustainable Aquaculture of Finfish. Stallard and Potter, Adelaide.

Munday, B.L., Hallegraeff, G.M., 1998. Mass mortality of captive southern bluefin tuna (*Thunnus maccoyii*) in April/May 1996 in Boston Bay, South Australia: A complex diagnostic problem. Fish Pathol. 33 (4), 343–350, 1998.10.

Nowak, B.F., Rough, K., Ellis, D., Crane, M., Cameron, A., Clarke, S., 2003. A Risk Assessment of Factors Influencing the Health of Southern Bluefin Tuna. FRDC 2001/253. Aquafin CRC, Adelaide.

Nowak, B.F., Evans, D., Ellis, D., Wilkinson, R., Porter, M., Crosbie, P., et al., 2010. Improving Husbandry and Performance of Southern Bluefin Tuna Through Better Understanding of the Relationship Between Fish Stress and Health. Aquafin CRC, Adelaide.

Nowak, B., Ellis, D., Kirchoff, N. (Eds.), 2013. Understanding Parasites And Investigating Ways to Mitigate their Influence on SBT Production. FRDC 2008/228. University of Tasmania.

Padula, D.J., Daughtry, B.J., Phua, S.T.G., Balshaw, S., 2007. Management of Food Safety Hazards in Farmed Southern Bluefin Tuna to Exploit Market Opportunities. Aquafin CRC, Adelaide.

Patterson, H., Begg, G., Curtotti, R., Vieira, S., 2009. Southern bluefin tuna fishery. In: Wilson, D.T., Curtotti, R., Begg, G.A. (Eds.), Fishery Status Reports 2009 – Status of Fish Stocks and Fisheries Managed by the Australian Government. Union offset printers, Canberra, A.C.T., Canberra, pp. 422–433.

Patterson, H., Begg, G., Hormis, M., 2010. Southern bluefin tuna fishery. In: Woodhams, J., Stobutzki, I., Vieira, S., Curtotti, R., Begg, G. (Eds.), Fishery Status Reports 2009 – Status of Fish Stocks and Fisheries Managed by the Australian Government. Union offset printers, Canberra, ACT, Canberra, pp. 359–367.

PIRSA, 2012. <http://www.pir.sa.gov.au/aquaculture/aquaculture_industry/tuna>. (viewed April, 2012).

PIRSA, 2015a. <http://pir.sa.gov.au/aquaculture/leasing_and_licensing/marine_license_holder_responsibilities> (viewed May 2015).

PIRSA, 2015b. <http://www.legislation.sa.gov.au/LZ/C/R/AQUACULTURE%20REGULATIONS%202005.aspx> (viewed May 2015).

PIRSA, 2015c. <http://www.legislation.sa.gov.au/LZ/C/A/AQUACULTURE%20ACT%202001/CURRENT/2001.66.UN.PDF> (viewed May 2015).

Plentex, 2015. Local demand for aquafeed. <http://www.plentex.com.au/irm/content/local-demand-for-aquafeed1.aspx?RID = 347> (viewed July, 2015).

Plevin, J.E., 2000. MFV Tacoma and the Haldane family, viewed January 18, 2012. <http://tacoma.org.au/documents/Tacoma_Haldane_History.pdf>.

Serventy, D.L., 1956. The southern bluefin tuna, *Thunnus thynnus maccoyii* (Castelnau), in Australian waters. Aust. J. Marine Freshw. Res. 7, 1–43.

Thomas, P., 2007. Maximising the Control of Quality in Farmed SBT. Aquafin CRC, Adelaide.

Thomas, P., Thomas, M., Schuller, K., D'Antignana, T., 2009. Application of the use of Dietary Supplements for Improving Flesh Quality Attributes of Farmed SBT. Aquafin CRC, Adelaide.

van Barneveld, R., Ellis, D., 2007. Practical Nutrition of Farmed Southern Bluefin Tuna Thunnus maccoyii. Aquafin CRC, Adelaide.

van Barneveld, R.J., Carter, C.J., Tivey, D.R., Brooker, J.D., 1999. Development of an Invitro Assay for the Assessment of Alternative Protein Sources for Use in Artificial Diets for Farmed Southern Bluefin Tuna (*Thunnus maccoyii*). Fisheries Research and Development Corporation Final Report, Project 95/068.

van Barneveld, R.J., Buchanan, J., Ellis, D., 2009. Optimisation of Farmed Southern Bluefin Tuna (Thunnus maccoyii) Nutrition to Improve Feed Conversion Efficiency and Reduce Production Costs. Aquafin CRC, Adelaide.

Wardle, C.S., Videler, J.J., Arimoto, T., Franco, J.M., He, P., 1989. The muscle twitch and the maximum swimming speed of giant bluefin tuna, *Thunnus thynnus L.* J. Fish Biol. 35, 129–137.

Young, J.W., Bradford, R.W., Lamb, T.D., Lyne, V.D., 1996. Biomass of zooplankton and micronekton in the southern bluefin tuna fishing grounds off eastern Tasmania, Australia. Mar. Ecol. Prog. Ser. 138, 1–14.

Young, J.W., Lamb, T.D., Le, D., Bradford, R.W., Whitelaw, A.W., 1997. Feeding ecology and interannual variations in diet of southern bluefin tuna, *Thunnus maccoyii*, in relation to coastal and oceanic waters off eastern Tasmania, Australia. Environ. Biol. Fish. 50, 275–296.

Chapter 10

Southern Bluefin Tuna Captive Breeding in Australia

Ben Nan Chen[1], Wayne Hutchinson[2] and Craig Foster[1]
[1]Clean Seas Tuna Ltd., Port Lincoln, SA, Australia, [2]South Australian Research and Development Institute, Urrbrae, SA, Australia

10.1 INTRODUCTION

Increasing demand for seafood has put significant pressure on the international fisheries for wild bluefin tuna species. Farming of southern bluefin tuna (SBFT, *Thuunus maccoyii*) is the largest aquaculture industry in South Australia producing 7486 tonnes valued at A$153.5 million in 2012–2013 (Stephan and Hobsbawn, 2014). Currently the volume of SBFT produced by the Australian tuna aquaculture industry is limited by the size of the wild catch quota agreed and reviewed annually by the Commission for the Conservation of the Southern Bluefin Tuna (CCSBT) who determine the management of this international fishery. CCSBT is comprised of representatives of voting member countries and regions such as Australia, Japan, New Zealand, Korea, Indonesia, Taiwan, and nonvoting cooperating nonmembers including the Philippines, South Africa, and the European Community. From 2007 until 2009 CCSBT set the Australian quota at 5265 tonnes of a global total allowable catch (TAC) of 11,810 tonnes. However, in 2009 there was a reduction of 23.4% to the Australian quota to 4016 tonnes for the 2010 and 2011 fishing seasons. This Australian quota was revised in 2011 allowing an increase of 32% over 3 years. In 2014, the Australian quota was 5193 tonnes of the global TAC of 12,449 tonnes and from 2015–2017 this will increase to 5665 tonnes of the global TAC of 14,647 tonnes. The fluctuation of the wild stock and quota limitation of the fishery supports the need to pursue propagation to allow further expansion of SBFT aquaculture industry in Australia. The development of the hatchery technology to provide propagated fingerlings is needed if SBFT aquaculture is to grow and to become truly sustainable. In response to this need, since 2006, Clean Seas Tuna Ltd (CST) has invested significantly in SBFT propagation R&D in collaboration

D.D. Benetti, G.J. Partridge, A. Buentello (Eds): Advances in Tuna Aquaculture.
DOI: http://dx.doi.org/10.1016/B978-0-12-411459-3.00009-6

with the Australian Seafood Cooperative Research Centre (SfCRC), and set the major objective to realize the commercial production of SBFT fingerlings from eggs spawned by captive broodstock.

Captive SBFT broodstock maintained and conditioned onshore by CST first spawned small numbers of viable eggs in 2008. In 2009, 51.8 million eggs and approximately 400 weaned 40 days after hatch (dah) SBFT fingerlings were produced. Similar numbers of SBFT fingerlings were produced in 2011 and 2012, and the fingerlings were transferred from nursery tanks into sea cages and a lined onshore pond. These fingerlings grew well but ongoing mortality occurred after transfer and no fingerlings survived when ambient seawater temperatures declined from late autumn. In the hatchery, SBFT larval survival rate was usually below 1% during the first 2–3 weeks of larval rearing. Mortality was mostly caused by surface adhesion during the first few days of feeding and sinking death at night. Reasons for poor survival were observed to involve poor swim bladder inflation that was recorded in all batches cultured. Inappropriate nutrition and less than optimal environmental conditions are also thought to contribute to slow growth during the early rearing stages when SBFT larvae are feeding on live foods.

Recently, larval survival has been improved significantly using improved upwelling devices and providing higher DHA enrichment of live foods during the first 2 weeks of feeding. Although survival of more than 20% of SBFT larvae until 13 dah has been achieved, the survival after this point declines quickly as larvae transfer to feeding on *Artemia* and yolk sac larvae. Subsequently, similar to Pacific bluefin tuna (PBFT, *Thunnus orientalis*), inappropriate weaning diets, cannibalism, wall collision, and handling stress following transfer to nursery tanks are other obstacles during later nursery stages. However, significant knowledge and hands-on experience toward closing the SBFT life cycle have been achieved in the larval rearing and nursery stages.

10.2 LIFE HISTORY OF SBFT

The highly migratory SBFT forage in the Indian Ocean and south-west Pacific Ocean between 30°S and 50°S (Basson et al., 2012). Like other bluefin tuna species, the cold tolerant SBFT are obliged to migrate to warm waters to spawn in an environment that is not best suited for their growth and maintenance. Shingu (1978) proposed that SBFT occupy different areas of the southern oceans at different stages while growing and feeding. The principal habitat of immature SBFT appears to be the surface layers while the habitat of adults is the sub-surface or mid-water layers. Caton (1991) suggests that juveniles migrate in the southward flowing Leeuwin current that originates near the SBFT spawning ground and eventually turns east into the Great Australian Bight (GAB) at Cape Leeuwin. Farley and Davis (1998) state that older juveniles move south along the continental shelf of Western Australia and first appear in the warm waters of the GAB as 1-year-old in summer and then disperse along the West Wind Drift in winter. The Australian summer surface fishery (November–March) for SBFT in the GAB is dominated by 2–4-year-old fish that are also caught off south eastern Africa and these juvenile fish then move into winter feeding grounds in the central Indian Ocean (Gunn et al., 2003). In Australian waters, fish of 3–4 years of age make seasonal migrations between the east and west and around Tasmania and New Zealand generally in the latter half of the year (Caton, 1991). Samples from the Taiwanese long-line fishery in the central Indian Ocean show that the dominant year classes are 4–6 years old comprised of immature fish less than 40 kg (140 cm). Very few SBFT older than 7–8 years are caught between 30–35°S suggesting that mature adult SBFT do not forage this far north and stay offshore until reaching maturity or they may make seasonal migrations to the Tasman Sea north of 40°S and to waters east of New Zealand during the southern winter (Olson, 1980; Caton, 1991).

The maximum lifespan of SBFT is estimated to be 40 years (Gunn et al., 2008) with fish achieving 200 kg and measuring up to 2 m in length (Caton, 1991). SBFT are a single population (Grewe et al., 1997) and mature fish migrate between feeding grounds across the Southern Ocean, along the West Wind Drift, to a staging ground in the eastern Indian Ocean off Western Australia and then move on to the spawning ground between Bali, Indonesia and Australia in an area between 7–20°S and 102–124°E (Davis et al., 1991). This is when and where the surface water is a combination of the strongest Pacific Warm Pool flow through from the Indonesian archipelago and the convergent Indian Ocean gyre circulation as functions of variable seasonal winds (Sharp, 2001). Spawning occurs in all months except July with the main spawning season extending from September to April with a peak in January–February (Caton, 1991; Farley and Davis, 1998). A study of the dynamics of spawning of SBFT concluded that spawning is not synchronized for the whole stock and it is suggested that there is a continual turnover of new spawners replacing post-spawning fish throughout the spawning season. A low number of post-spawning fish sampled on the spawning ground suggests that individuals leave the area as soon as they have completed spawning (Farley and Davis, 1998). It is possible that because of the wide distribution of SBFT across these feeding grounds that the cue to commence the spawning migration may occur at different times in different locations accounting for the extended spawning season due to the time required for these fish to migrate to the spawning grounds.

SBFT are multiple or batch spawners (Farley and Davis, 1998) and the mean size at 50% maturity is between 158–163 cm when they are between 11–12 years of age (Farley and Davis, 1998; Gunn et al., 2003, 2008). This is much older than PBFT which commence spawning at approximately 3–5 years of age (Masuma et al., 2008, 2011). The spawning frequency for SBFT is reported to be 1.6 days but if only those females in prime spawning condition are considered then the spawning interval reduces to 1.1 days suggesting that once they start spawning females will spawn daily (Farley and Davis, 1998). Relative batch fecundity of SBFT increases with body length and is estimated at 56.5 ± 16.1 oocytes per gram of body weight (Farley and Davis, 1998) equivalent to approximately 5.7 million oocytes per batch from a 100 kg female.

Gunn et al. (2006) suggested the possibility that SBFT are not obligate annual spawners with not all mature fish migrating and spawning each year. This suggestion is supported by the observation that only one of 46 fish to which pop-up archival satellite tags (PSATs) were deployed migrated to the spawning ground, although this number may have been higher had a greater number of tags remained attached for a longer period of time. More recently Evans et al. (2012) deployed PSATs on 90 large SBFT in waters off eastern Australia (2001–2006) and northern New Zealand (2007–2008) and categorized 23 fish with tags that remained attached for more than 160 days as "spawners" (n = 2) that migrated to the spawning ground; "likely spawners" (n = 7) that showed movement consistent with spawning migration but their tag detached prematurely; "likely non-spawners" (n = 12) that showed movement consistent with residency in the Tasman Sea; and "non-spawners" (n = 2) that stayed in the Tasman Sea for an annual cycle. These results suggest that matured SBFT may not spawn annually and may exhibit "skipped spawning." The authors suggest that changes in water temperature associated with annual changes in major water currents may trigger migration and hypothesize that resources may be invested in accelerated growth rather than spawning, in order to attain a larger size that may result in greater fecundity in later years. It is suggested that young matured SBFT are more likely to defer spawning, with this behavior declining with age as larger older matured fish expend less energy for growth. These phenomena may be attributed to the cold adapted physiology of SBFT that may result in significant thermal stress when inhabiting tropical spawning grounds (Evans et al., 2012).

10.3 CAPTIVE SPAWNING

Although the development of the technology for SBFT broodstock management and spawning is ongoing, spawning of SBFT broodstock has been achieved annually within the land-based holding system at CST Arno Bay Hatchery since 2008 (Table 10.1). This land-based broodstock holding system is approximately 3 million L in volume, and operates as a recirculating water treatment system that maintains water quality and allows control of water temperature and photoperiod to promote reproductive development and spawning of SBFT broodstock. Typically 15–25 fish between 100 and 150 kg are held within the broodstock tank and fish are added to the tank each year due to loss of fish from wall strikes attributed to chasing during spawning within the confined tank. Broodstock are fed to satiation once each

TABLE 10.1 Summary of Egg Production from Captive SBFT Broodstock held by Clean Sea Tuna Ltd at Arno Bay, South Australia between 2008 and 2014

Year	Total all Eggs (million)	Total Viable Eggs (million)	Spawning Events	Duration (days)
2008	0.6	0.1	3	3
2009	51.7	33.3	108	34
2010	46.5	26.9	99	93
Early 2011	25.5	16.7	28	69
Late 2011–Early 2012	10.8	6.5	47	35
Late 2012	1.5	1.0	15	67
2013	0.6	–	8	8
2014	5.2	2.9	27	115

day and are provided with a balance of protein and fat using local pilchards (*Sardinops sagax*) and Californian squid (*Loligo opalescences*) supplemented with vitamins. Feed intake is monitored and adjusted daily and follows a seasonal pattern with fish gouging during the "winter" period and consuming less during the simulated spawning period.

In 2009, spawning commenced in February and continued until April during which time 51.7 million eggs were spawned and 108 spawning events were recorded. During 2010, spawning commenced approximately 5 weeks earlier in January and continued until April during which time 46.5 million eggs were spawned and 99 spawning events were recorded (Table 10.1). Since then the number of eggs has declined progressively with 16.7 million, 6.5 million, 1 million, and 2.9 million viable eggs collected over the past four spawning periods (2011, 2011–2012, 2012–2013, 2014). During the best spawning years, fish were spawning between January and April with the frequency and duration of spawning varying between seasons. Egg production declined for the 2011–2012 and 2012–2013 spawning seasons when attempts were made to advance the spawning period to provide supply of SBFT larvae at a time that would allow fingerlings produced to be transferred to sea cages while warmer ambient sea water temperatures prevailed (i.e., January–February). Alternatively the decline in egg production may have been attributed to the death of one female in January 2012 that had spawned approximately 85% of the eggs produced during the 2009, 2010, 2011, and late 2011 spawning seasons.

Reproductive development of captive SBFT broodstock is controlled by manipulation of water temperature and photoperiod in combination with provision of high-quality nutrition. Studies show that in the wild matured SBFT prefer water temperatures between 19°C and 21°C and tend to remain in water temperature of less than 21.5°C as they migrate to the spawning ground (Patterson et al., 2008; Evans et al., 2012). While on the spawning ground, SBFT show a transition to surface-related behavior suggesting participation in spawning associated with surface temperatures greater than 24 °C (Farley and Davis, 1998; Patterson et al., 2008). Together with the thermal-photoperiod regulation, captive SBFT broodstock are treated with gonadotropin releasing hormone agonist (GnRHa) implants on a tag developed in Europe (Mylonas et al., 2007; De Metrio et al., 2010) to induce final maturation and spawning. The hormone dosage used is between 67–167 μg GnRHa kg^{-1} body weight that equates to between 8 mg and 20 mg hormone being delivered to a 120 kg fish. There is ongoing concern that the repeated implantation of individual fish has adverse effects with evidence of multiple tag heads remaining in fish associated with extensive muscle necrosis that may ultimately cause mortality of fish. Observations suggest that a water temperature shock timed to coincide with hormone implants may promote spawning of SBFT within the onshore system. It was observed that during spawning season, a number of males chase one female and most of the males released milt, while few of the females spawn eggs. This may imply that female sexual maturation and spawning (i.e., release of eggs) is a bottleneck for production of viable eggs from captive SBFT.

Spawning broodstock in sea cages as is described for PBFT by Masuma et al. (2011) and in Chapter 7, may be a viable option for SBFT if sites with suitable annual temperature profiles can be identified.

10.4 EGG INCUBATION

Eggs spawned and fertilized within the SBFT broodstock tank flow via a surface outlet into egg collectors where they accumulate until harvested. After collecting, rinsing, disinfecting, and counting, SBFT eggs are incubated either directly in larval rearing tanks or in separate incubators. When incubated directly in the larval rearing tanks, SBFT eggs are stocked at 5–10 eggs L^{-1} depending on the number of available eggs. The ambient conditions for incubation are the same as those used during early larval rearing and will be described below.

When hatched in incubators, SBFT eggs are stocked at no more than 1000 eggs L^{-1}. Observations suggest that aeration is unsuitable during incubation of SBFT eggs because the fine air bubbles cause larvae to stick to vertical screens and tank walls where bubbles break near these locations. This is not only a major cause of mortality in newly hatched larvae, but

FIGURE 10.1 Incubator system used by CST in the Arno Bay R&D hatchery to hatch SBFT eggs comprising one header tank and four egg incubators (450 L working volume each).

FIGURE 10.2 Central outlet screen and upwelling seawater device installed with each incubator in the CST R&D hatchery at Arno Bay. Newly hatched SBFT larvae are suspended in water column by the upwelling water current created within the incubator.

also in the early larval rearing stage. The incubators are 500–1000 L conical bottom fiberglass tanks with a seawater upwelling device installed to generate an upwelling water current within the incubators to suspend and homogeneously distribute eggs and larvae through the water column (Figures 10.1 and 10.2). During incubation, the daily water exchange rate is maintained at 20–25 turnovers (2000–2500%) with water temperature at 25–26°C, dissolved oxygen at 6.5–7.0 mg L^{-1} (95–98% saturation), pH at 8.0, and salinity at 32–37 g L^{-1}. Under these parameters, the viable SBFT eggs hatch in approximately 26–28 h, and hatching rate averages 80%. The size of newly hatched SBFT larvae is 2.85 mm standard length (SL). After siphoning dead eggs and removing egg shells, the larvae are transferred into larval rearing tanks using jugs or buckets.

FIGURE 10.3 SBFT larval rearing tank: 3000 L with 2 surface skimmers.

10.5 LARVAL REARING

If the incubation approach is applied, newly hatched (0–1 dah) SBFT larvae are typically stocked into larval rearing tanks at densities ranging from 5–20 larvae L^{-1} depending on the number of available larvae. The larval rearing tanks are conical bottomed round tanks with upwelling devices installed (as discussed in the previous section and below). The volumes of the larval rearing tanks used vary from 3000 L (Figure 10.3) to 13,000 L. Larger tanks can be used for SBFT larval rearing if upwelling devices are installed to create upwelling current to maintain the larvae within the water column. Both recirculating seawater treatment systems and flow-through systems have been applied for SBFT larval rearing. Incoming sea water is sand filtered and filtered down to 5 μm or less and then UV treated before being delivered into the larval rearing tanks, because larger suspended solids are not easily settled down in upwelling tanks.

During the larval rearing period, water temperature is maintained at 25°C but can go as high as 27°C which is similar to the water temperature experienced by SBFT larvae on the spawning grounds off the North West coast of Australia (Jenkins and Davis, 1990; Shiao et al., 2009). However, it is suggested that higher temperature may contribute to adverse microbial conditions in the upwelling larval tanks. Dissolved oxygen is controlled at 100% saturation in SBFT larval rearing tanks, as oxygen consumption of bluefin tuna is reported to increase rapidly at an early stage particularly as the gills begin to form (Miyashita et al., 1999). Other environmental parameters such as pH (8.0) and salinity (37 g L^{-1}) are maintained at the same levels as the incoming seawater supply to the CST's hatchery at Arno Bay. When using ozone in a recirculating system, the ORP reading of the sea water returning to larval tanks should be maintained at around 320 mV.

Photoperiod is controlled to provide 14 h of light and 10 h of darkness with a ramp period of 30 min at the beginning and end of each day to simulate natural light conditions. Multiple fluorescent lights suspended above each larval rearing tank (Figure 10.3) provide a light intensity at the water surface of 2000 lux (35 μmol s^{-1} m^{-2}) in the middle of the tank and 1000 lux (26 μmol s^{-1} m^{-2}) at the edge of the tank. However, in a recent study on the visual development of SBFT larvae, Hilder (2013) suggested that lower light intensity can be applied in the early stages of SBFT larval rearing. In this study, the proportion of 3 dah larvae feeding was low and unaffected by light intensity. However, 6 dah and 9 dah larvae showed a significant decrease in the proportion of feeding with increasing light intensity indicating increased feeding performance in lower light intensities with increasing age up to 9 dah. This suggested that lower light intensity could be applied in the early stages of SBFT larval rearing after first feeding is established.

SBFT larvae are negatively buoyant immediately post hatch and their body density increases with age, and regardless of swim bladder inflation (Woolley et al., 2013). Due to this negative buoyancy, larvae tend to sink to the tank bottom at night when they cease feeding and actively swimming. It is likely that this contributes to "sinking death" that is attributed to the adverse conditions to which aggregations of larvae are exposed to on the tank bottom. Vertical movement of SBFT larvae during darkness has also been observed in the wild (Davis et al., 1990). However, the natural upwelling system in the ocean can bring the larvae up to the surface waters during the day when feeding takes place. Therefore, to counteract larval sinking, upwelling devices are installed in SBFT larval rearing tanks to maintain larvae in suspension and the improved hydrodynamic conditions are considered to be critical for better survival during the first 2 weeks of culture.

Unlike PBFT larvae where aeration can be used to generate upwelling (Tanaka et al., 2009; Nakagawa et al., 2011), upwelling devices are installed in SBFT larval rearing tanks using water because fine air bubbles have been associated with mortality of early stage SBFT larvae. Therefore, no aeration at all is used with SBFT larval tanks in the first week after stocking. Strong turbulence in larval rearing tanks causes a low feeding incidence and consequently high mortality of PBFT larvae (Kato et al., 2008). On this basis, gentle upwelling is used to suspend larvae within the water column while allowing them to easily maintain their feeding position.

Swim bladder inflation is first observed in SBFT larvae at 4 dah, and using current rearing methods the incidence of larvae with inflated swim bladders varies from 7% to 27% over the first 10 days of larval rearing. This incidence is lower than PBFT larvae, reportedly as high as 54% (Kurata et al., 2011; Nakagawa et al., 2011) despite the use of surface skimmers during SBFT larval rearing from as early as 2–9 dah. It was reported that PBFT larvae have a narrow window of only 1 day at 3 dah to achieve primary swim bladder inflation (Kurata et al., 2013). For SBFT larvae, it is still unknown when the window available for achieving successful initial swim bladder inflation closes. However, the pneumatic duct was not visible in SBFT larvae at 8 dah, and the window for initial inflation closes once the connection to the oesophagus is lost (Woolley et al., 2013). The development of a rete mirabile provides evidence that the SBFT larvae develop into physoclistous adults using their swim bladder as a hydrostatic organ similar to PBFT.

Confounding attempts to promote initial swim bladder inflation is the observation that at this time surface adhesion (so-called surface death) causes high levels of mortality of SBFT larvae. During this period, SBFT larvae that need to break through the water surface and gulp air to inflate their swim bladders tend to become trapped by surface tension causing air exposure and consequently mortality. Similarly to PBFT, surface death of SBFT can be reduced significantly by applying surface oil (fish oil or krill oil) to larval tanks at 0.2–0.3 mL m^{-2} tank surface area to prevent larvae contacting the water surface (Kurata et al., 2013). However, adding oil to larval rearing tanks conflicts with the attempts to improve primary swim bladder inflation using surface skimmers which are operated during the same period of time to remove any oil film from the water surface that impede the ability of larvae to gulp air required to achieve initial swim bladder inflation. Recently it has been demonstrated that operation of surface skimmers for a 3 h period between 16:00 and 19:00 each day from 3 to 9 dah was sufficient to promote swim bladder inflation in >80% of PBFT larvae (Kurata et al., 2013). Further research toward understanding the timing of swim bladder inflation in SBFT will help to establish a practical procedure to both prevent surface death and enhance SBFT primary swim bladder inflation under hatchery conditions.

In larval rearing tanks, SBFT larvae are kept in clear water until the onset of first feeding at 3 dah when rotifers and microalgae (paste) are added. To maintain "green water" conditions, microalgal paste (Nanno 3600—Reed Mariculture, USA) is added to the larval rearing tanks during the 14 h light phase at a daily dosage of 20–40 ml KL^{-1} of tank water until the end of the rotifer feeding stage. In the wild, calanoid and cyclopoid copepod nauplii and cladocerans are the main prey of SBFT larvae until they approach 7.0 mm SL and quickly transition to feeding on smaller fish larvae (Young and Davis, 1990). Due to the difficulties associated with producing sufficient quantities of these prey for commercial marine fish culture, enriched L-strain rotifers (*Brachionus plicatilis*) are provided to SBFT larvae in the hatchery. Larvae are fed three times a day (0900 h, 1300 h, and 1730 h) at densities of 10–15 ind mL^{-1} from 3 to 5 dah, increasing to 15–20 ind mL^{-1} by 18–20 dah.

Newly hatched *Artemia* nauplii are provided to the larvae at 13 dah three times a day at 0.05–0.1 ind mL^{-1}. At 14 dah, enriched *Artemia* nauplii are provided to larvae three times daily at an initial density of 0.2–0.5 ind mL^{-1}, increasing to 1–2 ind mL^{-1} at 16 dah and remaining at this density until 24 dah. Yolk-sac larvae of yellowtail kingfish (*Seriola lalandi*) are fed to SBFT larvae from 18 dah at 20–25 ind L^{-1} until metamorphosed SBFT fingerlings are fully weaned onto formulated diets by 33–35 dah. It is

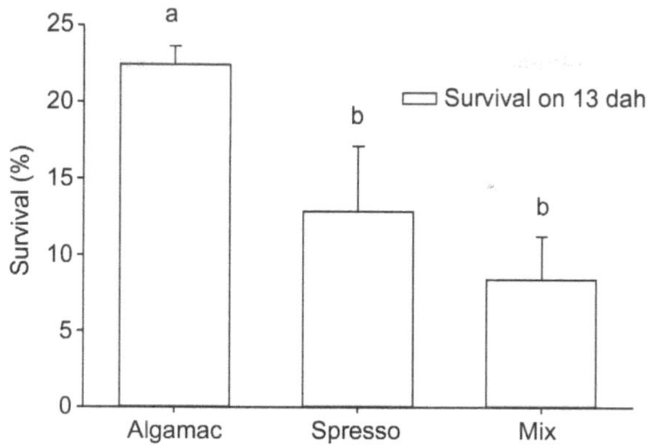

FIGURE 10.4 Comparison of SBFT larval survival at 13 dah using different enrichment products: AlgaMac, Spresso, and an equal mixture of both. Significant differences ($P < 0.05$) are represented using different letter notations (Chen, 2014).

suggested that smaller yolk sac larvae of other fish species such as mulloway (*Argyrosomus japonicus*), snapper (*Pagrus auratus*), and knifejaw (*Oplegnathus spp.*) may also be appropriate.

Tuna larvae require live foods with a high DHA content based on information such as the high DHA content (and DHA:EPA ratio) of ovarian tissue of female bluefin tuna at spawning (Mourente et al., 2002), the high DHA content of the natural food (cyclopoid and calanoid copepod nauplii) of wild SBFT larvae (Young and Davis, 1990), and high levels of DHA in the muscle fatty acid profile of wild caught SBFT (Nichols et al., 1998). Enrichment products that give a higher DHA content and a higher DHA:EPA ratio in the live foods are recommended for SBFT larval rearing. AlgaMac 3050 (Aquafauna Bio-Marine Inc, California, USA) is used at CST for both rotifer and *Artemia*, as a rotifer enrichment trial conducted showed that survival rates as high as 22.4% at 13 dah were achieved (Figure 10.4) (Chen, 2014). Further research is required to more fully understand the nutritional requirements of SBFT larvae as a guide to develop appropriate live food enrichment protocols as with other tuna species, particularly for *Artemia* enrichment (Biswas et al., 2006; Seoka et al., 2007, 2008).

10.6 WEANING AND NURSERY

The establishment of weaning and nursery protocols for SBFT juveniles is ongoing. Weaning usually starts around 20 dah when fish are still in the larval rearing tanks using a co-feeding approach that provides both weaning diets and prey fish larvae over a period of about 10–12 days. The weaning diets for SBFT juveniles can be any of the commercially available micro diets produced for tuna species, or moist diets prepared onsite by combining minced fresh local sardines and fishmeal. The weaning diets are hand fed to SBFT larvae every half hour during the light phase. Prey fish larvae are also added to the SBFT larval rearing tank several times during the day to provide a prey density of 20–25 ind L^{-1}. At approximately 33 dah, most of the SBFT juveniles are feeding successfully on the weaning diets and these fish quickly gain weight. At this time the young SBFT fingerlings that are feeding on weaning diets can be separated from the rest of the population and transferred to the nursery tanks, while the weaning process is continued with smaller individuals.

The size difference between individuals is observed to increase from approximately 18 dah, coinciding with the introduction of prey fish larvae. Particularly after the onset of weaning, some juveniles always express better growth rates which results in size variation in the tanks. Soon after these size differences are noticed, SBFT juveniles begin to show cannibalistic behavior. Mortality caused by cannibalism increases dramatically from this time, and over 50% of the population can be lost due to cannibalism in this period from 18 to 33 dah.

To minimize the mortalities caused by cannibalism, size grading (i.e., the separation of bigger fish from the smaller ones), is conducted frequently during the weaning period from 20 to 33 dah. When carrying out size grading, the water level of the larval or nursery tanks is reduced to a level that is convenient for technical staff to enter the tank. SBFT juveniles are slightly anesthetized using 2–3 ppm AQUI-S® Aquatic Anesthetic to allow bigger fish to be caught using jugs or buckets and moved to new tanks. Netting should be avoided at this stage to minimize the handling stress on the fish.

Juvenile SBFT begin to show schooling behavior from 30 dah and at this same time swimming speed increases significantly. Similar schooling behavior was also observed in PBFT juveniles (Sabate et al., 2010) and cannibalism eases at this time as the juveniles increase their swimming speed. However, similar to PBFT (Miyashita et al., 2000; Sawada et al., 2005), as soon as SBFT juveniles increase their swimming speed wall collision becomes the common cause of mortality. In PBFT, it is suggested that the incompatibility of the retinal adaptation with the change in the ambient light intensity at dawn may cause the visual disorientation of the fish that cannot control their high power swimming and thus collide with tank walls or nets (Masuma et al., 2001). A buffer net is usually set up in front of the tank wall to prevent collision. However, it is recommended to transfer SBFT juveniles to larger nursery tanks or sea cages as early as possible for better survival.

Developments of suitable weaning and nursery diets, as well as improvement of nursery husbandry techniques to minimize stress on larvae and juveniles are anticipated as critical steps for future commercial production of SBFT fingerlings.

Photos recording the SBFT larval development from 0 to 24 dah are displayed in Figures 10.5–10.9.

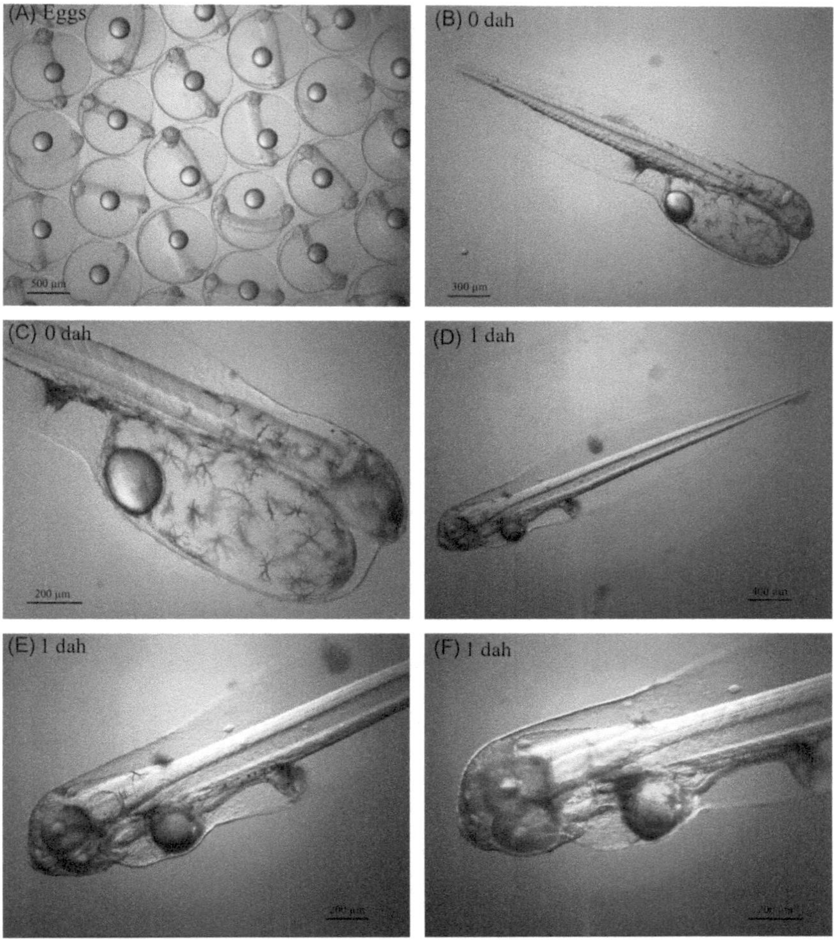

FIGURE 10.5 SBFT eggs and larvae: 0–1 dah.

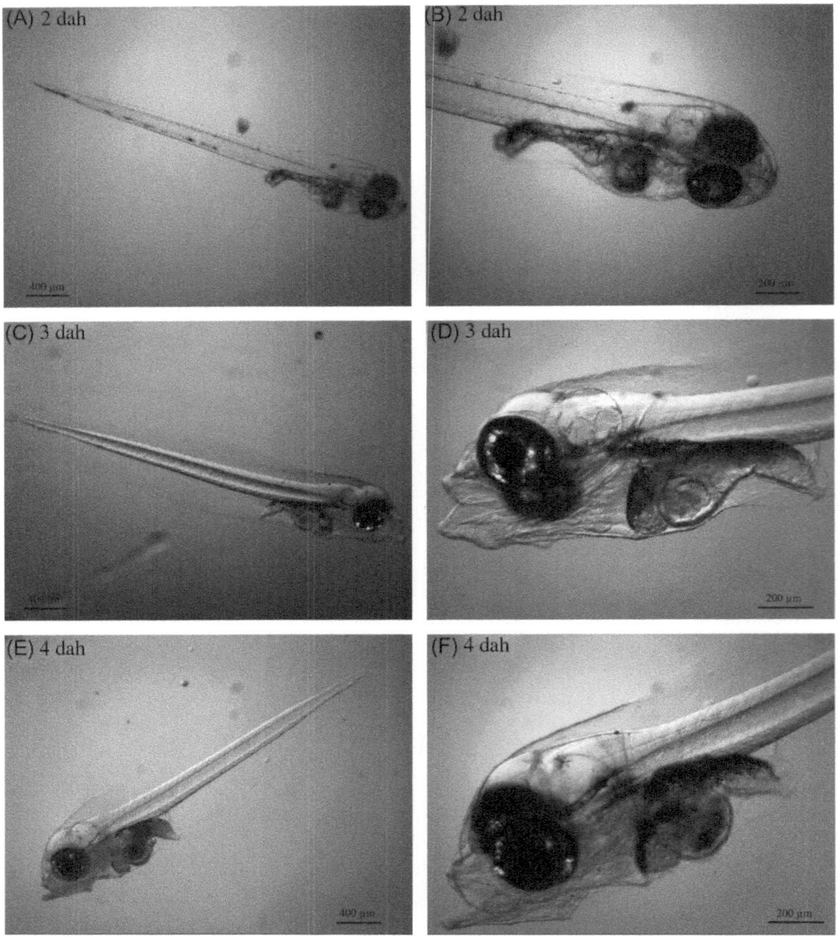

FIGURE 10.6 SBFT larvae: 2–4 dah.

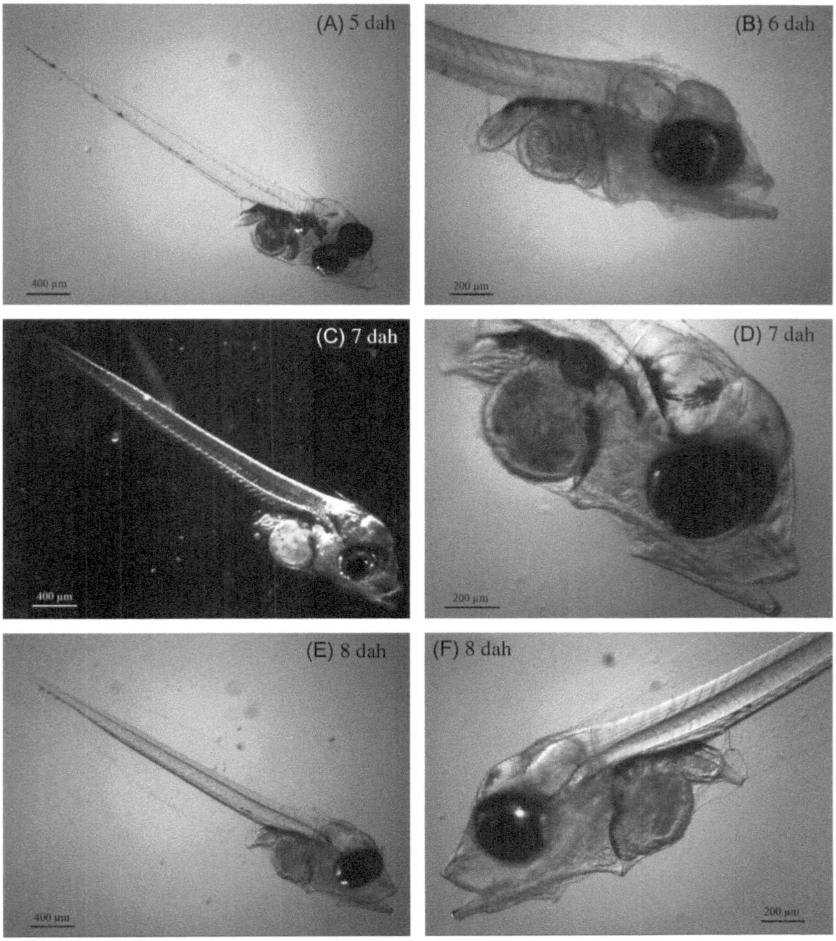

FIGURE 10.7 SBFT larvae: 5–8 dah.

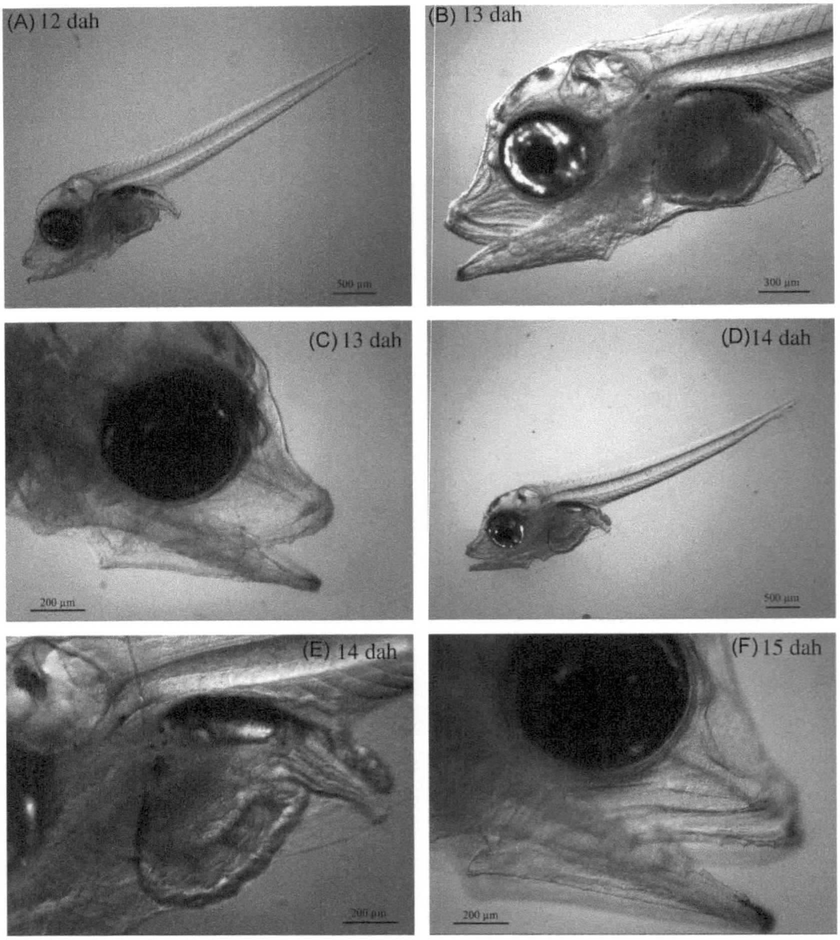

FIGURE 10.8 SBFT larvae: 12–15 dah.

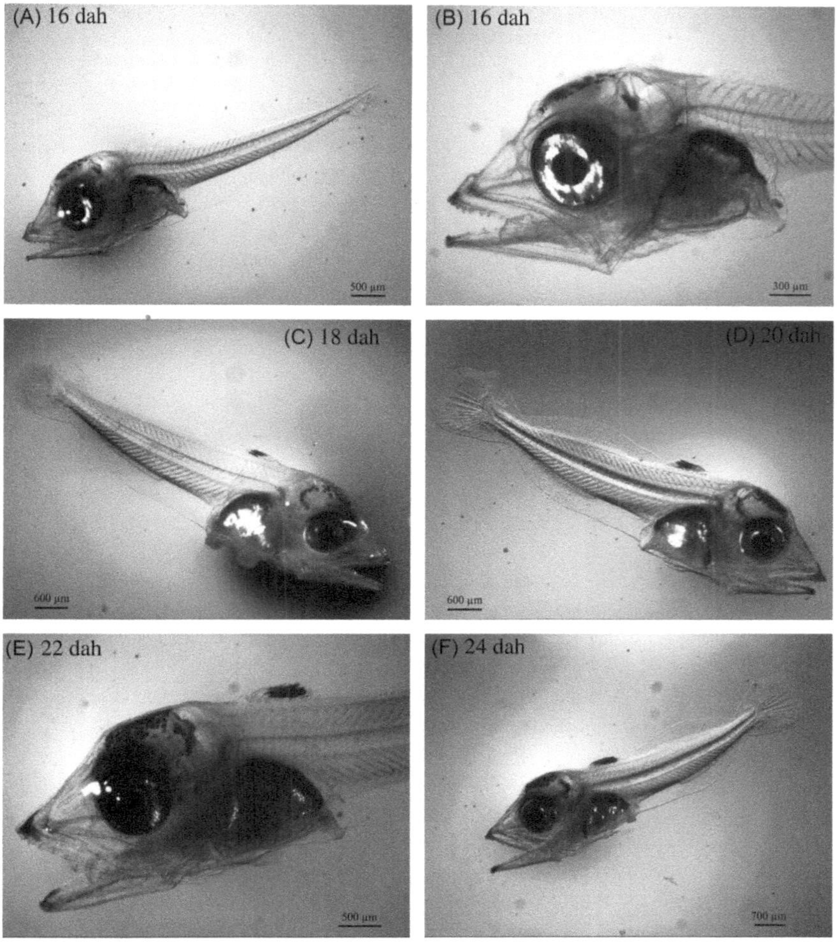

FIGURE 10.9 SBFT larvae: 16–24 dah.

REFERENCES

Basson, M., Hobday, A.J., Eveson, J.P., Patterson, T.A., 2012. Spatial interactions among juvenile southern bluefin tuna at the global scale: a large scale archival tag experiment. FRDC Report 2003/002.

Biswas, A.K., Nozaki, J., Kurata, M., Takii, K., Kumai, H., Seoka, M., 2006. Effect of *Artemia* enrichment on the growth and survival of Pacific bluefin tuna *Thunnus orientalis* (Temminck & Schlegel) larvae. Aquacult. Res. 37, 1662–1670.

Caton, A.E., 1991. Review of aspects of southern bluefin tuna biology, population and fisheries. In: Driso, R.B., Bayliff, W.H. (Eds.), Inter-American tropical Tuna Commission, Special Report 7, pp. 181–350.

Chen, B.N., 2014. Addressing causes of early mortality in hatchery produced southern bluefin tuna larvae. Final report (Project No. 2011/740) submitted to the Australian Seafood Cooperative Research Centre, Adelaide, Australia.

Davis, T.L.O., Jenkins, G.P., Young, J.W., 1990. Diel patterns of vertical distribution in larvae of southern bluefin *Thunnus maccoyii*, and other tuna in the East Indian Ocean. Mar. Ecol. Prog. Ser. 59, 63–74.

Davis, T.L.O., Lyne, V., Jenkins, G.P., 1991. Advection, dispersion and mortality of a patch of southern bluefin tuna larvae, *Thunnus maccoyii*, in the East Indian Ocean. Mar. Ecol. Prog. Ser. 73, 33–45.

De Metrio, G., Bridges, C.R., Mylonas, C.C., Caggiano, M., Deflorio, M., Santamaria, N., et al., 2010. Spawning induction and large-scale collection of fertilized eggs in captive Atlantic bluefin tuna (*Thunnus thynnus* L.) and the first larval rearing efforts. J. Appl. Ichthyol. 26, 596–599.

Evans, K., Patterson, T.A., Reid, H., Harley, S.J., 2012. Reproductive schedules in southern bluefin tuna: are current assumptions appropriate? PLoS ONE 7, e34550.

Farley, J.H., Davis, T.L.O., 1998. Reproductive dynamics of southern bluefin tuna, *Thunnus maccoyii*. Fish. Bull. 96, 223–236.

Grewe, P.M., Elliott, N.G., Innes, B.H., Ward, R.D., 1997. Genetic population structure of southern bluefin tuna (*Thunnus maccoyii*). Marine Biol. 127, 555–561.

Gunn, J., Farley, J., Hearn, B., 2003. Catch-at-age; age at first spawning; historical changes in growth; and natural mortality of SBFT: an integrated study of uncertainties in the population biology and dynamics of SBFT based on direct age estimates from otoliths. FRDC project No. 97/111.

Gunn, J., Evans, K., Patterson, T.A., Carter, T.I., 2006. Examining the movement and residency of adult southern bluefin tuna in the Tasman Sea and on their spawning grounds south of Indonesia using pop-up archival tags. Final report, Department of Agriculture, Fisheries and Forestry.

Gunn, J.S., Clear, N.P., Carter, T.I., Rees, A.J., Stanley, C.A., Farley, J.H., et al., 2008. Age and growth in southern bluefin tuna, *Thunnus maccoyii* (Castelnau): direct estimation from otoliths, scales and vertebrae. Fish. Res. 92, 207–220.

Hilder, P.I., 2013. Development of Vision and Larval Feeding Responses in Southern Bluefin Tuna and Yellowtail Kingfish. (PhD thesis). University of Tasmania, Hobart, Australia.

Jenkins, G.P., Davis, T.L.O., 1990. Age, growth rate, and growth trajectory determined from otolith microstructure of southern bluefin tuna *Thunnus maccoyii* larvae. Mar. Ecol. Prog. Ser. 63, 93–104.

Kato, Y., Takebe, T., Masuma, S., Kitagawa, T., Kimura, S., 2008. Turbulence effect on survival and feeding of Pacific bluefin tuna *Thunnus orientalis* larvae, on the basis of a rearing experiment. Fish. Sci. 74, 48–53.

Kurata, M., Seoka, M., Nakagawa, Y., Ishibashi, Y., Kumai, H., Sawada, Y., 2011. Promotion of initial swimbladder inflation in Pacific bluefin tuna, *Thunnus orientalis* (Temminck and Schlegel) larvae. Aquacult. Res. 43, 1–10.

Kurata, M., Seoka, M., Ishibashi, Y., Nakagawa, Y., Honryo, T., Katayama, S., et al., 2013. Timing to promote initial swimbladder inflation by surface film removal in Pacific bluefin tuna, *Thunnus orientalis* (Temminck and Schlegel) larvae. Aquacult. Res. 46, 1222–1232.

Masuma, S., Kawamura, G., Tezuka, N., Koiso, M., Namba, K., 2001. Retinomotor responses of juvenile bluefin tuna, *Thunnus thynnus*. Fish. Sci. 67, 228–231.

Masuma, S., Miyashita, S., Yamamoto, H., Kumai, H., 2008. Status of bluefin tuna farming, broodstock management, breeding and fingerling production in Japan. Rev. Fish. Sci. 16, 385–390.

Masuma, S., Takebe, T., Sakakura, Y., 2011. A review of the broodstock management and larviculture of the Pacific northern bluefin tuna in Japan. Aquaculture 315, 2–8.

Miyashita, S., Hattori, N., Sawada, Y., Ishibashi, Y., Nakatsukasa, H., Okada, T., et al., 1999. Ontogenetic change in oxygen consumption of bluefin tuna, *Thunnus thynnus*. Suisanzoshoku 47, 269–275.

Miyashita, S., Sawada, Y., Hattori, N., Nakatsukasa, H., Okada, T., Murata, O., et al., 2000. Mortality of northern bluefin tuna *Thunnus thynnus* due to trauma caused by collision during growout culture. J. World Aquacult. Soc. 31, 632–639.

Mourente, G., Megina, C., Diaz-Salvago, E., 2002. Lipids in female northern bluefin tuna (*Thunnus thynnus thynnus* L.) during sexual maturation. Fish Physiol. Biochem. 24, 351–363.

Mylonas, C.C., Bridges, C., Gordin, H., Rios, A.B., Garcia, A., De LA Gandara, F., et al., 2007. Preparation and administration of gonadotropin-releasing hormone agonist (GnRHa) implants for the artificial control of reproductive maturation in captive-reared Atlantic bluefin tuna (*Thunnus thynnus thynnus*). Rev. Fish. Sci. 15, 183–210.

Nakagawa, Y., Kurata, M., Sawada, Y., Sakamoto, W., Miyashita, S., 2011. Enhancement of survival rate of Pacific bluefin tuna (*Thunnus orientalis*) larvae by aeration control in rearing tank. Aquat. Living Resour. 24, 403–410.

Nichols, P.D., Virtue, P., Mooney, B.D., Elliot, N.G., Yearsley, G.K., 1998. Seafood the Good Food: the Oil (fat) Content and Composition of Australian Commercial Fishes, Shellfishes and Crustaceans. CSIRO Marine Research, Hobart, Australia.

Olson, R.J., 1980. Synopsis of biological data on the southern bluefin tuna, *Thunnus macoyii* (Castlenau, 1872). Inter-American Tropical Tuna Commission, Special Report 2, pp. 151–212.

Patterson, T.A., Evans, K., Carter, T.I., Gunn, J.S., 2008. Movement and behaviour of large southern bluefin tuna (*Thunnus maccoyii*) in the Australian region determined using pop-up satellite archival tags. Fish. Oceanogr. 17, 352–367.

Sabate, F.S., Sakakura, Y., Tanaka, Y., Kumon, K., Nikaido, H., Eba, T., et al., 2010. Onset and development of cannibalistic and schooling behavior in the early life stages of Pacific bluefin tuna *Thunnus orientalis*. Aquaculture 301, 16–21.

Sawada, Y., Okada, T., Miyashita, S., Murata, O., Kumai, H., 2005. Completion of the Pacific bluefin tuna *Thunnus orientalis* (Temminck et Schlegel) life cycle. Aquacult. Res. 36, 413–421.

Seoka, M., Kurata, M., Kumai, H., 2007. Effect of docosahexaenoic acid enrichment in *Artemia* on growth of Pacific bluefin tuna *Thunnus orientalis* larvae. Aquaculture 270, 193–199.

Seoka, M., Kurata, M., Tamagawa, R., Biswas, A.K., Biswas, B.K., Yong, A.S.K., et al., 2008. Dietary supplementation of salmon roe phospholipid enhances the growth and survival of Pacific bluefin tuna *Thunnus orientalis* larvae and juveniles. Aquaculture 275, 225–234.

Sharp, G.D., 2001. Tuna oceanography – An applied science. In: Block, B.A., Stevens, E.D. (Eds.), Tuna Physiology, Ecology and Evolution. Fish Physiology Series, vol. 19. Academic Press, New York, pp. 345–389.

Shiao, J.-C., Yui, T.-F., Hoie, H., Ninnemann, U., Chang, S.-K., 2009. Otolith O and C stable isotope compositions of southern bluefin tuna *Thunnus maccoyii* (Pisces: Scombridae) as possible environmental and physiological indicators. Zool. Stud. 48, 71–82.

Shingu, C., 1978. Ecology and stock of southern bluefin tuna. Japan Assoc. Fish. Res. Prot. Fish. Study 31, 81 (In Japanese. English translation in CSIRO Division Fisheries and Oceanography Report No. 131, pp. 79).

Stephan, M., Hobsbawn, P., 2014. Australian Fisheries and Aquaculture Statistics 2013, Fisheries Research and Development Corporation Project 2010/208. Australian Bureau of Agricultural and Resource Economics and Sciences, Canberra.

Tanaka, Y., Kumon, K., Nishi, A., Eba, T., Nikaido, H., Shiozawa, S., 2009. Status of the sinking of hatchery-reared larval Pacific bluefin tuna on the bottom of the mass culture tank with different aeration design. Aquacult. Sci. 57, 587–593.

Woolley, L.D., Fielder, S.D., Qin, J.G., 2013. Swimbladder inflation associated body density change and larval survival in southern bluefin tuna *Thunnus maccoyii*. Aquacult. Int. 21, 1233–1242.

Young, J.W., Davis, T.L.O., 1990. Feeding ecology of larvae of southern bluefin, albacore and skipjack tunas (Pisces: Scombridae) in the eastern Indian Ocean. Mar. Ecol. Prog. Ser. 61, 17–29.

Chapter 11

Diseases in Tuna Aquaculture

Jimena Balli[1], Ivona Mladineo[2], Sho Shirakashi[3] and Barbara F. Nowak[1]

[1]*Institute of Marine and Antarctic Studies, University of Tasmania, Launceston, TAS, Australia,*
[2]*Laboratory of Aquaculture, Institute of Oceanography & Fisheries, Croatia,*
[3]*Fisheries Laboratory, Kinki University, Wakayama, Japan*

11.1 INTRODUCTION

Successful health management of farmed fish is essential for sustainable aquaculture. The level of potential control of pathogens is related to the type of aquaculture system. Cage culture offers little control over waterborne pathogens and may contribute to free living organisms becoming parasitic (Nowak, 2007). Furthermore, stress level is harder to control in cage culture, in particular stress due to confinement or the presence of predators or extreme or unfavorable weather conditions. A disease outbreak is a result of the interaction between host, pathogen, and environment. In addition to the limited control over the presence of pathogens, cage culture may contribute to stress of the host and as a result immunosuppression. Risk of the outbreaks of the diseases can be assessed on the basis of the presence of the pathogens and parasites in the cage farming environment. This review focuses on three species of tuna: Atlantic bluefin tuna (ABFT), Pacific bluefin tuna (PBFT), and southern bluefin tuna (SBFT), all of which are ranched or farmed in cage culture. Different parasites emerged as health risks for ranched SBFT (Nowak, 2004) and PBFT (Nowak et al., 2006). Some of these parasitic infections have been associated with mortalities and reduced production (Polinski et al., 2013).

11.2 IMMUNE RESPONSE

Immune response recognizes and defends an organism against nonself for example a pathogen. Understanding immune response of the farmed species forms the basis for health management of the species, in particular

D.D. Benetti, G.J. Partridge, A. Buentello (Eds): Advances in Tuna Aquaculture.
DOI: http://dx.doi.org/10.1016/B978-0-12-411459-3.00008-4

development of vaccines and immunostimulants. Indirectly, knowledge of immune response improves understanding of risks to the health of the farmed species depending on the interactions between host, pathogen, and environment. Research on tuna immune response has included measures of the response at protein and gene expression levels. Some of the immune genes have been sequenced for all three species for example tumor necrosis factor (TNF)—both TNFα1 and TNFα2 were sequenced for PBFT (Kadowaki et al., 2009), SBFT (Polinski et al., 2013), and ABFT (Lepen Pleić et al., 2014).

Innate immune response develops first and is not specific for a particular pathogen. It can be classified as humoral and cellular. Studies of innate immune response in tunas have focused on humoral immune response, including lysozyme activity and complement activity (Watts et al., 2001; Kirchhoff et al., 2011a, 2012) and on gene expression (e.g., Kadowaki et al., 2009; Polinski et al., 2013; Lepen Pleić et al., 2014).

Lysozyme is an enzyme which has antibacterial activity. Patterns of lysozyme activity and alternate complement activity have been described for SBFT (Kirchhoff et al., 2011a,b, 2012). Overall, there is an increase in lysozyme activity in the first few weeks of ranching. Feeding SBFT a diet enriched with vitamin A and C resulted in 1.5-fold increase in lysozyme activity at week 8 post-transfer (Kirchhoff et al., 2011a). A significant negative correlation was observed between the number of adult *Cardicola forsteri* in heart and lysozyme activity as well as the number of blood fluke eggs in heart and lysozyme activity, however it was dependent on the SBFT cohort (Kirchhoff et al., 2012).

Complement is a group of proteolytic enzymes, the activation of which results in lysis or opsonization of the pathogen and induction of an inflammatory response. It is an essential component of the immune response which enhances the ability of phagocytic cells and antibodies to remove pathogens from an organism. Complement can be activated by classical (antibody complex), alternative (pathogen surface), or lectin (mannose binding lectin and mannose on pathogen surface complex) pathways. Most research on SBFT complement activity has focused on alternative pathway. Alternative pathway complement activity usually declined with ranching time (Kirchhoff et al., 2011a). However as the blood fluke infection usually increased with ranching time there could have been a confounding effect as in some cohorts there was a negative correlation between the *C. forsteri* and complement activity (Kirchhoff et al., 2012).

Cytokines are small proteins whose function is cell signaling. In the ABFT, three proinflammatory cytokines; TNFα1, TNFα2, IL1β, have been characterized and their potential role as health biomarkers was investigated during a 2-year rearing period (Lepen Pleić et al., 2014). In contrast to the results in PBFT, TNFα1 and TNFα2 in cage-reared ABFT had the same expression pattern, although the level of TNFα1 expression was higher during the same health conditions than of TNFα2, potentially related to the involvement of the former in intensive metabolism. Furthermore, contrary to other *in vivo* studies (Pelegrín et al., 2004; Pleguezuelos et al., 2000; Scapigliati et al., 2001; Zou et al., 2000), IL-1β was constitutively expressed even without stimulation and the liver was an important site of cytokine production during systemic inflammation. Inflammatory mediation through expression of IL-1β and TNFα was also localized at the site of *Didymosulcus katsuwonicola* (syn. *Didymocystis wedli*) infection (Mladineo and Block, 2010). Although constitutive expression of IL-1β and TNFα was observed in gills and skin implying a well-adapted innate immunity present at the barrier between the organism and environment, and up-regulation of both cytokines in didymozoid-infected gills, the lack of intensive cytokines response indicated the inability to successfully eliminate the parasite.

SBFT IgM has been characterized (Watts et al., 2001) and the SBFT IgM gene has been sequenced (Polinski et al., 2013). IgM and IgT from PBFT have been sequenced and their expression measured in secondary lymphoid organs (Mashoof et al., 2014). IgM expression was much greater in spleen and anterior kidney and only slightly greater in spleen and gills (Mashoof et al., 2014). Presence of specific antibodies against blood fluke *C. forsteri* in SBFT has been well documented (Watts et al., 2001; Aiken et al., 2006; Kirchhoff et al., 2012). There was a positive relationship between the antibodies and adult *C. forsteri* numbers with a time lag of 3 months (Aiken et al., 2006). Time in ranching had a significant effect on antibody titers and sero-prevalence which increased after transfer in March 2005 to reach a peak in December 2005 and then plateaued (Aiken et al., 2008). Overall, an increase in the antibody titers with time in ranching has been observed (Kirchhoff et al., 2012). Similarly in PBFT there was an increase in IgM at gene expression level in heart and gills with increasing blood fluke infection (Polinski et al., 2014a). The relative IgM transcription was correlated to the relative abundance of *Cardicola orientalis* DNA in gill samples (Polinski et al., 2014a). This antibody response most likely contributes to the improved survival of repeated blood fluke infections as the PBFT get older (Polinski et al., 2014a).

Immune response at the protein level was compared for SBFT from high (20 °C) and low (12 °C) temperature (Watts et al., 2002). However, the temperature was confounded with time in ranching (tuna from 12 °C were ranched for 7 months and tuna from 20 °C for 2.5 months). Highest mean serum total Ig concentrations were present in SBFT ranched in cold water whereas the warm water SBFT had low concentrations of total Ig (Watts et al., 2002). Serum lysozyme was higher in the SBFT from 12 °C (Watts et al., 2002). Similarly, mean alternative complement pathway activity was highest in the SBFT from 12 °C and mean classical pathway complement activity showed the same pattern (Watts et al., 2002).

In vitro analysis of immune gene expression of SBFT leucocytes incubated at 18 and 25 °C suggested that temperature affected the timing but not the degree of inflammatory response and that different cell populations showed different responses to temperature (Polinski et al., 2013). Heat shock protein 70 (Hsp70) transcription was induced in leukocyte but not in head kidney cell populations after 24 h incubation at 25 °C. Four potential inflammatory mediators TNFα2, IL-1β, IL-8, and Cox2 were upregulated in the head kidney leucocytes and in the peripheral blood leucocytes following LPS stimulations with the peak expression occurring faster in the cells incubated at 25 °C than 18 °C (Polinski et al., 2013).

Little is known of the ontogeny of immune response in tuna. Ontogeny of immune response of farmed PBFT was described using histology and the results suggested that the development of immune organs in tuna was more advanced relative to other marine teleosts (Watts et al., 2003). Kidney was the first organ to appear only on 0.5 days after hatch (dah), with undifferentiated stem cells in kidney present from 2 dah and lymphocytes from 7 dah. Thymus could be seen at 5 dah, lymphocytes were present from 7 dah and outer thymocytic zone and inner epithelioid zone from 15 dah (Watts et al., 2003). Spleen appeared at 2 dah and was erythroid until 30 dah which was the end of the study. On the basis of the presence of lymphocytes it was proposed that antibody response may be possible by 2 weeks dah (Watts et al., 2003).

11.2.1 Health of Ranched Atlantic Bluefin Tuna

Tuna aquaculture practice in the Mediterranean countries is mostly based on a 6-month period of ranching, except in Croatia, where juvenile tuna below 115 cm in fork length or 30 kg body weight, are caught and kept in cages for more than 1 year to reach marketable size. Consequently in the second case, aquaculture is characterized as farming rather than ranching (Miyake et al., 2003). In general, the number of infectious and noninfectious diseases reported in ABFT since the first facilities started to operate in early 1990s is negligible. At that time, mortalities were mainly related to sudden environmental changes (lightening, storm, heavy rains, forest fires) or inadequate husbandry and management (unbalanced diets, high density in cages) that after some empirical experience, have been minimized.

Initially, feeding imported frozen baitfish to cage-reared ABFT raised a question of potential dissemination and propagation of exotic viral agents, primarily viral hemorrhagic septicemia (VHS) from the herring (Jones et al., 1997; Marty et al., 1998) to both ranched and wild fish populations. After almost 25 years of tuna ranching in the Mediterranean, there was no evidence to support the existence of viral pathogens or clinical signs related to viruses in this species. However, only Spain has developed an epidemiological aquatic animal surveillance pilot program in inland Murcia region that is aiming to establish empirically the potential of viral diseases transmission to the local native wildlife as well as fish in aquaculture, propagated through baitfish fed to ABFT (Peñalver et al., 2012). Interestingly, after the first 4-year period of assessment of 470 samples from chub mackerel (*Scomber japonicus*), Atlantic mackerel (*S. scombrus*), European pilchard (*Clupea pilchardus*), and round sardinella (*Sardinella aurita*) no viral agents were detected. Nevertheless, isolation of betanodavirus (RGNNV genotype) in the larval mortalities of hatchery-reared PBFT (Nishioka et al., 2010), suggests potential susceptibility to encephalopathy and retinopathy virus infections.

Clinical signs of bacterial diseases are rare in ranched ABFT, although predisposing factors including capture trauma, dietary deficiency, high stocking density, and stress are all feasible in such intensive rearing conditions. One of the first reports of bacterial agents isolated from apparently healthy ABFT during the harvest was related to asymptomatic pasteurellosis that elicited multifocal chronic granulomatous changes in spleen and liver (Perić, 2002). Later, two extensive mortality outbreaks in juvenile and adult fish (2003 and 2004) in cages in the Adriatic Sea were caused by *Photobacterium damsela* subsp. *piscicida* (Mladineo et al., 2006), related to a sudden increase in water temperature and long-term feeding of low-quality baitfish that subsequently showed high concentrations of volatile amines (Šimat et al., 2009). No clinical signs were observed, except changed coloration, atypical swimming, and mortality within 1–2 days, while gross pathology included only general signs of septicemia. A smaller number of fish developed a chronic pasteurellosis form of infection, manifesting disseminated granulomas in kidney. Histopathology encompassed blood vessels congestion, hemorrhages, and lymphocytic infiltration in visceral organs and brain; focal coagulative necrosis of hepatocytes in liver accompanied by loss of adipocytes, and disseminated granulomas with a necrotic center, epithelioid cells and layers of connective tissue infiltrated by lymphocytes or macrophages in kidney. Mortalities subsided after the diet of frozen baitfish was changed to fresh local baitfish.

An outbreak of opportunistic bacterial infection including *Photobacterium damsela* subsp. *damselae*, *Vibrio* sp., and *Tenacibaculum* sp. was reported in hatchery-reared ABFT larvae (Gustinelli et al., 2011). Larvae displayed septicemia-related lesions in gills (hyperplasia, necrosis, teleangiectasiae), skin (epithelial sloughing, deep ulceration), muscle (hyaline degeneration, necrosis), and internal organs.

A few studies have attempted to isolate bacterial microflora from surfaces, mucosa, and organs of healthy ABFT, listing in total 161 isolates belonging to *Moraxella*, *Photobacterium*, *Brevundimonas*, *Weeksella*, *Klebsiella*, and *Pseudomonas* Gram-negative and *Staphylococcus xylosus* and *S. lentus* Gram-positive bacteria (Kapetanović et al., 2006). These authors suggested that the *K. pneumoniae* isolated from juvenile fish (6.5 to 20.5 kg)—being one among histidine decarboxylase bacteria (López-Sabater et al., 1994)—might represent a risk for histamine fish poisoning in humans (Kapetanović et al., 2011). However, it seems that in healthy hatchery-reared larvae there is a statistically significant difference between microbial communities colonizing the intestines between rearing groups (Gatesoupe et al., 2013), suggesting that opportunistic and potentially pathogenic bacteria might also greatly vary.

Although many parasitic taxa have been isolated from ranched ABFT in the Mediterranean, in general they have a tendency to decrease in number during ranching, fail to cause serious pathology (Mladineo et al., 2011), and vary significantly between ABFT populations that inhabit different geographic areas (Culurgioni et al., 2014). Only two taxa belonging to protozoan and myxozoa were reported, although no pathological changes were associated with their parasitization in the intestinal mucosa (*Microsporidium* sp.) or bile (*Ceratomyxa thunni*) (see Mladineo, 2006a; Mladineo and Bočina, 2006).

Hexostoma thynni (Polyophistocotylea: Hexostomidae) that induces localized lamellar fusion and hyperplasia at the attachment site in the gills of SBFT (Deveney et al., 2005), is rarely isolated in ABFT and without noticeable pathological effect. Aiken et al. (2007) have confirmed its cosmopolitan distribution among three geographically remote tuna species (*T. albacares*, *T. maccoyii*, and *T. thynnus*), suggesting that its pathogenicity toward tuna is species- and environment-specific.

Granulomatous changes most likely related to sanguinicolid blood fluke *C. forsteri* parasitization in tuna heart, have been reported in large fish in Spanish and Croatian facilities on rare occasions (Nowak et al., 2006). Adult *C. forsteri* was found in the heart of ranched ABFT from Spain (Aiken et al., 2007). Additionally, Llanos (2012) described typical hypertrophic lesions induced by *Cardicola* sp. eggs accumulated in gill capillaries of ranched tuna of 80–100 cm of tail length, which further supports the existence of the sanguinicolid in ranched ABFT in particular geographic areas that apparently does not greatly influence ABFT health and condition. Recently, the presence of *C. forsteri*, *C. orientalis*, and *C. opisthorchis* as well as undescribed species of *Cardicola* (eggs only) was reported from wild and ranched ABFT (Palacios-Abella et al., 2015). *C. opistorchis* was the most abundant species and more prevalent in the ranched (21%) than wild (6%) fish. It was found only in the heart, whereas *C. orientalis* was seen in the gills and *C. forsteri* was present in both heart and gills (Palacios-Abella et al., 2015).

In contrast, didymozoids (Digenea: Didymozoa) are a frequent and species-diverse group of digeneans, inhabiting a wide array of tuna organs (Mladineo and Tudor, 2004; Mladineo et al., 2010), but of limited etiological significance (Mladineo, 2007; Mladineo et al., 2008).

Although in juvenile ABFT, high numbers of some species of didymozoids can inflict acute inflammatory changes aggravated by secondary bacterial infections (Mladineo, 2007; Justo et al., 2009), they usually disappear in ranched animals after a couple of months in cages. Transcriptomic data of the most abundant gill didymozoid *Didymosulcus katsuwonicola* (syn. *Didymocystis wedli*) supported by electron transmission microscopy, suggested that innate immune response of the host, comprised of eosinophiles, mast, plasma, and rodlet cells in combination with encapsulation failed to remove the parasite, but succeeded to maintain its separation by encapsulation (Trumbić et al., 2015). Pathway analyses based on KEGG (Kyoto

Enciclopeadia of Genes and Genomes) sets showed the perturbations of components of innate immunity, complement, and coagulation cascades, as well as endocrine, digestive, and nervous functional pathways. The findings elucidated vivid cellular and molecular cross-talk between the host and the parasite. However, such balanced host–parasite interaction can be compromised by different factors encountered under intensive production conditions.

Oncophora melanocephala (Nematoda: Camallanidae) has been reported from ABFT. This parasite was deeply embedded by its buccal capsule in pyloric caeca mucosal layers, only eliciting local and negligible pathology in the form of small erosions accompanying punctuate hemorrhages after its spontaneous detachment (Mladineo et al., 2005). Larval stages of *Philometroides* sp. have been observed on rare occasions in ABFT from Italian and Adriatic facilities, although their high prevalence and abundance in broodstock could potentially lead to castration. Third larval stages of *Anisakis* spp. complex have been occasionally isolated from visceral subserosa in ABFT at varying infection levels, eliciting no pathological changes in the host, but representing a potential zoonotic hazard (Mladineo and Poljak, 2014). Interestingly, *Anisakis* spp. larvae infection levels seemed to decrease with the duration of ranching/rearing and often small blackish granuloma containing dead larvae were isolated from the visceral cavity (Mladineo et al., 2011).

Only a few specimens of the *Hepatoxylon trichiuri* (Cestoda: Trypanorhyncha) plerocercoid and of *Eutetrarhynchus* spp. (Cestoda: Eutetrarhynchidae) blastocysts have been isolated from the stomach and visceral cavity respectively in reared ABFT, inflicting no pathology (Mladineo, 2006b).

A low number of copepod taxa have been isolated from ranched/reared tuna that mainly exhibit a patchy distribution and tendency to accumulate in a small proportion of the host population. Likewise, a relatively large pennellid *Pennella filosa* is occasionally observed deeply embedded in the trunk musculature of larger specimens of ABFT. It elicits localized necrotic changes that do not affect general health and condition of the host, but might reduce the quality of the market product (Žilić and Mladineo, 2006). Similarly, other copepods isolated from ABFT gills (*Pseudocycnus appendiculatus*, *Euryphorus brachypterus*, and *Brachiella thynni*), did not induce major pathology. Recently Lepen Pleić et al. (2015) showed that *P. appendiculatus* attaching to the gill epithelium by clamping caused direct tissue disruption with underlying necrotic or apoptotic processes, and extensive proliferation of rodlet and goblet cells. An inflammatory reaction in the gills was supported by high levels of cytokine IL-1β but was localized at the attachment site. When the authors stimulated *in vitro* ABFT peripheral blood leukocyte culture by parasite total antigen extract, expression of IL-1β was observed only 12 h post-stimulation and showed relatively low-fold change. This supports the observation that in the ABFT, immune response to this copepod is relatively mild, probably modulated by the parasite's adaptive mechanisms that enable its survival within its host.

Neoplastic formations, as pathologies of unknown etiology have been sporadically observed in larger ABFT. One case related to a lipoma, proliferating in a mass of 1.8 kg in the dorsal musculature over the vertebral column in a 350 kg fish, with well-differentiated adipose cells (Marino et al., 2006). Nutritive imbalance caused by a diet with high lipid and low levels of antioxidants induced mortalities in Adriatic-reared ABFT by pansteatitis, reflected in severe hepatic necrosis and lipidosis (Roberts and Agius, 2008). Fish in excellent condition died, almost asymptomatically, except for lethargy, changed coloration, and altered swimming bahavior. Gross pathology included a large area of pyloric ceaca embedded in visceral fat tissue and bronze-colored liver, while histopathology revealed inflammatory infiltration of the abundant lipid tissues and hepatic necrosis.

11.2.2 Health of Farmed Pacific Bluefin Tuna

In Japan, PBFT are typically stocked into cages as wild caught juveniles of a few months of age (100–500 g) and raised for 2–3 years (30–50 kg). As previously described for Croatian ABFT, this extended period of grow-out is therefore referred to as farming, rather than just fattening or ranching. In recent years, advances in larval rearing techniques have resulted in the increasing use of hatchery-produced juveniles for farming. The majority of mortalities of these hatchery-reared fish occur before the juveniles reach 1 year of age, especially during the larval-rearing period and shortly after the fish are stocked into farming cages. Larval and juvenile deaths are predominantly unrelated to infectious diseases. Sinking and floating death, cannibalism, and collision death are the three main causes of mass mortalities in larvae and early juveniles (Sawada et al., 2005) and are described in detail in Chapter 8 of this book. However, in association with the increase in juvenile production, infectious diseases in farmed PBFT have become problematic in recent years.

Two viral diseases can cause significant losses in larval and juvenile PBFT. Red sea bream iridovirus (RSIV) outbreaks may occur among yearlings during the period of high water temperature (Kawakami and Nakajima, 2002; Sawada et al., 2005). Infected fish often become anorexic and appear darker. Mortalities can be significant in 3–4-month-old juveniles. Surviving juveniles become immune to the virus and no disease occurs in fish older than 1 year (Munday et al., 2003). The virus can be isolated from the spleen of infected fish which often displays hypertrophied cells. When RSIV occurs, the only practical countermeasure is to reduce feeding since no commercial vaccine for PBFT is currently available.

Viral nervous necrosis (VNN) infections also occasionally cause high mortalities in hatchery-reared larvae and young juveniles of typically less than a few months old (Nishioka et al., 2010). The virus can be transmitted vertically from broodstock or through larval feeds (Sugaya et al., 2009). Therefore, disinfection of eggs, feed, and larval-rearing water is important to prevent the disease (Higuchi et al., 2011). Infected fish show characteristic vacuolations in the central nervous system and retina, causing abnormal behavior such as whirling (Nishioka et al., 2010).

Vaccination may be effective for these viral diseases as in other farmed fish. However, vaccine development for PBFT is difficult because they are extremely vulnerable to handling and injection vaccination can result in mortalities. Thus, new injection techniques, immersion, or oral vaccination methods need to be developed.

Several bacterial diseases have been reported in juvenile PBFT during the hatchery period and shortly after transferring to sea cages. Pseudotuberculosis, vibriosis, and lactococcosis have been suspected on the basis of the signs and bacterial isolations, but they have not yet been confirmed.

There is no record of fungal diseases in farmed PBFT.

Various parasite species have been recorded in farmed PBFT and some are a significant threat for farming. A protozoan endoparasite *Ichthyodinium* sp. was observed in PBFT embryo and yolk sac larvae and has affected juvenile production (Ishimaru et al., 2012). The parasite multiplies within the yolk and eventually bursts out causing the fish larva to die. This parasite likely infects eggs through environmental water, thus the disinfection of rearing water is important to prevent the disease (Yuasa et al., 2007). To date, there are no other protozoan parasites of serious concern in farmed PBFT in Japan.

Whitish spindle-shaped cysts of a microsporidian parasite, *Microsporidium* sp. can be found in the trunk muscle of juvenile PBFT with the infection prevalence reaching over 90% (Zhang et al., 2010). Such cysts are not observed in larger fish and there are no external signs of infection. Not much is known about this microsporidian and there is no effective control measure against it.

To date, four species of kudoid myxozoans have been described from farmed PBFT. *Kudoa yasunagai* and *K. prunusi* form whitish cysts in the brain (Meng et al., 2011; Zhang et al., 2010). Infected PBFT show no apparent signs and the pathological effects of these *Kudoa* spp. are still unclear. Numerous oval-shaped cysts of *K. shiomitsui* are often found in the heart of juvenile PBFT, but fish appear to be healthy (Zhang et al., 2010). *K. hexapunctata*, hitherto recognized as *K. neothunni*, but which has recently been described as a new species, has been recorded from both wild and farmed PBFT (Yokoyama et al., 2014). This species infects skeletal muscle fibers as a form of subclinical pseudocyst. The effect of *K. hexapunctata* on PBFT is unclear but this species is suspected as a causative agent of human food poisoning cases after consumption of wild juvenile PBFT sashimi (fresh raw fish) (Suzuki et al., 2015). The infection of *K. hexapunctata* tends to be more severe among juveniles and infection level (prevalence and intensity) tends to be low in larger PBFT (Suzuki et al., 2015). The infection in harvest-sized farmed PBFT appears to be minor, if any. Another multivalvulid myxozoa *Unicapsula* sp., has also been found in the muscle of farmed

PBFT, but its detailed identity and effects on the host are still unknown. Life cycles of all of these myxozoans remain unknown and there is no effective countermeasure to prevent and cure the infections in farmed tuna.

There are records of several monogeneans from wild tuna, but not from farmed PBFT in Japan. In Mexico, a skin fluke *Capsala* sp. has been reported from farmed PBFT (Aiken et al., 2007).

Blood fluke infection is the most problematic parasitic disease in Japanese PBFT farming (Ogawa, 2015). In PBFT, two species of blood flukes, *C. orientalis* and *C. opisthorchis* have been described. The former species resides in gill arteries while the latter species predominantly infects the heart (Ogawa et al., 2010, 2011). These two species may co-infect the same individuals and produce numerous eggs which accumulate in the host gills, causing the fish to suffocate to death (Shirakashi et al., 2012a). The eggs can be also found in the heart where sometimes they cause nodule formation. Blood fluke infections are a problem predominantly among 0-year-old fish (occasionally in 1-year-old), thus farmers need to treat juveniles with the oral administration of praziquantel which effectively eradicates the adult flukes (Shirakashi et al., 2012b). Reinfections of blood flukes occur after the first drug treatment, thus repeated treatments are required to minimize the mortality. Recently, terebellid polychaetes have been identified as the intermediate hosts of *C. orientalis* and *C. opisthorchis* (Table 11.1, Sugihara et al., 2014, Shirakashi unpublished data).

Didymozoan infections may occur, mostly in young PBFT of wild origin. There was a high infection rate of *Didymocystis weldi* in juvenile PBFT caught from the Sea of Japan, but the parasite disappeared within a year of farming (Takebe et al., 2013). There are some reports of *Didymocystis* sp. infections in larger farmed PBFT, but they are not very common. This suggests that the didymozoan infections occur mainly in the wild probably through food or at particular geographical locations. In general, didymozoans pose no threat to host health.

Several parasitic crustaceans infect farmed PBFT, but their taxonomic identities are not well studied. Parasitic copepods are often observed on the skin, particularly on the lower abdomen or the posterior dorsal area, during the period of low water temperature. Infected fish often display some skin lesions, but the infection is not fatal. *Caligus macarovi* has been identified from farmed PBFT, but *Euryphorus* spp. may also be present (Nagasawa, 2011). Cymothoid isopods, possibly *Nerocila* sp., have also been spotted on the body surface, but they do not appear to cause serious harm to the fish.

11.2.3 Health of Ranched Southern Bluefin Tuna

SBFT ranching started in 1991 in Port Lincoln, South Australia (Hayward et al., 2008; Kirchhoff et al., 2011). Schools of 2- to 4-year-old wild SBFT are captured, transferred, and fattened in near-shore cages. This process can

TABLE 11.1 *Cardicola* **spp. Intermediate and Definitive Hosts and Effects on the Tuna Industry**

Cardicola Species	Intermediate Host	Definitive Host	Effect on Industry	Reference
C. forsteri	*L. modestus*	SBFT	Mortality	Cribb et al. (2011)
		ABFT	None	
C. orientalis	*Nicolea gracilibranchis*	SBFT	Mortality	Polinski et al. (2013), Shirakashi et al. (2012a), Palacios-Abella et al. (2015)
		PBFT	Mortality	
		ABFT	None	
C. opisthorchis	*Terebella* sp.	PBFT	Mortality	Sugihara et al. (2014)
		ABFT	None	

Mortality occurs in SBFT and PBFT if the fish are not treated.

be stressful for the fish, potentially leading to immunosuppression (Deveney et al., 2005; Kirchhoff et al., 2011a,c). The first few months of ranching had an effect on the health and condition of SBFT with changes in hemoglobin concentration, lysozyme activity, and alternative complement activity (Kirchhoff et al., 2011b).

There are no records of viral or fungal infections in SBFT. Some opportunistic bacteria have been linked to infections. *Aeromonas* spp. infections have been associated with parasitic trauma, especially with *C. chiastos*, which caused damage to the eye making tuna more likely to be affected by mechanical collisions and further damage and leading to secondary infection and loss of condition (Munday et al., 2003). *Vibrio* spp. and *Aeromonas* spp. have been isolated from internal organs, especially from individuals which presented external wounds (Munday et al., 2003; Valdenegro-Vega et al., 2013). *Photobacterium damselae* subsp. *damselae* was isolated from gills, spleen, and kidney of SBFT and from spleen of moribund or dead fish. Subclinical infection with *P. damselae* subsp. *piscicida* was reported from spleen of moribund SBFT (Valdenegro-Vega et al., 2013).

Scuticociliate *Uronema nigricans*, is considered to be the causative agent of swimmer syndrome in SBFT. Scuticociliates are free-living marine protists, which feed on suspended particulate matter but under certain circumstances can become opportunistic parasites (Lee et al., 2004; Moustafa et al., 2010; Munday et al., 1997). These ciliates can become histophagous and cause systemic infection (Lee et al., 2004; Moustafa et al., 2010).

Swimmer syndrome usually occurs when the water temperature is below 18 °C, the ciliate initially parasitizes the olfactory rosette, then invades the olfactory nerves, and eventually the brain, causing locomotory dysfunction followed by the death of the infected fish (Munday et al., 1997, 2003; Nowak, 2007). Changed husbandry practices significantly reduced the incidence of swimmer syndrome with very few confirmed cases reported in the last 12 years.

Blood flukes *C. forsteri* and *C. orientalis* have recently emerged as serious pathogens of SBFT (Aiken et al., 2009; Cribb et al., 2011; Polinski et al., 2013). Blood flukes have a two-host life cycle including an intermediate invertebrate host, a polychaete, in which they undergo asexual reproduction, and a definitive host, tuna, where sexual reproduction takes place (Table 11.1) (Aiken et al., 2009; Cribb et al., 2011; Kirchhoff et al., 2012). *Cardicola forsteri* is considered to be a significant problem in farmed SBFT; death of the fish has been attributed to severe bronchitis (Colquitt et al., 2001; Aiken et al., 2009; Cribb et al., 2011). Once the cercaria penetrates the host, it migrates to the final site (e.g., heart, in the case of *C. forsteri*), maturing there into an adult (Aiken et al., 2009). Adults lay eggs within the heart, the eggs are then transported to the gills through the blood stream. The eggs develop in the gills until they hatch and miracidia emerge disrupting the gills. Infectious free swimming miracidia search for an intermediate host to penetrate and undergo asexual reproduction. Eggs of *C. forsteri* can also be found in the spongy layer of the ventricle or in the afferent filamental arteries (Aiken et al., 2009; Kirchhoff et al., 2012). SBFT can be parasitized with both *C. forsteri* and *C. orientalis* in concurrent infections (Polinski et al., 2013, 2014a; Shirakashi et al., 2013). *Cardicola forsteri* adults are found mostly in the heart and their eggs in the gills of SBFT while *Cardicola orientalis* has tropism for the gill arteries (Polinski et al., 2013; Shirakashi et al., 2013). *Cardicola orientalis* can cause physiological problems in the fish, adults may cause clogging of the gill arteries while eggs can affect the gill integrity or block blood spaces in the gill lamellae leading to mortalities (Polinski et al., 2013; Shirakashi et al., 2013).

Some antihelminthics have been used to treat fish against different parasitic diseases, with praziquantel being one of the first options against monogeneans, digeneans, and intestinal cestodes. It has been used to treat helminthic parasites of human and domestic animals and is commercially available for both ornamental and farmed fish (Hardy-Smith et al., 2012; Shirakashi et al., 2012b). Praziquantel was effective against adult *C. forsteri* in experimental treatments, though repeated treatments may be needed due to its rapid clearance from fish (Hardy-Smith et al., 2012). Treated fish showed a reduction in the number of eggs in the gills and heart due to either the reduction of the presence of adult parasites or to a lower egg production (Hardy-Smith et al., 2012). In vitro, praziquantel induced T-cell receptor and IL-8 transcriptional expression (Polinski et al., 2014b). Expression of the inflammatory cytokines including Il-1β was upregulated in blood cell cultures after exposure to praziquantel (Polinski et al., 2014b). This suggests that praziquantel has immunomodulatory activity in fish (Polinski et al., 2014b).

There is variability in the level of infections between SBFT companies and the companies can explain variability in prevalence and abundance of the infection (Aiken et al., 2015). This may be because of differences in husbandry, or due to different average sizes of SBFT farmed by each of the companies, or due to the location of the operations (Aiken et al., 2015). SBFT farmed further offshore in deeper water had no blood fluke infections (Kirchhoff et al., 2011c). This suggests that management strategies can be successful in reducing blood fluke infections.

Epizootics of parasitic copepods *P. appendiculatus* and *C. chiastos* have been reported from ranched SBFT (Hayward et al., 2009, 2011). Gill copepod *P. appendiculatus* infection intensity and prevalence peaked in the winter months, but then declined during the second winter. This decline was attributed to the host immune response (Hayward et al., 2008). *Caligus chiastos* prevalence declined when the water temperature dropped to the average of 14.6 °C, peaking again when the water temperature increased to 20.5 °C (Hayward et al., 2009). Prevalence and intensity of infestations with *C. chiastos* were correlated with two stress indicators: plasma cortisol and glucose. There was a correlation with the number of *Caligus* spp. present on an individual tuna and the damage to the eye, and the severity of the damage of the eye was associated with reduced condition index of the fish (Hayward et al., 2011). Degan's leather jacket (*Thamnaconus degeni*) was identified as a reservoir for *C. chiastos* (see Hayward et al., 2011). This species was attracted to SBFT ranching areas by uneaten feed. Improved feed management and an increase in the distance between the ocean floor and the bottom of the cage eliminated sea lice infections in SBFT.

Neoplastic lesions are rarely seen in ranched SBFT. Two lipomas and a neurofibrosarcoma were identified in two ranched individuals showing visible nodular lesions (Johnson et al., 2008). A lipoma was reported previously from wild SBFT (Lester and Kelly, 1983). These types of lesions usually occur only in the larger SBFT.

11.3 FUTURE CHALLENGES

While there are relatively few significant health problems in farming and ranching of tunas, intensification of farming which may also lead to increased biomass on sites may result in other health issues. Currently PBFT is the only species with any commercial hatchery production. The production of all other species (and still the majority of PBFT production in Japan and all in Mexico) is based on the capture of wild juveniles and fattening them in cages. Closing the life cycle for other species and moving from ranching farming will present new challenges typical for marine hatcheries including health issues such as viral diseases affecting tunas during hatchery stage. Hatchery-reared juveniles will be naïve to marine pathogens which they will encounter once moved to cages. This will contribute to disease outbreaks. Hatchery-reared PBFT are significantly affected by blood flukes and have to be treated with praziquantel. There will be a need for development of new treatments and vaccines to reduce the impact of infectious diseases.

Fundamental knowledge of tuna immunity and how it is affected by farming and ranching will contribute to our ability to use immunomodulation to prevent disease outbreak. Development of vaccines is characteristic of any mature aquaculture industry and will be essential for farming of the species with closed life cycles. Effective vaccines are available against most viral and bacterial diseases affecting Atlantic salmon and research is continuing on development of vaccines against fish parasites. Due to the characteristics of the species, oral delivery will be the preferred method for tuna.

Some pathogens, for example nodavirus, have been detected in farmed tuna but the susceptibility of the species to those pathogens is unknown. It will be essential to investigate this further to understand the risk factors which may contribute to these pathogens causing disease outbreaks. As moving more to hatchery production will increase risk of viral diseases there is a need for development of permanent cell lines to enable research on viral pathogens of tuna.

Understanding life cycles and reservoirs of pathogens and parasites allows implementation of effective control measures. While we have identified some of the intermediate hosts of blood flukes, there is no information on the intermediate hosts for other blood flukes from some geographical areas and some other parasites, including myxosporeans. This is a significant knowledge gap which limits management strategies for those parasites.

While there are some species-specific differences in health of the three species of tuna and ranching/farming practices differ in different regions and for different species, there are many common health issues or risks which suggest that collaborative research would be most effective. We are facing the same global challenges, for example climate change which will affect aquaculture, in particular cage farming and ranching. This is due to the limited control of cage farming conditions and reliance on the availability of wild fish for ranching.

REFERENCES

Aiken, H., Hayward, C., Cameron, A., Nowak, B., 2009. Simulating blood fluke, *Cardicola forsteri*, infection in farmed southern bluefin tuna, *Thunnus maccoyii*, using stochastic models. Aquaculture 293, 204–210.

Aiken, H., Hayward, C., Nowak, B., 2015. Factors affecting abundance and prevalence of blood fluke, *Cardicola forsteri*, infection in commercially ranched southern bluefin tuna, *Thunnus maccoyii*, in Australia. Vet. Parasitol. 210, 106–113.

Aiken, H.M., Hayward, C.J., Nowak, B.F., 2006. An epizootic and its decline of a blood fluke, *Cardicola forsteri*, in farmed southern bluefin tuna, *Thunnus maccoyii*. Aquaculture 254, 40–45.

Aiken, H.M., Bott, N.J., Mladineo, I., Montero, F.E., Nowak, B.F., Hayward, C.J., 2007. Molecular evidence for cosmopolitan distribution of platyhelminth parasites of tunas (*Thunnus* spp.). Fish Fish. 8, 167–180.

Aiken, H.M., Hayward, C.J., Crosbie, P.B., Watts, M., Nowak, B.F., 2008. Serological evidence of an antibody response in farmed southern bluefin tuna naturally infected with the blood fluke *Cardicola forsteri*. Fish Shellfish Immunol. 25, 66–75.

Colquitt, S.E., Munday, B.L., Daintith, M., 2001. Pathological findings in southern bluefin tuna, *Thunnus maccoyii* (Castlenau), infected with *Cardicola forsteri* (Cribb, Daintith & Munday, 2000) (Digenea: Sanguinicolidae), a blood fluke. J. Fish Dis. 24, 225–229.

Cribb, T.H., Adlard, R.D., Hayward, C.J., Bott, N.J., Ellis, D., Evans, D., et al., 2011. The life cycle of *Cardicola forsteri* (Trematoda: Aporocotylidae), a pathogen of ranched southern bluefin tuna, *Thunnus maccoyii*. Int. J. Parasitol. 41, 861–870.

Culurgioni, J., Mele, S., Merella, P., Addis, P., Figus, V., Cau, A., et al., 2014. Metazoan gill parasites of the Atlantic bluefin tuna *Thunnus thynnus* (Linnaeus) (Osteichthyes: Scombridae) from the Mediterranean and their possible use as biological tags. Folia Parasitol. 61, 148–156.

Deveney, M.R., Bayly, T.J., Johnston, C.J., Nowak, B.F., 2005. A parasite survey of farmed southern bluefin tuna, *Thunnus maccoyii* (Castelnau). J. Fish Dis., 279–284.

Gatesoupe, F.-J., Covès, D., Ortega, A., Papandroulakis, N., Vadstein, O., de la Gándara, F., 2013. A spatiotemporal study of bacterial community profiles associated with Atlantic bluefin tuna larvae, *Thunnus thynnus* L., in three Mediterranean hatcheries. Aquacult. Res. 44, 1511–1523.

Gustinelli, A., Campana, M., Caggiano, M., Florio, D., Caffara, M., Quaglio, F., et al., 2011. Observations on health problems in hatchery reared tuna (*Thunnus thynnus*) in Italy. 15th EAFP International Conference on Diseases of Fish and Shellfish. September 12–16, 2011 Split, Croatia. European Association of Fish Pathologists.

Hardy-Smith, P., Ellis, D., Humphrey, J., Evans, M., Evans, D., Rough, K., et al., 2012. *In vitro* and *in vivo* efficacy of anthelmintic compounds against blood fluke (*Cardicola forsteri*). Aquaculture 334–337, 39–44.

Hayward, C.J., Aiken, H.M., Nowak, B.F., 2008. Epizootics of metazoan gill parasites did not threaten feasibility of farming southern bluefin tuna (*Thunnus maccoyii*) in a trial extending over summer months. Vet. Parasitol. 154, 122–128.

Hayward, C.J., Bott, N.J., Nowak, B.F., 2009. Seasonal epizootics of sea lice, *Caligus* spp., on southern bluefin tuna, *Thunnus maccoyii* (Castelnau), in a long-term farming trial. J. Fish Dis. 32, 101–106.

Hayward, C.J., Svane, I., Lachimpadi, S.K., Itoh, N., Bott, N.J., Nowak, B.F., 2011. Sea lice infections of wild fishes near ranched southern bluefin tuna (*Thunnus maccoyii*). Aquaculture 320, 178–182.

Higuchi, K., Eba, T., Tanaka, Y., Kumon, K., Nishi, A., Nikaido, H., et al., 2011. Large-scale disinfection of eggs of spangled emperor *Lethrinus nebulosus* to prevent viral nervous necrosis in seed production of Pacific bluefin tuna *Thunnus orientalis*. J. Fish. Technol. 4, 15–20.

Ishimaru, K., Iida, N., Okada, T., Miyashita, S., 2012. *Ichthyodinium* infection in the embryos and yolk sac larvae of Pacific bluefin tuna *Thunnus orientalis*. Fish Pathol. 47, 143–146.

Johnson, C.J., Deveney, M.R., Bayly, T., Nowak, B.F., 2008. Gross and histopathological characteristics of two lipomas and a neurofibrosarcoma detected in aquacultured southern bluefin tuna, *Thunnus maccoyii* (Castelnau), in South Australia. J. Fish Dis. 31, 241–247.

Jones, J.B., Hyatt, A.D., Hine, P.M., Whittington, R.J., Griffin, D.A., Bax, N.J., 1997. Special topic review: Australasian pilchard mortalities. World J. Microbiol. Biotechnol. 13, 383–392.

Justo, M.C.N., Tortelly, R., Menezes, R.C., Kohn, A., 2009. First record in South America of *Koellikerioides internogastricus* (Digenea: Didymozoidae) with new host record and pathological alterations. Vet. Parasitol. 161, 158–161.

Kadowaki, T., Harada, H., Sawada, Y., Kohchi, C., Soma, G.-I., Takahashi, Y., et al., 2009. Two types of tumor necrosis factor-a in bluefin tuna (*Thunnus orientalis*) genes: molecular cloning and expression profile in response to several immunological stimulants. Fish Shellfish Immunol. 27, 585–594.

Kapetanović, D., Kurtović, B., Vardić, I., Valić, D., Teskeredžić, Z., Teskeredžić, E., 2006. Preliminary studies on bacterial diversity of cultured bluefin tuna (*Thunnus thynnus* from the Adriatic Sea. Aquacult. Res. 37, 1265–1266.

Kapetanović, D., Vardić Smrzlić, I., Valić, D., Teskeredžić, E., 2011. Bacteria in the culture of bluefin tuna *Thunnus thynnus*. In: Abstract Book of the 15th International Conference on Diseases of Fish and Shellfish, September 12–16, 2011 Split, Croatia.

Kawakami, H., Nakajima, K., 2002. Cultured fish species affected by red sea bream iridoviral disease from 1996 to 2000. Fish Pathol. 37, 45–47.

Kirchhoff, N.T., D'Antignana, T., Leef, M.J., Hayward, C.J., Wilkinson, R.J., Nowak, B.F., 2011a. Effects of immunostimulants on ranched southern bluefin tuna *Thunnus maccoyii*: immune response, health and performance. J. Fish Biol. 79, 331–355.

Kirchhoff, N.T., Leef, M.J., Ellis, D., Purser, J., Nowak, B.F., 2011b. Effects of the first two months of ranching on the health of southern bluefin tuna *Thunnus maccoyii*. Aquaculture 315, 207–212.

Kirchhoff, N.T., Rough, K.M., Nowak, B.F., 2011c. Moving cages further offshore: effects on southern bluefin tuna, *T. maccoyii*. PLoS One 6, 6–13.

Kirchhoff, N.T., Leef, M.J., Valdenegro, V., Hayward, C.J., Nowak, B.F., 2012. Correlation of humoral immune response in southern bluefin tuna, *T. maccoyii*, with infection stage of the blood fluke, *Cardicola forsteri*. PLoS One 7, 1–14.

Lee, E.H., Kim, S.M., Kwon, S.R., Kim, S.K., Nam, Y.K., Kim, K.H., 2004. Comparison of toxic effects of nitric oxide and peroxynitrite on *Uronema marinum* (Ciliata: Scuticociliatida). Dis. Aquat. Organ. 58, 255–260.

Lepen Pleić, I., Secombes, C.J., Bird, S., Mladineo, I., 2014. Characterization of three pro-inflammatory cytokines, TNFα1, TNFα2 and IL-1β, in cage-reared Atlantic bluefin tuna *Thunnus thynnus*. Fish Shellfish Immunol. 36, 98–112.

Lepen Pleić, I., Bušelicć, I., Trumbić, Ž., Bočina, I., Šprug, M., Mladineo, I., 2015. Expression analysis of the Atlantic bluefin tuna (*Thunnus thynnus*) pro-inflammatory cytokines, IL-1β, TNFα1 and TNFα2 in response to *in vitro* stimulations and gill parasite infections. Fish Shellfish Immunol. 45, 946–954.

Lester, R.J.G., Kelly, W.R., 1983. Tumour-like growths from southern Australian marine fish. Tasmanian Fish. Res. 25, 27–32.

López-Sabater, E.I., Rodríguez-Jerez, J.J., Roig-Sagués, A.X., Mora-Ventura, M.A.T., 1994. Bacteriological quality of tuna fish (*Thunnus thynnus*) destined for canning: effect of tuna handling on presence of histidine decarboxylase bacteria and histamine level. J. Food Prot. 4, 276–344.

Marino, F., Monaco, S., Salvaggio, A., Macri, B., 2006. Lipoma in a farmed bluefin tuna, *Thunnus thynnus* (L.). J. Fish Dis. 29, 697–699.

Marty, G.D., Freiberg, E.F., Meyers, T.R., Wilcock, J., Farver, T.B., Hinton, D.E., 1998. Viral hemorrhagic septicemia virus, *Ichthyophonus hoferi*, and other causes of morbidity in Pacific herring *Clupea pallasi* spawning in Prince William Sound, Alaska, USA. Dis. Aquat. Organ. 32, 14–40.

Mashoof, S., Pohlenz, C., Chen, P.L., Deiss, T.C., Gatlin III, D., Buentello, A., et al., 2014. Expressed Ig μ and τ transcripts share diversity segment in ranched *Thunnus orientalis*. Dev. Comp. Immunol. 43, 76–86.

Meng, F., Yokoyama, H., Shirakashi, S., Grabner, D., Ogawa, K., Ishimaru, K., et al., 2011. *Kudoa prunusi* n. sp. (Myxozoa: Multivalvulida) from the brain of Pacific bluefin tuna *Thunnus orientalis* (Temminck & Schlegel, 1844) cultured in Japan. Parasitol. Int. 60, 90–96.

Miyake, P.M., De la Serna, J.M., Di Natale, A., Farrugia, A., Katavić, I., Miyabe, N., et al., 2003. General review of bluefin tuna farming in the Mediterranean area. Coll. Vol. Sci. Pap. ICCAT 55, 114–124.

Mladineo, I., 2006a. *Microsporidia* sp. in Atlantic bluefin tuna (*Thunnus thynnus*). Bull. Eur. Assoc. Fish Pathol. 26, 152–155.

Mladineo, I., 2006b. *Hepatoxylon trichiuri* (Cestoda: Trypanorhyncha) plerocercoid in cage-reared northern bluefin tuna, *Thunnus thynnus* (Osteichthyes: Scombridae). Acta Adriat. 47, 79–83.

Mladineo, I., 2007. Histopathology of five species of *Didymocystis* spp. (Digenea: Didymozoidae) in cage-reared Atlantic bluefin tuna (*Thunnus thynnus* thynnus). Vet. Res. Commun. 30, 475–484.

Mladineo, I., Tudor, M., 2004. Digenea of Adriatic cage-reared bluefin tuna *Thunnus thynnus* thynnus. Bull. Eur. Assoc. Fish Pathol. 24, 144–153.

Mladineo, I., Bočina, I., 2006. *Ceratomyxa thunni* sp. n. (Myxozoa: Ceratomyxidae) in Atlantic northern bluefin tuna (*Thunnus thynnus*) caught in the Adriatic Sea, Island of Jabuka. Zootaxa 1224, 59–68.

Mladineo, I., Block, B.A., 2010. Expression of cytokines IL-1β and TNF-α in tissues and cysts surrounding *Didymocystis wedli* (Digenea, Didymozoidae) in the Pacific bluefin tuna (*Thunnus orientalis*). Fish Shellfish Immunol. 29, 487–493.

Mladineo, I., Poljak, V., 2014. Ecology and genetic structure of zoonotic *Anisakis* spp. from Adriatic commercial fish species. Appl. Environ. Microbiol. 80, 1281–1290.

Mladineo, I., Čanković, M., Ščekić, V., Žilić, J., 2005. *Camallanus* sp. infection in cage-reared bluefin tuna (*Thunnus thynnus*). P.9.1, Page 119 in Book of Abstracts. EAFP 12th International Conference, September 11–16, Copenhagen, Denmark.

Mladineo, I., Miletić, I., Bočina, I., 2006. *Photobacterium damselae* subsp *piscicida* outbreak in cage-reared Atlantic bluefin tuna *Thunnus thynnus*. J. Aquat. Anim. Health 18, 51–54.

Mladineo, I., Žilić, J., Čanković, M., 2008. Health survey of Atlantic bluefin tuna, *Thunnus thynnus* (Linnaeus, 1758), Reared in Adriatic cages from 2003 to 2006. J World Aquacult. Soc. 39, 281–289.

Mladineo, I., Bott, N.J., Nowak, B.F., Block, B.A., 2010. Multilocus phylogenetic analyses reveal that habitat selection drives the speciation of Didymozoidae (Digenea) parasitizing Pacific and Atlantic bluefin tunas. Parasitology 137, 1013–1025.

Mladineo, I., Šegvić, T., Petrić, M., 2011. Do captive conditions favor shedding of parasites in the reared Atlantic bluefin tuna (*Thunnus thynnus*)? Parasitol. Int. 60, 25–33.

Moustafa, E.M.M., Naota, M., Morita, T., Tange, N., Shimada, A., 2010. Pathological study on the scuticociliatosis affecting farmed Japanese flounder (*Paralichthys olivaceus*) in Japan. J. Vet. Med. Sci 72, 1359–1362.

Munday, B., O'Donoghue, P., Watts, M., Rough, K., Hawkesford, T., 1997. Fatal encephalitis due to the scuticociliate *Uronema nigricans* in sea-caged, southern bluefin tuna *Thunnus maccoyii*. Dis. Aquat. Organ. 30, 17–25.

Munday, B.L., Sawada, Y., Cribb, T., Hayward, C.J., 2003. Review diseases of tunas, *Thunnus* spp. J. Fish Dis. 26, 187–206.

Nagasawa, K., 2011. *Caligus macarovi* (Copepoda, Caligidae) from Pacific bluefin tuna, *Thunnus orientalis*, cultured in Japan. Crustaceana 84, 1145–1147.

Nishioka, T., Mori, K., Sugaya, T., Tezuka, N., Takebo, T., Imaizumi, H., et al., 2010. Involvement of viral nervous necrosis in larval mortality of hatched-reared Pacific bluefin tuna *Thunnus orientalis*. Fish Pathol. 45, 69–72.

Nowak, B.F., 2004. Assessment of health risks to southern bluefin tuna under current culture conditions. Bull. Eur. Assoc. Fish Pathol. 24, 45–51.

Nowak, B.F., 2007. Parasitic diseases in marine cage culture—An example of experimental evolution of parasites? Int. J. Parasitol. 37, 581–588.

Nowak, B., Mladineo, I., Aiken, H., Bott, N., Hayward, C., 2006. Results of health surveys of two species of farmed tuna: southern bluefin tuna (*Thunnus maccoyii*) in Australia and northern bluefin tuna (*Thunnus thynnus*) in the Mediterranean. Bull. Eur. Assoc. Fish Pathol. 26, 38–42.

Ogawa, K., 2015. Diseases of cultured marine fishes caused by Platyhelminthes (Monogenea, Digenea, Cestoda). Parasitology 142, 178–195.

Ogawa, K., Tanaka, S., Sugihara, Y., Takami, I., 2010. A new blood fluke of the genus *Cardicola* (Trematoda: Sanguinicolidae) from Pacific bluefin tuna *Thunnus orientalis* (Temminck & Schlegel, 1844) cultured in Japan. Parasitol. Int. 59, 44–48.

Ogawa, K., Ishimaru, K., Shirakashi, S., Takami, I., Grabner, D., 2011. *Cardicola opisthorchis* n. sp. (Trematoda: Aporocotylidae) from the Pacific bluefin tuna, *Thunnus orientalis* (Temminck & Schlegel, 1844), cultured in Japan. Parasitol. Int. 60, 307–312.

Palacios-Abella, J.F., Rodriguez-Llanos, J., Mele, S., Montero, F.S., 2015. Morphological characterization and identification of four species of *Cardicola* Short 1953 (Trematoda: Aporocotylidae) infecting the Atlantic bluefin tuna *Thunnus thynnus* (L.) in the Mediterranean Sea. Syst. Parasitol. 91, 101–117.

Pelegrín, P., Chaves-Pozo, E., Mulero, V., Meseguer, J., 2004. Production and mechanism of secretion of interleukin-1β from the marine fish gilthead seabream. Dev. Comp. Immunol. 28, 229–237.

Peñalver, J.G., Pedrero, E.M.-D., Tafalla, C., Bermúdez, L., Gómez, O., Viuda, E., et al., 2012. Assessment of risk transmission virus diseases across of food used to feed bluefin tuna (*Thunnus thynnus*). An. Vet. Murcia 28, 15–21.

Perić, Z., 2002. Morphological and histological changes of the parenchimatous organs of blue fin tuna, Thunnus thynnus (Linnaeus, 1758). In: Proceedings of First International Symposium Domestication of the Blue Fin Tuna *Thunnus thynnus thynnus* (DOTT). Cartagena, Spain, February 3–8; 2002.

Pleguezuelos, O., Zou, J., Cunningham, C., Secombes, C.J., 2000. Cloning, sequencing and analysis of expression of a second IL-1β gene in rainbow trout (*Oncorchincus mykiss*). Immunogenetics 51, 1002–1011.

Polinski, M., Hamilton, D.B., Nowak, B., Bridle, A., 2013. SYBR, TaqMan, or both: highly sensitive, non-invasive detection of *Cardicola* blood fluke species in southern bluefin tuna (*Thunnus maccoyii*). Mol. Biochem. Parasitol. 191, 7–15.

Polinski, M., Shirakashi, S., Bridle, A., Nowak, B., 2014a. Transcriptional immune response of cage-cultured Pacific bluefin tuna during infection by two *Cardicola* blood fluke species. Fish Shellfish Immunol. 36, 61–67.

Polinski, M., Bridle, A., Neumann, L., Nowak, B., 2014b. Preliminary evidence of transcriptional immunomodulation by praziquantel in bluefin tuna and Atlantic salmon in vitro cultures. Fish Shellfish Immunol. 38, 42–46.

Roberts, R.J., Agius, C., 2008. Pan-steatitis in farmed northern bluefin tuna, *Thunnus thynnus* (L.), in the eastern Adriatic. J. Fish Dis. 31, 83–88.

Rodríguez Llanos, J., 2012. Parasites of the Atlantic bluefin tuna (*Thunnus thynnus*) in fattening farms in the Spanish Mediterranean. Pathologies and their use as biomarkers. Master thesis, Master interuniversitario de acuicultura, Valencia, pp. 43. (In Spanish).

Sawada, Y., Okada, T., Miyashita, S., Murata, O., Kumai, H., 2005. Completion of the Pacific bluefin tuna *Thunnus orientalis* (Temminck et Schlegel) life cycle. Aquacult. Res. 36, 413–421.

Scapigliati, G., Bonocore, F., Bird, S., Zou, J., Pelegrin, P., Falasca, C., et al., 2001. Phylogeny of cytokines: molecular cloning and expression analysis of sea bass *Dicentrarchus labrax* interleukin-1. Fish Shellfish Immunol. 11, 711–726.

Shirakashi, S., Kishimoto, Y., Kinami, R., Katano, H., Ishimaru, K., Murata, O., et al., 2012a. Morphology and distribution of blood fluke eggs and associated pathology in the gills of cultured Pacific bluefin tuna, *Thunnus orientalis*. Parasitol. Int. 61, 242–249.

Shirakashi, S., Andrews, M., Kishimoto, Y., Ishimaru, K., Okada, T., Sawada, Y., et al., 2012b. Oral treatment of praziquantel as an effective control measure against blood fluke infection in Pacific bluefin tuna (*Thunnus orientalis*). Aquaculture 326, 15–19.

Shirakashi, S., Tsunemoto, K., Webber, C., Rough, K., Ellis, D., Ogawa, K., 2013. Two speceis of *Cardicola* (Trematoda: Aporocotylidae) found in southern bluefin tuna *Thunnus maccoyii* ranched in South Australia. Fish Pathol. 48, 1–4.

Šimat, V., Maršić-Lučić, J., Tudor, M., Mladineo, I., 2009. Long-term storage influence on volatile amines (TVB-N and TMA-N) in sardine and herring utilized as food for tuna fattening. J. Appl. Ichthyol. 25, 766–770.

Sugaya, T., Mori, K., Nishioka, T., Masuma, S., Oka, M., Mushiake, K., et al., 2009. Genetic heterogeneity of betanodaviruses in juvenile production trials of Pacific bluefin tuna, *Thunnus orientalis* (Temminck & Schlegel). J. Fish Dis. 32, 815–823.

Sugihara, Y., Yamada, T., Tamaki, A., Yamanishi, R., Kanai, K., 2014. Larval stages of the bluefin tuna blood fluke *Cardicola opisthorchis* (Trematoda: Aporocotylidae) found from *Terebella* sp (Polychaeta: Terebellidae). Parasitol. Int. 63, 295–299.

Suzuki, J., Murata, R., Yokoyama, H., Sadamasu, K., Kai, A., 2015. Detection rate of diarrhoea-causing *Kudoa hexapunctata* in Pacific bluefin tuna *Thunnus orientalis* from Japanese waters. Int. J. Food Microbiol. 194, 1–6.

Takebe, T., Saeki, Y., Masuma, S., Nikaido, H., Ide, K., Shiozawa, S., et al., 2013. Prevalence and transmission capability of *Didymocystis wedli* (Digenea; Didymozoidae) in cage-reared young Pacific bluefin tuna *Thunnus orientalis* in the Amami area of Japan. Nippon Suisan Gakkaishi 79, 214–218.

Trumbić, Ž., Taggart, J.B.T., Bron, J.E., Bočina, I., Mladineo, I., 2015. Ultrastructural and transcriptomic aspects of the Atlantic bluefin tuna *Thunnus thynnus* (Linnaeus, 1758) gill epithelium infected with *Didymosulcus katsuwonicola* (Pozdnyakov 1990). Ninth International Symposium of Fish Parasites, Valencia, Spain.

Valdenegro-Vega, V., Naeem, S., Carson, J., Bowman, J.P., Tejedor del Real, J.L., Nowak, B., 2013. Culturable microbiota of ranched southern bluefin tuna (*Thunnus maccoyii* Castelnau). J. Appl. Microbiol. 115, 923–932.

Watts, M., Munday, B.L., Burke, C.M., 2001. Isolation and partial characterisation of immunoglobulin from southern bluefin tuna *Thunnus maccoyii* Castelnau. Fish Shellfish Immunol. 11, 491–503.

Watts, M., Munday, B.L., Burke, C.M., 2002. Investigation of humoral immune factors from selected groups of southern bluefin tuna, *Thunnus maccoyii* (Castelnau): implications for aquaculture. J. Fish Dis. 25, 191–200.

Watts, M., Kato, K., Munday, B.L., Burke, C.M., 2003. Ontogeny of immune system organs in northern bluefin tuna (*Thynnus orientalis*, Temminck and Schelgel 1844). Aquacult. Res. 34, 13–21.

Yokoyama, H., Suzuki, J., Shirakashi, S., 2014. *Kudoa hexapunctata* n. sp (Myxozoa: Multivalvulida) from the somatic muscle of Pacific bluefin tuna *Thunnus orientalis* and re-description of *K. neothunni* in yellowfin tuna *T. albacares*. Parasitol. Int. 63, 571–579.

Yuasa, K., Kamaishi, T., Mori, K.I., Hutapea, J.H., Permana, G.N., Nakazawa, A., 2007. Infection by a protozoan endoparasite of the genus *Ichthyodinium* in embryos and yolk-sac larvae of yellowfin tuna *Thunnus albacares*. Fish Pathol. 42, 59–66.

Zhang, J.Y., Meng, F., Yokoyama, H., Miyahara, J., Takami, I., Ogawa, K., 2010. Myxosporean and microsporidian infections in cultured Pacific bluefin tuna *Thunnus orientalis* in Japan. Fish. Sci. 76, 981–990.

Zou, J., Holland, J., Pleguezuelos, O., Cunningham, C., Secombes, C.J., 2000. Factors influencing the expression of interleukin-1 β in cultured rainbow trout (*Oncorhynchus mykiss*) leukocytes. Dev. Comp. Immunol. 24, 575–582.

Žilić, J., Mladineo, I., 2006. Effect of *Pennella filosa* (Pennellidae, Copepoda) on meat quality of bluefin tuna (*Thunnus thynnus*). Meso 1, 42–45. (In Croatian).

Chapter 12

Nutrition of Cultured Tuna Species

Alejandro Buentello[1], Manabu Seoka[2] and Jorge Suarez[3]
[1]*Ichthus Unlimited, LLC, West Des Moines, IA, USA*, [2]*Skretting ARC/Japan, Fukuoka, Japan*,
[3]*Rosenstiel School of Marine and Atmospheric Science, University of Miami, Miami, FL, USA*

12.1 INTRODUCTION

Proper nutrition is vital to the development of any new candidate species for commercial aquaculture, playing key roles in reproductive success, including the effective establishment of broodstock, the production of high-quality eggs and larvae and ultimately, the cost-effective production of nutritious seafood.

Important clues to the nutritional requirements of any animal can be obtained by looking at the natural food items for that animal, in this instance the natural prey species of scombrid fish, and also by determining the composition of wild-caught animals, both prey and predator. The latter certainly being a source of information in terms of lipids and fatty acid requirements. The somewhat limited data available to date from short feeding trials also can add to the overall perspective.

In this chapter, we will try to go beyond this approach and gather relevant information from those few institutions that have had access to either juvenile fish or have been able to muster the significant resources necessary to conduct nutritional research with commercial-size fish. In doing so, we will attempt to provide an updated overview of what is known about the nutrition of the large scombrids and also, importantly, place the problems of their feeding and nutrition in a global perspective.

12.2 COMPOSITION OF WILD SCOMBRID PREY

Today, information about nutrient requirements of scombrid fish is fragmentary at best. Therefore, establishing a baseline level of knowledge utilizing

D.D. Benetti, G.J. Partridge, A. Buentello (Eds): Advances in Tuna Aquaculture.
DOI: http://dx.doi.org/10.1016/B978-0-12-411459-3.00012-6

feeding habits and stomach contents of wild tuna follows a nutritional model that has been used successfully for the initial domestication of new aquaculture species.

The tropical oceans conditions are advantageous for the reproduction of tuna because of the relatively stable physical conditions throughout the year compared to the temperate areas (Tanabe, 2001). However, tropical areas provide lesser nutritional sufficiency (due to specific trophic conditions) for larval tuna than temperate areas do. As a result, some researchers have identified the early ichthyophagi (piscivorous behavior) exhibited by several scombrids, together with their extremely fast growth and development of adult-like digestive capabilities, as adaptation strategies of early juvenile tuna to these environments (Kaji et al., 1996, 1999; Kaji, 2000; Morote et al., 2008; Buentello et al., 2011). This notion appears corroborated by the observation of an increased dependence on fish larvae as prey organisms from the larval stage through to the early juvenile stage of tuna (Torres et al., 2011; Satoh et al., 2013).

Early scombrid stages typically range from 35°N to 35°S in the western Pacific, where surface water temperatures are warmer than 25°C (Tanabe, 2001) and at depths between 0 and 100 m (mixed layer above the thermocline); thus, the tropical ocean is important not only as the spawning site but also as nursery grounds of larvae and juvenile tuna. The feeding is mainly diurnal and significant differences are found in stomach contents between larvae sampled during the day or night (Richards and Dove, 1971). In the Mediterranean, *Thunnus thynnus* larvae of ~6 mm standard length exhibit positive selection for cladocerans over other prey; whereas, copepod nauplii dominate the diets of earlier stages (Catalan et al., 2011). Well after the post flexion stage (~7 mm in total length) and based on frequencies of occurrence, juvenile scombrids (≥10 mm) appear to rely most heavily on larval fish (~93%) and much less on the classes cephalopoda (~2%), euphausiacea (~1%), copepoda, and amphipoda (negligible) as well as on other unidentified prey taxas (~4%) (Tanabe, 2001; Torres et al., 2011).

For juvenile ABFT, studies conducted at the Spanish Institute of Oceanography (De la Gandara, 2012) pointed out that young ABFT consume relatively large prey swallowing them whole and thus, allowing for a high identification rate. However, as the digestion rate is not the same for all prey, the relative importance of rapidly digested items may be underestimated. Analysis of stomach content indicated that, by weight, the composition of prey consisted mostly of cephalopods (65%), especially the short-fin squid, *Illex coindetii* (61%), followed by fish (33%). Small crustaceans were also present in the sampled stomachs in smaller percentage. This study also identified a high degree of variation in results obtained even from consecutive years. It is clear that the abundance of specific prey items plays a major role in the composition of stomach content in juvenile ABFT identifying them as opportunistic rather than selective predators.

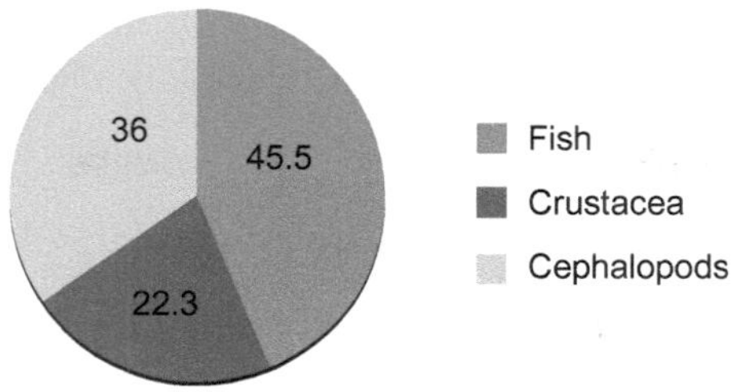

FIGURE 12.1 Average proportion (%) of prey items by weight found for yellowfin tuna from several regions in the Pacific Ocean. *Adapted from Buckley and Miller (1994).*

As tuna age, they become highly adaptable foragers of fishes, squids, or large zooplankton. Analyses of stomach contents of adult yellowfin tuna (YFT) in the eastern tropical Pacific obtained by net hauls indicate that an ommastrephid squid (e.g., *Dosidicus gigas*) together with the portunid crab *Euphylax dovii* are some of the most important food items in terms of the volume and number ingested. In addition to the squid and crab, the small scombrid *Auxis* sp. and epipelagic exocoetids such as *Oxyporhamphus micropferus* are also important prey items for YFT in this area (Perrin et al., 1973). Similar analyses in the same region (Alverson, 1963) have found that the major components in the diet of adult YFT ($\sim$1 m fork length) are fish (47% of total volume) followed by crustaceans (45.4%) with cephalopods accounting for only 7.6% of the total volume. More recent studies (Fiedler and Bernard, 1987) reveal that albacore, *Thunnus alalunga* and skipjack, *Katsuwonus pelamis* caught off California feed on juvenile northern anchovy, *Engraulis mordax*, other epipelagic fishes and planktonic crustaceans. Likewise, similar size ABFT prey on a wide variety of schooling fish, cephalopods, crustaceans, and benthic invertebrates (Logan et al., 2011) in trophic migration throughout most of the year in preparation for the annual reproductive migration to well-established spawning grounds in the Gulf of Mexico or the Mediterranean Sea (Rooker et al., 2007).

Adult tuna are quasi-cosmopolitan inhabitants of the world's oceans and highly mobile fishes, which adapt their feeding habits to what is available in the different ecosystems they occupy at a particular time. As an illustration, tuna from the eastern Pacific Ocean prey mostly on other fish followed by squid (Figure 12.1). Crustaceans are negligible in the diet of New Caledonia tuna; whereas they represent a large proportion of the diet of tuna in Papua New Guinea and Polynesia (Allain, 2005).

Although the analysis of common prey items for tuna species may be useful in generating preliminary knowledge about their nutritional needs, it is important for the reader to be aware of the potential limitations of this approach. For instance, adult tuna have a significant capacity for lipid storage for use in short- or long-term energy demands (Saito and Mrata, 1998). Because the time period during which large-sized tuna turn over their whole-body fatty acid trophic signature may span from days to months, the scombrid fatty acid composition will never exactly match that of its prey (Budge et al., 2006).

12.3 COMPOSITION OF WILD SCOMBRIDS

The composition of wild scombrids can give some indication of possible dietary requirements, at least in respect to nutrients such as lipid and fatty acids. Obviously, significant variation can be expected in carcass lipid levels due to the seasonal availability of prey and location of the foraging efforts. Significantly, the protein composition has been observed to be less variable. The very strong inverse relationship between body fat and water in highly migratory tuna species indicates that the fish obtain energy for their journeys from muscle lipid reserves (Clay, 1988). The fact that flesh lipid levels can vary so widely has important consequences for farming. Clearly the level of fat in the flesh will be highly dependent upon dietary fat levels but seasonal factors affecting lipid metabolism in fish may also be important. The potential benefits of a highly functional lipid metabolism including rapid growth may have to be balanced such that potential deleterious effects of reduced product quality and concomitant consumer acceptance are avoided.

European scientists under the SELFDOTT initiative analyzed the chemical composition of juvenile wild-caught ABFT (800 ± 133 g; 34 ± 2 cm; $n = 30$) finding the carcass to be composed of 80.7% CP, 3.8% CL, and 11.3% ash, on average (De la Gandara, 2012). Carcass amino acids (AA) analysis indicated (as % of dry sample) that Glu, Asp, Lys, Leu, His, Arg, Ala, Val, Gly, Thr, Ile, Ser, Phe, Pro, Tyr, Met, Tau, and Cys accounted for 76.3% of the total AA content in the following percentages: 10.2, 7.2, 6.9, 6.1, 6.1, 4.9, 4.5, 4.0, 3.9, 3.7, 3.6, 3.2, 3.1, 2.8, 2.7, 2.4, 0.7, and 0.4%, respectively. Contrasting these values with those of other teleost fish, the study found that juvenile ABFT generally have a higher AA content with the most significant differences being in His, Met, Lys, Leu, and Asp.

Similarly, fatty acid analysis of liver and muscle (Table 12.1) compared with other marine teleosts indicated that juvenile ABFT has significantly higher EPA and DHA concentrations than various other marine species. This notion of ABFT having an enriched profile in *n*-3 polyunsaturated fatty acids was recently corroborated in a study by Mourente et al. (2015) which identified trophic linkages between the ABFT fatty acid composition and diatomaceous food webs from the temperate and subtropical Atlantic Ocean.

Proximate compositions of tuna muscle tissue differ by fish origin and yearly season and also by the particular muscle group being analyzed (Morishita et al., 1987; Date and Yamamoto, 1988; Nakamura et al., 2005). In addition, lipid content of tuna fattened in cages is generally much higher than that of wild fish (Shimizu et al., 1973; Hirano et al., 1980; Saeki and Kumagai, 1984).

TABLE 12.1 Fatty Acid Composition of Juvenile ABFT

Fatty Acid	Liver	Muscle
	% of Total Fatty Acids	
Saturates	28.4	34.2
Monosaturates	12.5	8.1
n−9	10.3	6.7
n−6	5.7	8.9
n−3	48.4	47.5
EPA + DHA	47.1	45.5
n−3/*n*−6	9.3	5.4
20:4 *n*−6	2.3	5.2
20:5 *n*−3	5.5	6.9
22:6 *n*−3	41.6	38.5

EPA = Eicosapentaenoic acid = 20:5 *n*−3
DHA = Docosahexaenoic acid = 22:6 *n*−3
Arachidonic acid = 20:4 *n*−6.
Adapted from De la Gandara (2012).

A comparative analysis of four regions of ordinary muscle (permutations of front and rear with dorsal and ventral) of PBFT revealed that only the belly flaps differ significantly in terms of muscle tissue proximate composition as most scombrid species including PBFT tend to accumulate fat in the ventral region (Nakamura et al., 2005). According to this work, the average muscle tissue of 13-kg PBFT reared in cages is composed of 68% moisture, 1% ash, 23% CP, and about 8% CL. In contrast, belly flaps may accumulate up to 38% fat in lieu of reduced protein (~16%) and moisture (45%) contents, with ash staying about the same 1%.

One of the main objectives of tuna farming, as it is practiced today, is not necessarily to increase fish size but instead to increase the fat content of the captive biomass. The driver of this practice is the Japanese market, which considers that most wild-caught tuna have less than desirable lipid contents, even in the ventral sections. A study aimed at evaluating the effects of farming over specific morphometric relationships in ABFT held in cages for 5 months in Murcia, Spain, documented that farmed tuna became significantly heavier and their somatic condition also increased. Apparently, somatic differences were more evident above 180 cm in fork length (Aguado-Gimenez and Garcia-Garcia, 2005). The researchers concluded that below this length, smaller fish were still growing rapidly and their higher metabolic rate prevented them from becoming fatty.

Similarly, a study conducted in the vicinity of the Favignana Island, Sicily, Italy, contrasting the carcass proximate composition of wild-caught ABFT versus that of similar-sized, cage-fattened fish in the same area after 6 months in captivity documented significant decreases in moisture (71% to 63%, wild vs farmed, respectively), CP (22% to 20% wild vs farmed, respectively) with a concomitant increase of lipid content (6% to 15% wild vs farmed, respectively) (Roncarati et al., 2008). These reports are in agreement with those of Saito et al. (1995) and Tsukamasa et al. (2007), which also documented high lipid contents (13%) for cultured PBFT of about 50 kg in contrast with that of even larger animals (7.7% and ~70 kg) of the same species but obtained from the wild.

In utilizing this information for development of tuna diets, nutritionists may need to also consider that the exceedingly high fat contents found in cultured tuna is likely due to the lack of prolonged swimming activity in cultured fish as well the availability of high caloric food on a daily basis. Once feeding artificial feeds become a common practice in tuna grow out operations, balanced formulations will no doubt take into account the ideal physiological status of these large migratory fishes and present to the market an optimized product.

12.4 CURRENT FEEDING PRACTICES

To supply the well-established global markets for sushi/sashimi grade tuna, most tuna farming today is practiced in a non-sustainable manner, for the most part, as it still relays on wild caught fish. Prevalent feeding practices are also disputable because during the fattening period tuna are fed sardine/squid rations until they grow to a size and acquire a flesh composition agreeable to buyers. This is not only impractical and largely unsustainable but it may pose ecological risks that require immediate attention. For example in northwest Mexico, sardine availability is generally limited to April–September and frozen product is more expensive and less nutritious. Importantly, the baitfish composition is highly variable over the course of a farming/growing season and is likely to affect production efficiency.

The daily feeding of large quantities of untreated fresh or frozen fish/squid results in unreasonably high feed conversion rates (22.6:1 to 17.8:1; Ottolenghi et al., 2004; Estess et al., 2014).

In addition, it is our experience that although sardine/squid diets provide acceptable nutrition to adult tuna in the short term, prolonged feeding with these items is insufficient to meet the metabolic needs of tuna. Repeated freezing and thawing cycles destroy the biological activity of key vitamins and thus the practice of coating baitfish with a various mixtures of these nutrients embedded in a gelatin matrix has gained some popularity, despite the obvious and significant leaching potential.

TABLE 12.2 Comparison of Gross Nitrogen and Phosphorus[a] Waste Output on a Daily Basis (mg of nutrient/kg fish/day) in Three Different Mediterranean Species

Species	Particulate N	Particulate P	Dissolved N	Dissolved P	Reference
Thunnus thynnus	49.4	70.0	694.2	51.7	Aguado-Gimenez et al. (2006)
Sparus aurata	51.6	31.3	182.4	11.3	Lupatsch and Kissil (1998)
Dicentrarchus labrax	62.3	21.6	382.9	14.4	Lemarie et al. (1998)

[a]*Total nitrogen and total phosphorus.*

Perhaps of greater concern is the fact that current feeding strategies facilitate the proliferation of pathogens (Ward et al., 2001). This situation resulted in a sudden collapse of the sardine (*Sardinops sagax*) fishery in southern Australia in 1995, the yield falling from 165,000 MT to 37,000 MT (Ward et al., 2001).

These considerations and the need to provide sustainability and permanence to tuna aquaculture are the main driving factors for the development of commercially manufactured tuna feeds (Smullen, 2009).

12.4.1 Waste and Composition of Waste from Fattening Tuna Operations

The qualitative and quantitative composition of waste generated by captive scombrids compared to other cultured species provides information regarding their metabolism and growth (Table 12.2). The lower total nitrogen (TN) particulate waste output of tuna is a consequence of the greater digestibility of the food protein by tuna, and the higher TN dissolved waste output because of their relatively lower growth and higher metabolic cost (Brill, 1987).

Likewise, tuna makes better use of protein as an energy source than other carnivorous fish including sea bream and European sea bass, as reflected by the higher degree of protein hydrolysis in the scombrid digestion process (Essed et al., 2003). Total phosphorus (TP) waste output depends to a great extent on the phosphorus content of the diet, plasma levels (Phillips, 1962), and phosphorus requirements, so that phosphorus retention efficiency declines when intake exceeds the requirements of the fish (Ruohonen et al., 1999). The fact that TP requirements and digestibility are quite similar for all three species and that the TP content is higher in tuna food (defrosted fish vs feedstuffs) explains why TP waste production was higher in tuna than in gilthead sea bream or sea bass (Aguado-Gimenez et al., 2006).

Wild-caught tuna are typically reluctant to feed on commercially prepared dry diets and moist or semi-moist diets. In addition, as moist diets contain significant amount of water the resulting feed conversion ratio is high due to limitations on energy content in the feed. In Australia, the poor acceptance and/or significant mortality of fish fed these "research diets" has made the SBFT industry reluctant to switch from feeding industrial baitfish to pelleted diets.

12.5 TOWARD FORMULATED DIETS FOR CULTURED TUNA

12.5.1 Digestive Physiology

In contrast with the majority of marine fish larvae cultured to date, a fully developed digestive tract, including gastric digestion, is acquired by scombrids in a comparatively short period (days vs weeks or months) after the onset of exogenous feeding. The capacity to degrade and absorb dietary nutrients by the larval tuna is sufficient to support fast growth by digesting prey naturally available in the sea.

12.5.1.1 Characterization of Scombrid Digestive Enzymes

Tuna appear to be characterized by a unique digestive physiology that is well adapted to their rapid growth, strong piscivorous habits, and endothermy. However, there are still few studies on this subject area. Tuna larvae exhibit a very early nutritional point-of-no-return, particularly at high water temperatures (Margulies, 1993). Because first-feeding larvae typically have minimal nutrient reserves in combination with very demanding needs to support high levels of locomotion and an extremely rapid growth rate, a thorough understanding of their digestive capacity is absolutely essential to enable the provision of nutrient particles suitable for their developing gut to digest and assimilate in an efficient manner (Izquierdo et al., 2001; Yufera and Darias, 2007).

Studies on larval PBFT fed enriched rotifers from 3 dph indicate that the gastric glands and pyloric ceca first appear during the flexion phase, supporting the hypothesis of early functionality of the digestive system (Kaji et al., 1996). Similarly, in this species, the activities of proteolytic enzymes exhibited peaks at 3, 14, and 25 dph which coincides with first feeding, flexion phase, and transformation phase to juvenile, respectively. Likewise, synthesis of enteric pepsinogen starts just prior the flexion phase with activation of this zymogen progressing concurrently with larval development (Kaji, 2000). Comparable enzymatic activities were documented by Buentello et al. (2011) for YFT, with chymotrypsin, aminopeptidase, amylase, and both acid and alkaline phosphatases being similar to that in PBFT but more than two orders of magnitude higher than those evaluated for California halibut (*Paralichthys californicus*; Alvarez-Gonzalez et al., 2006) of a similar size (YFT larvae 30 day after hatch ~27 mm TL); in the case of amylase activity, it was up to 70-fold the level found in yellowtail, *Seriola lalandi*, of similar size (Chen et al., 2006). Taken together these results suggest not only that the digestive capacity of larval and early juvenile scombrids are superior to that of other marine fish but, the flexion phase (~12 dph, 6–7 mm) appears to mark the onset of such superior digestive capabilities.

Larval PBFT start growing more rapidly after 20–25 dph with fast differentiation and development of the stomach and pyloric ceca. This development is accompanied by a drastic increase in trypsin-like activity enabling them to eat larval fish of other species and/or their siblings (cannibalism) to maintain their rapid somatic growth (Miyashita et al., 1998). A similar ontogenetic pattern has been found in YFT larvae (Kaji et al., 1996). It has been pointed out that the activities of several gastrointestinal enzymes are greater in the YFT than in similar-age/size marine fish larvae of other species, and that these enzymes display a precocious increase in specific activity concurrent with somatic growth (Buentello et al., 2011). In young and adult tuna, moreover, it has been shown that the temperature of the peritoneal cavity affects digestive process; for instance, once PBFT reach a body weight of 20 kg, they are able to maintain high and rapid digestibility even at a low water temperatures. This is related to heat generation caused by muscular movement and from the chemical digestive process of ingested food in the thermally isolated viscera (Kubo, 2013). This thermal strategy allows tuna to eat frequently when their food is abundant (Carey et al., 1984) and subsequently to maintain their rapid somatic growth. In addition, scombrid species appear to employ unique strategies for rapid meal digestion such as the maintenance of low stomach pH found in a recent study on ranched SBFT fed a modified baitfish diet (Leef et al., 2012).

12.5.2 Nutrient Requirements

Diet development for commercially fattened tuna species is particularly difficult for several reasons. Among them, the cost of research with advanced juvenile specimens of these species is extremely high, having to maintain operations on numerous net-pens, and the high costs of crews, boats, and related infrastructure. In addition, the price tag of each organism used for research, depending on its size could reach close to US$1000. Additionally, the commercial size at which these fish are sent to the market is significantly larger than most other cultured aquatic species—usually, ranging above 20–30 kg and beyond 1 m in fork length. Until the complete life cycle for PBFT was closed at Kinky University in 2002 (Sawada et al., 2005), there was no availability of juvenile fish for nutritional research and even then access to laboratory-produced animals has been quite limited to a single institution.

As discussed above, research with animals of close-to-commercial size is very expensive and is compounded even more by the lack of sufficient animals necessary to limit the variance, making it extremely difficult to design statistically sound experiments (Glencross et al., 2002a,b).

12.5.2.1 Proteins and Amino Acids

Protein requirements vary among species with carnivorous juvenile fish typically needing more dietary protein such as 45% for Atlantic salmon, *Salmo salar*; 55% for yellowtail *Seriola* sp., red sea bream, *Pagrus major*, and malabar grouper, *Epinephelus malabaricus*; and 50% for bullseye puffer fish, *Sphoeroides annulatus*, compared to some other more omnivorous species such as 32–36% for channel catfish, *Ictalurus punctatus*, and 30% for Nile tilapia, *Oreochromis niloticus*, for growth and maintenance (Biswas et al., 2009a).

Generally, increasing dietary protein can lead to improved fish production, especially for carnivorous fish (Lee et al., 2002). However, excess dietary protein will be metabolized for energy and will increase excretion of nitrogenous waste material into the aquatic environments (Catacutan and Coloso, 1995; Tibbetts et al., 2000), which may be detrimental to fish growth and the culture environment, as well further increasing the high operational costs associated with the feed.

The optimum dietary protein content needed for good fish performance will depend on the energy content of the feed (Cowey, 1979). Providing the optimum energy levels in diets for fish is important because low dietary energy will result in the utilization of dietary protein for energy rather than for protein synthesis. Thus, energy-yielding nutrients such as lipids and carbohydrates can theoretically reduce the oxidation of protein to energy and hence improve the utilization of dietary protein for growth and tissue repletion having a protein-sparing effect (Wilson, 1989).

Integral to the development of a feed for any species is the identification of their CP, CL, and energy requirements. Protein is required to provide the fundamental units (AA) for growth, while dietary lipid provides both essential fatty acids and some of the energy needed for the metabolic processes of growth. Further, energy can also be derived from the metabolism of protein and some dietary carbohydrates.

Due to the aforementioned complexities, several alternative and/or indirect methods of measuring protein utilization and amino acid requirements have been examined in PBFT and SBFT (Table 12.3; Van Barneveld et al., 1997; Carter et al., 1998, 1999; Glencross et al., 1999a).

Studies with SBFT have examined growth rates as a function of protein intake, considering the effect of water temperature and dietary energy, and results obtained suggest that optimal daily protein intake for growth is about 10 g protein/kg/day (Glencross et al., 1999a,b, 2002a), while

TABLE 12.3 Essential Amino Acid Composition of Key Muscle Types (g/kg on a dry-matter basis) of SBFT and a Proposed Ideal Amino Acid Balance (%), Relative to Lysine

Amino Acid	White Muscle	Red Muscle	Ideal Balance (%)
Methionine	10.9	11.0	35
Threonine	17.2	18.2	56
Valine	19.2	20.4	63
Isoleucine	17.1	18.7	57
Leucine	27.6	29.3	90
Phenylalanine	14.9	15.1	47
Lysine	30.7	32.5	100
Histidine	30.3	30.5	96
Arginine	22.3	23.9	73

Takii et al. (2005) determined the daily protein requirement for maintenance in PBFT to be 5.5 g/kg body weight. Recently, a study was conducted by Biswas et al. (2009a), with the objective to determine the optimum dietary protein and lipid levels for developing a formulated diet for PBFT juveniles. Five test diets were prepared with varying proportions of protein/lipid including 72.8/9.2, 66.8/14.8, 61.9/17.9, 57.2/21.9, and 53.3/27.0, using defatted enzyme treated fishmeal (FM) and salmon egg oil. The diet with 61.9% CP and 17.9% CL showed the highest growth performance of PBFT in terms of final body weight, specific growth rate, and feed efficiency.

The determination of AA requirements for tuna species has not been conducted extensively for different reasons linked to the difficulty of maintaining such species in captivity, but there exists an interesting study on AA absorption with the technique of everted sleeve (Ingham and Arme, 1977) with a small part of intestine. Due to the relative shortness of the tuna's intestinal length, it follows that the absorption of AA and the capacity of the pyloric ceca to also uptake AA must be high. This notion appears to be supported by the study of Martínez-Montaño et al. (2013), which evaluated the interaction between lysine and arginine in the proximal intestine of PBFT using the everted intestine method. Scombrid species have a relatively short digestive tract characterized by a small ratio between intestinal length and fork length which is typical of carnivorous fish but more pronounced in the case of tuna. The high AA absorption capacity, therefore, compensates for the reduced intestinal length.

12.5.2.2 Lipids and Fatty Acids

Dietary lipid and fatty acid requirements of tuna seem to vary depending on their developmental and growth stages as in other fish species and also possibly even among the tuna species. Dietary lipids perform four main functions in the body: (i) provide energy, (ii) provide essential fatty acids (EFA), (iii) serve as structural components, and (iv) serve in regulatory functions (eicosanoids, leukotrienes, thromboxanes, prostaglandins, etc.). As with protein, lipid requirements can be broken down into three main categories: gross lipid requirement, qualitative, and quantitative EFA requirements (Mourente and Tocher, 2009). Although some studies have been conducted with PBFT and ABFT, very little is still known about the detailed lipid and fatty acid requirements of tuna. However, the fact that tuna are highly migratory and piscivorous fish species with rapid growth, relatively high levels of DHA (docosahexaenoic acid, 22:6n-3), and a high ratio of DHA:EPA (eicosapentaenoic acid; 20:5n-3), compared to most other species, indicates they have unique lipid and fatty acid requirements (Mourente and Tocher, 2009).

Larval PBFT exhibits good growth when given yolk-sac larvae of marine fish species containing around 20% lipid on a dry-weight basis (DW; Seoka et al., 2007a). Furthermore, best growth in juvenile PBFT was achieved with an artificial feed containing about 18% lipid and about 62% CP on a DW basis (Biswas et al., 2009a), although there is still no data for young and adult tuna. On the other hand, larval PBFT fail to grow when continuously fed nutritionally enriched *Artemia* which are widely used for larval production of other marine fish without serious problems (Seoka et al., 2007a). Increasing the level of DHA in *Artemia* further improves growth and survival of larval PBFT, which suggests that DHA is a more critical essential fatty acid for PBFT compared with other marine fish. However, this improvement is not so significant when compared to the excellent growth and survival attained when feeding of yolk-sac larvae of surrogate fish species to larval tuna (Seoka et al., 2007b).

Growth failure has also been found in larval PBFT given an artificial diet containing neutral lipid (generally from anchovy oil) as the main dietary lipid source (Seoka et al., 2010). However, this negative impact can be countered by partly replacing the neutral lipid with high inclusions of phospholipids from salmon roe in the artificial feed (Seoka et al., 2008). Phospholipids in enriched *Artemia* are low in DHA; whereas salmon roe phospholipids are rich in DHA, suggesting that larval PBFT strongly rely on dietary phospholipid-bound DHA to maintain and promote growth and survival (Seoka et al., 2008), as also shown in other marine fish species (Tocher, 2003; Cahu et al., 2009). Furthermore, it has been reported that use of a commercial fish oil rich in phospholipid-bound DHA in an artificial feed improves growth performance of PBFT juveniles (Biswas, 2010). As is the case in other marine fish (Tocher et al., 2008), larval and juvenile tuna seem to have a specific requirement for preformed phospholipids in their diet.

Feeding *Artemia* enriched with choline, a precursor molecule for the most abundant phospholipid (phosphatidylcholine) in the body (Seoka et al., 2007a), showed no positive effect on growth of PBFT larvae (Biswas et al., 2006). In contrast, dietary inclusion of a high level of soybean phospholipids (lecithin) combined with a high level of DHA from neutral lipids of salmon roe in an artificial feed can enhance the growth of PBFT larvae and juveniles, although a small amount of phospholipid from salmon row was present in the neutral lipids (Seoka et al., 2008). Therefore, both qualitative and quantitative impacts of dietary phospholipids on tuna's growth and survival need to be studied further in order to understand their lipid and fatty acid requirements. Also, as dietary phospholipids play an important role in the early development and functionality of the digestive system (Cahu et al., 2003, 2009), a better understanding of the specific dietary requirements for phospholipids in tuna would support further developments in larval rearing.

The polyunsaturated fatty acid (PUFA) composition of most marine fish is dominated by the *n*-3 highly unsaturated fatty acids (HUFA), EPA, and DHA. Compared with other species, the fatty acid composition of tuna appears unique and characterized by relatively high levels of DHA (25–36% in total fatty acids) and DHA:EPA ratio (3.4–5.8) (Mourente and Tocher, 2009). The lipid biochemistry underpinning the high level of DHA and the high DHA:EPA ratio in tuna is still unclear, although the biosynthetic capacity for these particular fatty acids has been reported to be very limited in scombrids (Morais et al., 2011) and, in consequence, it has been suggested that tuna larval diets should reflect this, containing high levels of *n*-3 HUFA with a high DHA:EPA ratio (Mourente and Tocher, 2009). Indeed, the DHA content and DHA:EPA ratio of striped beakfish, *Oplegnathus fasciatus*, yolk-sac larvae, an ideal live food for tuna larvae, are around 20 mg/g DW (around 20% in total fatty acids) and 3.3, respectively (Seoka et al., 2007b). Feeding enriched *Artemia* containing 24 mg DHA/g DW (about 12% in total fatty acid) but with a DHA:EPA ratio of 1:1 resulted in stunted growth and poor survival of PBFT larvae, emphasizing the importance of a nutritional balance between dietary *n*-3 HUFAs in tuna feeds (Seoka et al., 2007b; Mourente and Tocher, 2009). A recent study revealed that PBFT have five copies of the green-sensitive opsin gene and this number is the highest among all fish studied to date (Nakamura et al., 2013). The remarkable accumulation of DHA in neural tissues and organs including retina has previously been found in ABFT, PBFT, and YFT (Ishihara and Saito, 1996; Saito et al., 1996; Mourente et al., 2002; Roy et al., 2010). It is therefore assumed that DHA has a unique role in the retina to secure the sharpest vision required for a top predator like tuna.

12.5.2.3 Carbohydrates

Data on the utilization of carbohydrates in tuna is limited. This section presents relevant information and findings from existing studies on the use of carbohydrates in balanced diets for scombrid species and other marine carnivorous fish.

Carbohydrates are a good source of energy at a low cost. This makes them attractive for the development of fish feed. However, dietary carbohydrates must be used with care as their use in high amounts remains controversial (Bou et al., 2014). Dietary carbohydrates are used by carnivorous fish in a limited manner (Moon, 2001; Hemre et al., 2002). They affect growth but the growth level produced is determined by many factors, which include source, concentration, level of food intake, and digestibility (Brauge et al., 1994).

An acute dose of carbohydrates, for example, glucose, can cause an intense and prolonged state of hyperglycemia in many fish species (Booth et al., 2006; Moon, 2001). The intolerance in fish can be caused by a higher sensitivity of insulin to amino acids rather than glucose (Hemre et al., 2002; Moon, 2001), and also by inefficiencies in the utilization and absorption of peripheral glucose, by inadequacies in regulation of homeostatic glucose and by imbalances of endogenous versus exogenous glucose sources (Booth et al., 2013).

Dietary carbohydrates are used less in feed for aquatic farmed species than in feed for poultry and mammals. Their level of digestibility depends greatly on source, composition, processing, and level of inclusion. The digestibility tends to be lower than that of protein and fat sources (NRC, 2011).

Matus de la Parra et al. (2007) evaluated the digestive enzymes of six growing PBFT using freshly dissected digestive tracts (24.15 ± 7.58 kg and 1.07 ± 0.13 m), and categorizing enzyme activities according to pH and temperature optima in crude extracts of stomach; cecal mass; and proximal, middle, and distal intestine. The broader range of activity was exhibited by the α-amylase activity, with higher activity over the alkaline pH values and higher temperature.

A dietary carbohydrate level suitable for juvenile PBFT was investigated by Biswas et al. (2009b), who also sought to verify the performance of formulated diet (Table 12.4) compared to a diet of raw sand-lance (*Ammodytes* sp.). The purpose of the study was to find a suitable dietary sugar level, substituting with a mixture of 68% enzyme treated fish meal and 8% salmon egg oil, for getting optimal growth of juvenile PBFT. The results showed that suitable diets for juvenile fish of this species should contain 60% protein, 16% lipid, and 13% carbohydrate. This diet produced optimal growth and efficient utilization of carbohydrates.

From these results, Biswas et al. (2009b) concluded that carbohydrates can replace dietary protein and lipid to a certain extent without negative effects on growth performance or carcass quality. In addition, there was an improvement in the physical properties of pellets with the use of carbohydrates, up to 12.8%, and improvement in growth and feed utilization, with no negative effect on the metabolism of carbohydrate by juvenile PBT.

TABLE 12.4 Experimental Diet Used by Biswas et al. (2009b) to Optimize Carbohydrate Levels in Tuna Diets

Ingredient	% of Diet
Enzyme-treated fish meal[a]	63.8
Salmon egg oil[b]	7.5
α-Starch	8.0
Vitamin mixture[c]	5.0
Mineral mixture[c]	5.0
Soybean lecithin	1.9
Taurine	2.0
Feeding stimulants[d]	0.5
Cellulose	3.9
Wheat gluten	2.3
APM[e]	0.1

[a]*Profish S.A, Santiago, Chile, defatted with* n-Hexane.
[b]*Sujiko oil®, Nisshin Marinetech Co. Ltd., Yokohama, Japan.*
[c]*Halver's vitamin and mineral mixtures (Halver, 1957).*
[d]*The mixture of Glutamic acid, 8.5; L-Histidine HCl H_2O, 232.8; and Inosine-5′-monophosphate 2Na, 200.9 mg.*
[e]*L-Ascorbyl-2-monophosphate magnesium^{2+} salt.*

12.5.2.4 Minerals

Minerals may be ingested by fish both from the diet and from the surrounding water. Minerals are needed within close ranges of specific concentrations and in efficient forms for normal metabolic, cellular, and tissue processes to occur (Watanabe et al., 1997).

Determining mineral requirements can be challenging. The difficulty stems from having to determine the potential contribution of minerals from the water, the possibility of minerals leaching from the diet prior to consumption, identifying test diets with a low concentration of the target mineral, and the lack of information on mineral bioavailability to fish species (NRC, 2011).

The gills can absorb many minerals, helping meet metabolic requirements. Osmoregulatory processes, reacting to the salinity of the water, help in the absorption of minerals from the diet or water, and affects the urine and feces produced. Macro-minerals and micro-minerals are important elements in the diets, as they are important factors in a variety of body processes. The body requires macro-minerals in reasonably high concentrations for the formation of skeletal structures and other hard tissues, the transfer of electrons, control of the acid:base balance, the production of membranes, and osmoregulation. The most commonly recognized macro-minerals are calcium, chlorine, magnesium, phosphorus, potassium, and sodium (NRC, 2011).

Micro-minerals (or trace minerals) are important components, although generally required at lower concentrations, as components of hormones and enzymes. They also function as cofactors and/or activators of many enzymes and biochemical processes. The most common trace minerals are chromium, copper, iodine, iron, manganese, selenium, and zinc.

Diets for fish require particular attention to meeting the requirements of the targeted species. One set of them is the diets for tuna. Care must be taken to avoid low levels of certain minerals as they may cause deficiencies. The diets may be used to adequately supplement critical minerals, which the tuna can absorb from drinking sea water, but which they need for healthy growth at higher concentrations. Tuna may need supplementation of the following critical minerals: iron, selenium, phosphorus, calcium, and magnesium (Masumoto, 2002).

12.5.2.5 Vitamins

For scombrid species, as for other animals, vitamins are necessary for normal physiological function. In the marine environment, wild tuna may find these compounds in sufficient amounts as natural components of their various prey items, likely concentrated by the food chain. However, captive tuna, which are most often fed frozen baitfish, often find the provision of these organic compounds in lesser concentrations than what is necessary to meet their metabolic needs or with a reduced/deteriorated biological activity, due to product handling and repeated cycles of freezing and thawing.

Assumptions could be made that scombrid species require the same range of water-soluble and fat-soluble vitamins that have been identified as being necessary in the diets of all fish studied to date (NRC, 2011). However, the quantitative requirements for each vitamin vary slightly between species; therefore, determinations for the main scombrid species being cultured to date (PBFT, SBFT, YFT, and ABFT) will have to be determined to optimize formulations. As a first approach, a generalized vitamin premix for salmon (Halver, 1957) has been quasi-appropriate for pioneering nutritional trials conducted with PBFT at Kinki University and has proven sufficient thus far to prevent vitamin deficiency symptoms of PBFT in short-term trials.

Nonetheless, some modifications in vitamin premixes may be necessary when considering the unique scombrid physiology. For instance, there are several reports of an uncommonly high level of DHA in the carcass composition of large scombrids, including PBFT juveniles which display a very high DHA:EPA ratio (Sawada et al., 1993). For most marine fish in both Northern and Southern Hemispheres, this ratio is typically lower than 2

(Ackman, 1980). This is clearly not the case for PBFT, which exhibits DHA:EPA ratios of up to 6, with muscle DHA levels ranging from 25% to 35% (Ishihara and Saito, 1996). Similarly, Medina et al. (1995) found DHA:EPA ratios exceeding 7 in ABFT. Because these fatty acids are particularly prone to oxidation, a closer look at vitamins with proven antioxidant function, such as vitamin E (tocopherol) and vitamin C (ascorbate), should result in the fortification of these compounds in tuna diets. For instance, although tank collisions made for an extremely short feeding trial (2 weeks), the determined requirement for early juvenile PBFT was 1200 mg/kg of L-ascorbyl-2-monophosphate (Biswas et al., 2013). This level exceeds 22-fold that of the highest vitamin C requirement (similar monophosphate) determined for teleost species (cobia, *Rachycentron canadum*, up to 54 mg/kg, NRC, 2011) and may partially explain the DHA/EPA-related increased need for vitamin C.

In addition, scombrids have often been categorized as "high-performance" fish. Similar to ergogenic aids in high-performance human athletes (Clarkson, 1996), the well-documented superior metabolic performance of scombrids (Dickson, 1995) including hyper-stamina and recovery from anaerobic bouts necessitates an increased supply of key vitamins including thiamin, riboflavin, vitamin B_6, B_{12}, and folate. As an example, Moyes (1996) used cardiac metabolic rates and other physiological indicators to discriminate between skipjack tuna (*K. pelamis*) and common carp (*Cyprinus carpio*). In this case, the high-performance status was obtained by the scombrid metabolism occurring at much higher rates than all other teleost fishes. For instance, the skipjack tuna ventricular respiratory capacity is up to 21-fold that of the less metabolically active carp.

Another case in point is the discovery in 1978 (Guppy and Hochachka, 1978) that tuna white muscle has the highest lactate dehydrogenase activity found in nature (5700 units/g), which powers the rather phenomenal anaerobic bursts of *Euthynnus pelamis* (synonym with *K. pelamis*). This is still true today as the volume density of mitochondria within a scombrid muscle fiber is much higher than mammalian species and even superior to that found in hummingbirds (Dymowska et al., 2012). This enables scombrid species not only to swim at sustained speed of up to 5 body lengths/s but, to accelerate to 20 body lengths/s (40 mph) and maintain that speed for up to 8 min, with 30 s rests between intervals (Guppy and Hochachka, 1978). Because several of the water-soluble vitamins have primarily coenzyme functions (vitamin B complex) in metabolic pathways and other water-soluble (choline, inositol, and vitamin C) and lipid-soluble vitamins (A, D, E, and K) have functions other than coenzymes, but are all extremely important for the intermediary metabolism of vertebrates including fish, it follows that the nutrient requirements for vitamins in tuna species may be higher than those of all other teleost fish.

The notion of a potentially higher vitamin requirement for scombrid species appears to be corroborated by European research under the SELFDOTT initiative, which analyzed the concentration of vitamins C and E in muscle and liver tissues in juvenile, wild-caught ABFT (800 ± 133 g; 34 ± 2 cm; $n = 30$). For vitamin C, muscle and hepatic tissue vitamin C contents (in μg/g DW) were 51 and 632, respectively; whereas vitamin E contents in the same tissues were 1.3 and 180, respectively. These levels were found to be significantly higher than those obtained from other marine organisms including squid (*Sepia elegans*), mackerel (*Trachurus mediterraneus*), sea bream (*Sparus aurata*), and dentex (*Dentex dentex*) (De la Gandara, 2012).

The quantitative requirements for most of the vitamins have been established for several fish species including Atlantic salmon, Pacific salmon (*Oncorhynchus* spp.), rainbow trout (*Oncorhynchus mykiss*), common carp, rohu (*Labeo rohita*), channel catfish (*Ictalurus punctatus*), and tilapia (*Oreochromis* spp.), while only some of the requirements are known for red sea bream (*Pagrus major*), hybrid striped bass (*Morone saxatilis* x *M. chrysops*), red drum (*Sciaenops ocellatus*), barramundi (*Lates calcarifer*), cobia (*Rachycentron canadum*), European sea bass (*Dicentrarchus labrax*), Japanese flounder (*Paralichths olivaceus*), grouper (*Epinephelus* spp.), and yellowtail (*Seriola* spp.) (NRC, 2011). To our knowledge, no vitamin requirement has been determined for tuna species except that of early juvenile PBFT (25 days post hatching) for vitamin C in a rather short feeding trial. The nutritional requirement for vitamin C in this trial was estimated to be 1200 mg/kg of diet of L-ascorbyl-2-monophosphate-Mg which is equivalent to 454 mg/kg of ascorbic acid (Biswas et al., 2013).

Although feeding trial studies oriented at defining optimal levels for other nutrients in scombrid fish have used vitamin premixes recommended for salmon species (e.g., Biswas et al., 2009a), Table 12.5 takes into account the above-mentioned physiological considerations and may more closely approach the vitamin requirements of metabolically demanding scombrid species without incurring toxicity. That is, these estimated requirements may suffice until studies are conducted to define these requirements for optimal growth, health, and enhancement of defense mechanisms.

12.5.2.6 Energy Utilization

The successful culture of tuna species depends in part upon the provision of diets containing adequate levels of energy, as well as an appropriate balance of nutrients to support optimal growth and to maintain health.

Bioenergetics—the study of the balance between dietary energy supply, metabolic expenditure, and gain—faces unique challenges in tuna research. In theory, the definition of metabolic requirements for energy in scombrid fish should be as simple as that for all living organisms: the energy available for growth is equal to the feed intake energy minus all metabolic costs, including heat (NRC, 2011). In practice, however, the approaches utilized to define energy requirements are exceedingly difficult to apply to large size and very mobile tuna species, either in the wild or in oceanic net pens.

TABLE 12.5 Suggested Vitamin Levels for Scombrid Species

Vitamins	Suggested Level	Units
Water-Soluble Vitamins		
Thiamin	20	mg/kg
Riboflavin	18	mg/kg
Vitamin B_6	30	mg/kg
Pantothenic acid	60	mg/kg
Niacin	230	mg/kg
Biotin	2	mg/kg
Vitamin B_{12}	0.04	mg/kg
Folic acid	3	mg/kg
Choline	2000	mg/kg
Myo-inositol	600	mg/kg
Ascorbic acid	454	mg/kg
Fat-Soluble Vitamins		
Retinyl palmitate (A)	3.5	mg/kg
Cholecalciferol (D)	60	μg/kg
α-Tocopherol (E)	150	mg/kg
Menadione (K)	20	mg/kg

Similarly, methodologies for assessing the energy cost of growth in fish have been long available (e.g., Ege and Krogh, 1914) and perfected over time such that comparisons of dietary energy intake with fecal and other waste outputs, together with tissue deposition yield fairly reasonable estimates (Einen and Roem, 1997). However, these studies require sufficient animals (statistical power) of the right size (~5 to 1000 g) to thrive over an extended period of time (over 10 weeks) to arrive at accurate estimates. This has not been the case for scombrid species even in Japan where the short duration of the nutritional trials (e.g., Ohnishi et al., 2014) conducted thus far and poor survival have been the chief limitations.

Likewise, traditional respirometry techniques, which are often used to determine bioenergetic requirements of small size fish, are exceedingly difficult to conduct with these ram ventilators which must swim continuously (Graham and Laurs, 1982), and no institution has to date suitable respirometry chambers to provide adequate husbandry conditions for animals larger than 20 kg, which is the target capture size for traditional tuna ranching. Notwithstanding, there are some very creative efforts using large, sea-based, polypropylene respirometry chambers to overcome the difficulties of working with large pelagic fish (Fitzgibbon et al., 2007, 2008).

Shipboard research has also provided some respirometry data on tuna specimens soon after capture with the limitation that these attempts are often unable to differentiate between metabolic rates and the stress of capture itself (Estess et al., 2014). To add complexity to the determination of bioenergetic requirements for tuna, each cultured species often differs in mobility, cardiovascular function, thermal strata, and endothermic capacity (Blank et al., 2004), and body size has profound effects on their ability to thermo-regulate (Graham and Dickson, 2001; Kitagawa et al., 2010).

The Tuna Research and Conservation Center (TRCC) in Pacific Grove, California, was the first institution to measure PBFT and YFT routine metabolic rates in a swim tunnel at a temperature of 20°C (Blank et al., 2007a,b). Oxygen consumption of PBFT (7.1–9.4 kg) swimming at 0.9 or 1.8 body lengths/s ranged from 235 to 498 mg O_2/kg/h, respectively. For other marine carnivorous fish (salmonids), it has been determined that that the consumption of 1 g of oxygen leads to a release of 13.6 kJ of energy (Cho and Bureau, 1995).

As an approximation, one may choose to use this factor to compute the amount of energy a 7.1–9.4 kg PBFT would need to derive from dietary components (3.2–6.8 kJ/kg of body weight/h) to support swimming, maintain internal optimal temperature, and perform all other metabolically vital functions including growth. Expressed for a 24 h period (76.7–162.6 kJ/kg/day), these energetic needs are of similar magnitude as the one estimated by Glencross et al. (2002b) for adult SBFT in net pens at 15°C and under the assumption that growth had ceased (150 kJ/kg/day). These values are also similar to the maximum energetic regime of 176 kJ/kg/day allocated for both YFT and PBFT at TRCC and considered optimal (Estess et al., 2014). The reader should be mindful that these are only approximations and data derived from restrained (in tanks or in a respirometry chamber) fish or fish forced to swim (as in a swim tunnel) may not always be representative of fish under culture conditions (net pens).

The previous estimations are useful, however, in highlighting the fact that tuna species have increased metabolic requirements as compared to other fish (Swimmer et al., 2004), and thus their energy needs are also peculiarly high (Korsmeyer and Dewar, 2001; Ohnishi et al., 2014). Maximum energy requirements to gain 1 kg of mass (in MJ/kg) for Atlantic salmon (26.8), rainbow trout (26.8), European sea bass (*Dicentrarchus labrax*, 26.2), and Asian sea bass (*Lates calcifer*, 27.8) (NRC, 2011) are but a fraction of those calculated for tuna species –56 MJ of gross energy (intake energy, at 16°C; Glencross et al., 2002a)—although most estimates are only valid for the specific set of conditions that generated them, they do provide an opportunity to understand the striking energetic differences between tuna and other teleost fish. These differences are consistent with physiological (Korsmeyer et al., 1996; Swimmer et al., 2004) and modeling evidence (Glencross et al., 1999b).

Wild tuna may also have different metabolic needs as compared to those kept captive. When a free-ranging tuna dives deep or when it cruises through a low-temperature zone, muscular work provides metabolic heat that, while resisting the thermal gradient, places an enormous energetic demand on these animals (Carey et al., 1984). Although the preferred temperature for cultured scombrid species ranges from 18°C to 25°C, they can endure temperatures as low as 3°C at 2000 m deep or briefly as high as 30°C, when spawning (Patterson et al., 2008).

As for the partitioning of this energy, from an initial 2900 kJ/kg/day gross energy intake, juvenile (39 d old, 5 cm fork length and 1.1 g average body weight) PBFT held at 26.5°C lose about 23% of this energy in fecal and branchial losses; whereas standard metabolism, retained energy, and voluntary activity account for 41%, 21%, and 15%, respectively (Ohnishi et al., 2014). These values are in contrast with those obtained at 20°C by Estess et al. (2014) for the same species but of larger size (11.4 kg), which indicate that only 12.4% (as opposed to 21%) is converted into tissue or retained. These differences are largely due to the significant size differences, different caloric intake, levels of activity, and environmental temperature to which the animals were subjected. Despite these differences, the efficiency of energy retention of tuna species appears to be much lower than for other species of cultured fish. Therefore, much effort is necessary to improve formulations and implement feeding strategies such that energy dissipation can be reduced in tuna aquaculture.

12.5.2.7 Digestibility and Availability

The terms "digestibility" and "availability" both refer to the digestion of complex feedstuffs so that nutrients contained therein are made available for uptake by the body. Those components not absorbed are excreted in feces. The nutrients that are digested presumably contribute to growth and metabolism (NRC, 2011). Digestibility studies examine the digestion and absorption of nutrients of a particular species, providing central knowledge in aquaculture for the accurate determination of the nutritional requirements of that species. These studies involve a difficult, often impossible task for many aquatic species: the collection of fecal matter with minimal leaching (Bureau et al., 1999). Additionally, these studies, if they can be carried out, need to hold the experimental fish in tanks which is not an option for some species, be it because of their pelagic behavior or their physical size. As an alternative, an *in vitro* methodology that includes physiological digestion and absorption processes can be implemented and thus provide the necessary digestibility information for the accurate development of compound feeds (Martínez-Montaño et al., 2010).

Using three enzyme systems based on purified commercial, Atlantic salmon or SBFT digestive enzymes, Carter et al. (1999) compared *in vitro* crude protein and dry matter digestibility of ingredients and feeds, finding that CP digestibility was higher than dry matter digestibility and values were highest using commercial enzymes than using crude extracts from Atlantic salmon and lowest using crude extracts from SBF. In this study, Atlantic salmon was used as a surrogate species to compare apparent *in vivo* and *in vitro* digestibility values finding positive and significant correlations ($P < 0.001$) between animal-based protein meals for all the enzyme systems.

Also, Aguado-Gimenez et al. (2004) estimated the digestibility of total nitrogen and phosphorus, when these elements produce most of the organic wastes with environmental impact, and to compare two fecal collection methods in studies of the utilization of nutrients by ABFT. In this study, the digestibility of total nitrogen and phosphorus was studied under common fattening conditions for the purpose of determining nutrient utilization and generating tools to estimate waste output. Animals were fed *ad libitum* once a day with a mixture of mackerel, herring, pilchard, and gilt-sardine. Sampled tuna weighed between 150 and 350 kg. Feces were collected by scuba divers in the cages by dissecting the distal intestine with a manual sieve after slaughtering. Acid insoluble ash was used as the endogenous inert marker.

Nutrient utilization and/or waste production in cultured species have been widely studied. However, only a few studies exist on scombrid fish (Bruce, 1997; Carter et al., 1999; Aguado-Gimenez et al., 2004; Martínez-Montaño et al., 2010). The development of feeds for large scombrids, although an important mariculture species, has been limited due to the lack of detailed information on their nutrition. Pelleted feeds (moist) have been manufactured but at a slow rate. The difficulty stems from the fact that trials on feeding are difficult, time-consuming, and very expensive. A strategy to accelerate the manufacturing of viable formulated feed for tuna has been the evaluation of *in vitro* digestibility of ingredients and nutrients. However, *in vitro* studies provide preliminary approximations at best and the need persists to screen digestibility coefficients of feed ingredients in tuna diets. Also, long-term feeding trials are still necessary to be able to develop viable formulated diets for tuna. Although this is the most dependable method, feeding trials with tuna are difficult, require a lot of time, and are costly as only a few variables can be tested during each growing cycle (Carter et al., 1999).

A comparison of apparent digestibility coefficients (ADC) for total-N using two types of collection methods: dissection of the intestine and direct collection yielded the following results:

- The ADC was lower when feces were collected by dissection of the intestine (DI, 65.8%) than when by direct collection (DC, 93.7%).
- Total-P ADC was lower when feces were directly collected (DC: 48.1%) than when collected by DI (66.1%).

Tuna showed nitrogen digestibility similarly high to that of other carnivorous fish species. Phosphorus digestibility, on the other hand, was low, just as in many other fishes, due to, it seems, dietary phosphorus excess. DC results appear more consistent and appropriate for the estimation of waste output. The need for a more accurate nutritional assessment and formulated artificial diets was demonstrated.

Martínez-Montaño et al. (2010) evaluated AA *in vitro* in the proximal intestinal region of the PBFT. To do so, an AA pool from fresh sardine hydrolyzed with a pancreatic enzyme extract from the PBFT was used. In this study, a comparison was made of different AAs to determine their absorption rates. The everted intestine technique with fresh proximal intestinal sections from the PBFT was used for that purpose. Hydrolyzed sardine muscle was obtained using porcine pepsin first and an alkaline pancreatic PBFT extract as a second step. The hydrolyzed extract was tested with samples at three concentrations (0.14, 0.09, and 0.06 mg of soluble protein/mL). Assimilation of the nonessential amino acids (NEAAs; mainly Ala, Glu, and Gly) was higher than that of the essential amino acids (EAAs). Certain AAs, including Arg and Lys, were observed to have more affinity with the cationic transporters than did His. The glucogenic AA Glu always displayed double the absorption rate than that of Asp, probably due to an energy requirement of the cells. The highest absorption rates were exhibited by the neutral AAs, Ala and Gly, at all concentrations.

12.5.3 Practical Diets

The future of tuna aquaculture will depend on how well it meets its specific challenges. Among them, growing tuna species on diets that are not only nutritionally adequate for these unique species but also economically feasible, sustainable, and environmentally responsible is of paramount importance.

12.5.4 Ingredients

The advent of nutrition research for scombrid species arrives at a time during which there is a wider range of alterative ingredients to FM than what has been prevalent before. Many of these ingredients are more complex than FM and require thorough evaluations in order to determine their nutritional value and appropriate use levels in prospective tuna diets. In the evaluation of specific ingredients, ideally, the science of nutrition should endeavor to gain knowledge on the nutritional implications of use of those ingredients and, once this knowledge is gained then to seek applications to commercial feed formulations.

Today, economic and ecological sustainability considerations are creating a significant motivation to reduce levels of FM in marine fish feeds. Many studies are carried out each year to test the effect of incorporating various protein sources in feeds for different aquaculture species including tuna. These studies encompass a great diversity of ingredients, diet formulations, and experimental designs. For some fish, a number of studies have shown that a large proportion of FM could be replaced by plant ingredients without a deleterious effect on growth rate and feed efficiency provided the high-quality ingredients are used (Hua and Bureau, 2012).

Following this global trend, a study was conducted at the Fisheries Laboratory of Kinki University, Wakayama, Japan, to investigate whether enzyme-treated FM (EFM) could be partially replaced by soybean meal (SBM) or enzyme-treated SBM (ESBM) without compromising the growth of laboratory-raised PBFT juveniles (Biswas et al., 2011). Authors concluded that an inclusion level of 10% SBM in substitution of EFM was possible without adversely affecting growth performance in juvenile PBFT. Enzyme-treating SBM did not improve the outcome and was actually slightly detrimental to fish. As this study lasted only 12 days and the etiology of SBM-induced enteritis has been documented to take longer (Baeverfjord and Krogdahl, 1996; Urán-Carmona, 2008) in other fish species (e.g., Atlantic salmon) even when SBM incorporated at 33% of diet, this assertion needs to be verified in longer-term trials.

Similarly, because early juvenile fish typically do not have a well-developed immune system and some of the allergic reactions to SBM do not become manifest until Atlantic salmon is adapted to sea water (100 g) (Sahlmann et al., 2015); the same may be true for PBFT and thus the above-mentioned study needs to be repeated with larger-size PBFT and for a considerably longer period (~10–15 weeks).

While research on plant protein sources as FM alternatives has gained considerable interest in recent years (Gatlin et al., 2007; Barrows et al., 2008), possible inclusion levels are often limited by species-specific digestive limitations and by the presence of anti-nutritional factors (ANFs; Francis et al., 2001) in plant-derived proteins. Harsh processing methods and solvent extraction are commonly used to reduce ANFs-related negative effects. However, the use of extreme heat is known to elicit detrimental changes in the processed products. These changes may negatively affect nutritional and organoleptic attributes which may render the new materials unacceptable to scombrid species. Several vitamins quickly degrade under heat treatments as do color and flavor compounds. Texture is also negatively affected; frequently vegetable tissues soften and chemical compounds need to be added to regain firmness, binding power, and structure (San Martin et al., 2002). All these changes result in products that are far from similar to the original fresh or unprocessed products.

Alternatively, relatively new methods of removal of antigenic materials such as fermentation or other microbial-based processes together with breeding programs oriented at reducing ANF may prove fundamental in improving the nutritive value of vegetable proteins without denaturing desirable elements in these ingredients. As an example, a vast array of products can be derived from soybeans and are found nowadays in more than 20,000 items in the food shelves of super markets worldwide.

12.5.5 Feeding Stimulants

Ji et al. (2007) conducted an experiment to identify substances in horse mackerel, *Trachurus japonicas*, one of the main forage fish for PBFT farming in Japan that may stimulate feeding of juvenile PBFT. A casein-based diet devoid of these substances was supplemented with mackerel extracts separated in three different fractions, namely, amino acids, nucleotides, and organic nitrogenous bases. The diet containing all three fractions resulted in a significant stimulatory feeding effect in PBFT compared to fish fed the diet containing only organic nitrogenous bases. However, the dietary exclusion of nucleotides and/or amino acid fractions also resulted in lower PBFT feeding activity, indicating that these fractions—but not the organic nitrogenous base—contain feeding stimulants that positively affect juvenile PBFT feeding behavior. Further characterization of the nucleotide and amino acid profiles of each fraction revealed that major components were inosine-5′ monophosphate (IMP) and histidine, alanine, taurine, lysine, and glutamate, respectively. In addition, histidine, glutamate, and IMP synergized to produce the strongest feeding stimulant activity than any other possible combinations of amino acids and nucleotides. Further studies by Haga et al. (2011) corroborated the strong stimulatory action excreted by IMP upon larval PBFT feeding activity when presented with a casein peptide-based micro diet. Non-scombrid species such as yellowtail, *Seriola quinqueradiata*, horse mackerel, and marbled rockfish, *Sebastes marmoratus*, also experience enhanced feeding activity when presented with diets containing this nucleotide (Hosokawa et al., 2001; Ikeda et al., 1988; Takaoka et al., 1990).

The results obtained by Ji et al. (2007) are in contrast to those reported by Kohbara et al. (2006) which document that only IMP, but not histidine or glutamate, stimulates the gustatory response of juvenile PBFT in an electrophysiological study. This contradiction may be partly related to the fact that the electrophysiological study did not test potential synergies between IMP and histidine or glutamate. It also should be pointed out that the electrophysiological study was not designed to distinguish between substances which stimulate feeding and those which inhibit it.

The identification of attractant compounds that stimulate feeding activity in fish is compounded by the source of natural or synthetic extracts. For example, the nucleotide fraction demonstrated a strong enhancement of feeding activity in yellowtail, *S. quinqueradiata*, when a synthetic extract from horse mackerel was used (Ji et al., 2007); whereas amino acid fractions have a similar effect when using a synthetic extract from krill (Hosokawa et al., unpublished data).

The identification of histidine as feeding stimulant for PBFT is of interest because this amino acid also stimulates feeding behavior in prominent fresh water species such as rainbow trout, *Oncorhynchus mykiss*, and Japanese eel, *Anguilla japonica*, but it had not been reported as feeding stimulant for marine fish species (Takii and Ji, 2005). Histidine is an essential amino acid present in high concentration in the scombrid muscle tissue and has been associated with a possible buffering mechanism which helps maintain pH homeostasis in tuna (Suyama et al., 1986). The higher concentration of histidine in tuna muscle is suggestive of an elevated dietary requirement which may help explain in part the preference of PBFT for feeds supplemented with this amino acid.

One of the main objectives of tuna farmers is the efficient conversion of either bait or compounded feeds into live fish, with minimum waste, and clearly any way of increasing the acceptability or palatability of a fish diet is advantageous. In the case of PBFT, IMP-supplemented micro diets significantly improve larval survival and growth (Haga et al., 2011), improving the outcome of larval rearing efforts in this challenging species. The rationale of supplementing feeding stimulants also may be expanded as these compounds enhance acceptability of diets in which FM has been replaced with plant protein sources (Papatryphon and Soares, 2000). However, cost-effectiveness and other practical issues including those of regulatory nature need to be considered when evaluating the possible inclusion of particular stimulants. For example, in Japan crystalline glutamate is listed as a feed additive which can be commercially used in fish feeds but histidine and IMP are not. As an alternative, the utilization of natural ingredients which also stimulate feeding activity in fish may prove useful in developing compound tuna feeds with high palatability. Squid meal is one such ingredient which supports optimal growth when incorporated in diets for juvenile PBFT (Seoka et al., 2010). A research group in Japan has recently reported that the addition of an extract made from squid by-products to a compounded diet improves feeding actively in juvenile PBFT. Thus, the identification of natural ingredients and/or synthetic additives which promote feeding activity in tuna species appears to be a promising field for nutritional research.

12.5.6 Substitution of Fishmeal with Plant Protein Sources

Although many attempts have been made to develop moist feeds for scombrids that could support similar performance as that obtained for bait-fed tuna, today there is only limited success attained as described in Section 12.5.8. Most of these efforts have been based on muscle and/or natural prey species compositions resulting in a high-protein, high-energy, low-carbohydrate diet as befits a carnivore and opportunistic predator.

Previous studies with PBFT have demonstrated that while there are no major problems with the amino acid balance of normal Chilean FM, the digestibility of nutrients and energy from this protein source are low in early juvenile PBFT (Biswas et al., 2009a). In an effort to develop a suitable artificial diet for early juvenile stages of this species, recent studies discussed by Mourente and Tocher (2009) have been conducted which indicate the usefulness of utilizing enzyme-treated FM in formulated diets for tuna. The original study (Ji et al., 2008) estimated the availability of some enzyme-treated FM, in place of conventional FM, as a main dietary protein source for juvenile PBFT. Only a diet utilizing enzyme-treated Chilean FM was able to sustain similar growth and other performance indicators to those PBFT juveniles fed with the surrogate fish sand-lance *Ammodytes personatus*. Although these results identified enzyme-treated Chilean FM as a suitable dietary protein source for PBFT, animals fed with this experimental diet had higher carcass lipid content and lower accumulation of Ω-3 LC-PUFA than the sand-lance-fed fish (Mourente and Tocher, 2009).

There are few studies on FM replacement in diets for scombrids. However, the digestibility of alternative proteins has been investigated using a soft, dry pelleted diet in the yellowtail, a marine pelagic species, obtaining digestibility values of 86% in diets containing from 30% to 50% soybean meal (Watanabe et al., 1992; Watanabe, 2002). The study conducted on PBFT by Biswas et al. (2011) was aimed at investigating whether the expensive enzyme-treated FM (EFM) can be replaced partially by normal FM and enzyme-treated soybean meal (ESBM) or normal soybean meal (SBM) in an attempt to establish low cost diet for PBFT juveniles without compromising growth. The results demonstrated that 20% SM (26% EFM replacement) can be included in the diet of PBT juveniles without compromising growth performance at least under the experimental design used in this study.

12.5.7 Larval Diets

Basic food items in tuna larval production are almost the same as in larval production of other marine fish species including algae, rotifers, *Artemia*, and artificial diets. However, in tuna marine fish yolk-sac larvae have been widely used after the rotifer and *Artemia* feeding periods due to the strong piscivorous character of tuna larvae and difficulty of weaning tuna larvae to an artificial diet. Positive progress has recently been made in developing artificial diets for tuna larval production in Japan and Europe (Biswas, 2010; SELFDOTT, http://www.repositorio.ieo.es/e-ieo/handle/10508/1119; TRANSDOTT, http://www.transdott.eu/transdott/).

12.5.7.1 Embryogenesis and Yolk-Sac Stage Larvae

Tuna larvae hatch in 24–36 h after fertilization depending on water temperature (Miyashita, 2002). As with other fish species, they rely on endogenous nutrients until the digestive tract from mouth to anus completely opens and the fish can start to ingest the first exogenous nutrients. A remarkable decrease in free AA (FAA) have been observed in eggs and yolk-sac larvae of PBFT with embryonic development, which suggests an important role of FAA as sources for energy production and protein synthesis to support the early development of PBFT (Takii et al., 1997). Triacylglycerol (TAG) was also suggested to be another important energy source for the development of PBFT eggs (Seoka et al., 1998) but a different and independent trial suggested only a minor role of this lipid class as an energy source for egg development (Seoka, unpublished data); whereas after hatching PBFT larvae probably consume TAG as shown in ABFT larvae (Morais et al., 2011). During the endogenous nutritional stage of ABFT, the level of DHA is maintained but the proportion of EPA declines coinciding with a great increase in gene expression of Δ6 fatty acyl desaturase and elongase 5, which could both be involved in the conversion of EPA to DHA (Morais et al., 2011). Contents of protein and phospholipids during the embryonic and larval development of PBFT and ABFT remain relatively constant or increase slightly, so these nutrients appear to be utilized for formation of the embryonic and larval body (Takii et al., 1997; Morais et al., 2011).

12.5.7.2 First Feeding and Early Digestive Capabilities

Nutritionally enriched rotifers are normally the first diet used in tuna larval production, as is the case in the larval production of other marine fish (Miyashita, 2002). Larval PBFT begin ingesting rotifers after opening of the mouth and anus which usually occurs 2.5–3 days after hatching (Miyashita, 2002). Differentiation of the liver, pancreas, and intestine starts before the first ingestion of rotifers in PBFT yolk-sac larvae, which indicates that the larvae are ready for first feeding even at the endogenous stage (Miyashita et al., 1998). During the flexion phase (11 to 17 dph), the functional jaw teeth can be found and the blind sac, gastric glands, and pyloric ceca begin to form in PBFT larvae (Miyashita et al., 1998). Trypsin- and pepsin-like digestive enzyme activities increase remarkably with the formation of the gastric glands and pyloric ceca (Miyashita et al., 1998).

This dramatic formation and activation of the teeth, digestive organs, and enzymes indicate improved ability to digest more complex food items. Therefore, it is suggested that the phase from flexion to the early post flexion is the right time to switch from rotifer to other feed items such as *Artemia*, fish yolk-sac larvae, and/or artificial diet in PBFT larvae. In fact, feeding of enriched *Artemia* and then yolk-sac larvae of other marine fish starts in PBFT larval production because tuna show preference for larger food items and display a stronger piscivorous behavior starting at this phase (Miyashita, 2002). Similar findings have been obtained for YFT (Kaji et al., 1996). High activity of digestive enzymes such as chymotrypsin, aminopeptidase, amylase in larval YFT at this developmental stage would be in agreement with the current understanding on evolutionary adaptations of YFT larvae which exhibit early piscivory and a growth rate remarkably faster than most other marine fish (Buentello et al., 2011).

12.5.7.3 Live Prey

Wild PBFT larvae catch copepods as live prey at the early growth stage (Uotani et al., 1990); and attempts to utilize copepods in tuna larval production have been made in Japan and Europe. However, most protocols apply a typical rotifer-*Artemia* feeding requiem also for tuna until they start requiring yolk-sac larvae of other marine fish (Miyashita, 2002; Masuma et al., 2008).

The lorica length (LL) of rotifers positively selected by PBFT larvae ranges from 100 μm to 239 μm from 3 to 7 dph, and the lorica width (LW) ranges from 80 to 179 μm, but within this size range there is a shift toward selection of rotifers with larger sizes of LL and LW correlated to time after hatching (Sawada et al., 2000). Rotifer density at the commencement of feeding is in general around 5–10 individuals/mL of rearing water, which is similar or somewhat higher than in other marine fish larval production, and the density is usually maintained until the onset of *Artemia*-feeding (Miyashita, 2002). Feeding S-type rotifer or a combination feeding of S- and L-type rotifers is typical in tuna larval production. The rotifers are commonly enriched nutritionally using commercial product(s) in order to improve contents of essential fatty acids and micro-nutrients (Masuma et al., 2008). No study has been reported on the nutritional requirements, including EPA and DHA requirements, of larval tuna during the rotifer feeding period. However, examples of nutrient profiles in enriched rotifers and enriched *Artemia* as well as yolk-sack larvae used in a commercial PBFT larval production can be found in pertinent literature (Seoka et al., 2007a).

Artemia are very common live feeds in larval production of most marine fish species including tuna larval production. However, continuous feeding of *Artemia* leads to stunted growth and poor survival of PBFT, possibly also in other tuna, even if *Artemia* are nutritionally enriched before feeding (Seoka et al., 2007b). Therefore, enriched *Artemia* are used mainly to reduce fish size variations in PBFT larval production, and the feeding period is very short, only for 3–4 days, to avoid growth failure (Miyashita, 2002; Seoka et al., 2007b). The reason for this negative impact of *Artemia* on larval growth remains largely unknown but appears to be of nutritional nature. Although DHA enrichment of *Artemia* significantly improves larval growth and survival of PBFT, this improvement is negligible when compared with the excellent growth of larvae fed yolk-sac larvae of surrogate fish species (Seoka et al., 2007b). No remarkable differences have been found in proximate composition between *Artemia* and yolk-sac larvae except for a higher content of carbohydrate and lower contents of phosphatidylcholine and sterol ester in enriched *Artemia* (Seoka et al., 2007a). In addition, enriched *Artemia* can be characterized by a lower DHA:EPA ratio and lower DHA level in phospholipids compared with yolk-sac larvae (Seoka et al., 2007a,,b). Also, protein content in lipid-enriched *Artemia* may become sub-optimal for good growth. Such nutritional inferiority of enriched *Artemia* to yolk-sac larvae may partly be responsible for the growth failure of PBFT larvae fed enriched *Artemia*. Alternatively, the relatively high contents of shell proteins including chitin, arthropodin, sclerotin, and other hard-to-digest proteins in the *Artemia* exoskeleton (Gagne and Blaise, 2004) may explain in part the limited nutritional value of *Artemia* as compared to yolk-sac larvae of marine fish.

Yolk-sac larvae of marine fish species have been recognized as an ideal live prey after the rotifer-*Artemia* feeding period. Yolk-sac larvae of Japanese knife jaw *Oplegnathus fasciatus* (Miyashita, 2002), spangled emperor *Lethrinus nebulosus* (Masuma et al., 2010), and Japanese herring *Sillago japonica* (no reference available) are typical in PBFT larval production in Japan while yolk-sac larvae of gilthead seabream *Sparus aurata* are used in the Mediterranean for ABFT production (SELFDOTT, http://www.repositorio.ieo.es/e-ieo/handle/10508/1119). These fish species have been selected as "yolk-sac larvae suppliers" because they spawn a large amount of eggs to produce the yolk-sac larvae, continuously. Also, yolk-sac larvae of the tuna species being produced can be given when the tuna brood fish gives a surplus of eggs and/or there are no other yolk-sac larvae available in the hatchery. The nutritional profile of the yolk-sac larvae changes significantly with development through active metabolism of endogenous nutrients and it is also influenced by the nutritional status of the brood fish. It is, therefore, complicated to make a nutritional comparison among yolk-sac larvae of different species. However, yolk-sac larvae of the above-mentioned species share some biological characteristics such as hatching from buoyant eggs, similarly sized (2–3 mm in total length), having a single oil droplet in the yolk, and the first feeding occurring a few dph. This type of yolk-sac larvae has a similar nutrient profile and can be characterized by high protein, free amino acid, and lipid contents, but low carbohydrate content, as shown in previous studies (Rønnestad et al., 1999, 2003). Furthermore, these yolk-sac larvae contain the necessary nutrients for their own development and very importantly, these nutrients are present in a water-soluble form; that is, lipoproteins (Ohkubo and Matsubara, 2002). Digestion and absorption of ingested lipoproteins appear to be easy even in tuna larvae not yet possessing a fully developed digestive capacity.

Although much improvement is still needed, high quality protein and lipid sources have been incorporated into balanced feeds for early juvenile PBFT with some degree of success (Ji et al., 2008; Biswas et al., 2009a,b).

12.5.8 Feeding Juvenile and Grow-Out Stages

Feeding experiments with juvenile ABFT (~2.8 kg) were conducted at the Spanish Institute of Oceanography, Mazarron, Spain in 2010 (De la Gandara,

2012). Following an unsuccessful attempt in 2009, researchers consulted with personnel affiliated with Skretting – Australia and the Hellenic Centre for Marine Research (Aghios Kosmas) and modified the so-called Aurora diet. Extruded and dried pellets were formulated to contain 50% crude protein, 17% lipid, and an undisclosed supplementation level of vitamins with special emphasis on vitamin C. Pellets were produced at Skretting ARC (Feed Technology Plant) in Stavanger, Norway, extruding the raw ingredients through a 2-mm die and were 2 mm in length with a 50:50 mixture of floating and sinking pellets to try to emulate sinking behavior of thawed baitfish. There were ~ 40, 2.8-kg fish in each of two oceanic net pens. Pellets were offered at ~6% BW/day or to apparent satiation in a single daily meal. To entice fish to begin consuming dry extruded pellets a soft diet of minced sardinella (*S. aurita*) mixed with wheat gluten as binder (AquaSoft) was offered first and then gradually replaced by the extruded pellet. A number of flavor and color attractants were also tested in the hope to improve acceptance including Candurin® Silver Sparkle (pharmaceutical colorants), green- and yellow-colored pellets, paprika, chili powder, and Seagarden natural seafood powders (*Calanus*, blue muscle and shrimp). Pellets were also soaked in sea water in an effort to improve texture.

The 2010 feeding trial at the Spanish Institute of Oceanography also resulted in failure as fish did not consume the formulated diets or, if they did, they would immediately spit out the pellet particle. None of the attractants appeared to worked and juvenile ABFT fed only on sardinella, indicating that new tuna feeds need to be developed focusing more on texture and taste so that the critical starting point—feed consumption—is attained.

Similar experiences were reported in the commercial cultivation of YFT sub-adults (off the coast of La Paz, Mexico) when offered formulated extruded diets manufactured by two of the most important aquafeed producers in North America. A first approach tested a dry pellet intended as supplemental nutrition. This diet had an undisclosed over fortification of nutrients which are typically unavailable from thawed baitfish and was meant to be used in conjunction with baitfish/squid feeding. However, ~30 kg YFT accurately discriminated the pelleted diet which went to waste/pollution (Morris, 2006, personal communication). A second strategy used a complete formulation for salmon which was intended as total replacement of baitfish feeding.

The thought was that, if baitfish feeding was discontinued long enough, starving fish would be persuaded into consumption of the artificial diet. Indeed, fish went with no food as long as they could possibly withstand it (~20 days) only to resemble tuna specters unsuited for any market. As tuna ranching is conducted in oceanic net pens, schools of fish sporadically swim through the large-mesh netting in which case starved YFT would go into a feeding frenzy, colliding with the net as they tried to obtain some nourishment from the passing baitfish school. This again rendered the product unsuitable for the Japanese market which does not pay for scratched fish not to mention low fat content and malnourished. The second aquafeed manufacturer provided a quadrangular pouch-pellet, ~1.5 inches per side with the instructions of soaking the pellet for 30 min before making it available to the fish. Results were fairly similar.

Only when the in-house nutritionist prepared a soft-textured, moist diet did feed consumption resume at a similar rate to that in baitfish-fed net pens. This proprietary formulation provided both flavor and texture characteristics amenable to wild-caught YFT accustomed to the consumption of live prey items. This moist diet was sufficiently nutrient-dense such that with a daily feeding rate of only 3–4% adequately fulfilled the nutritional requirements of YFT grown in a sub-tropical environment. Nutrient density and availability appears to be a crucial aspect of tuna nutrition as according to their age, the average tuna stomach capacity is somewhat low compared to other species, at around 0.3% of BW. That is, for a 1-m long YFT, which weighs approximately 20 kg, the stomach contents will be about 60 g, or a little larger than a hen's egg. In addition, tuna have relatively shorter retention times in the gastrointestinal tract as compared to other teleost species and, although free ranging scombrids are able to improve digestion efficiency by warming up their visceral cavity (Gunn et al., 2001), this is of little use when confined in net pens, and the shorter retention time of the intestinal chyme requires increased density and availability of nutrients.

Despite the comparatively better acceptance, manufacturing tuna moist diets at a scale (e.g., commercial) that can satisfy the needs of a typical size tuna ranch (30–50 metric tons/day) is impractical and requires on-site cold storage capabilities which are seldom present in tuna ranching operations. Typically, these moist pellets are composed of minced bait fish, FM, fish oil, binders, and vitamins (Table 12.6) and produced on small-scale experimental equipment and often cannot be reproduced on a commercial scale.

Therefore the challenge today for a formulated tuna diet is to have good acceptance and mouth feel, be nutrient-dense enough so as to produce superior growth and condition with reduced mortality, and most importantly, produce fish that are well accepted by the Japanese market. An acceptable shelf-life and ease of handling are also essential. In Australia and Japan, a few companies including Pivot Aquaculture (Gibsons), Ridley Aqua-Feed Pty. Ltd., Skretting, Maruha Nichiro Group, and Hayashikane Sangyo Feed Company have undertaken this challenge with limited success on SBFT and PBFT. At the tuna aquaculture session of the 2014 World Aquaculture Meeting in Adelaide Australia (Jun 8–12, 2014), a representative from the company Nissui (T. Goto, Hachioji, Tokyo, Japan) indicated that significant advances had been made in terms of color, smell, texture, and shape, describing a double concentric sausage PBFT feed with 38% CP, 28% CL, and 9% moisture with a relatively high caloric content (4.2 kcal/g)

TABLE 12.6 Typical Composition of a Moist Pellet Diet

Ingredients	Control	1	2	3
Choline chloride	–	0.2	0.2	0.2
Lecithin	–	1.0	1.0	1.0
Vitamins & mineral premix	–	0.3	0.3	0.3
Stay-C	–	0.5	0.5	0.5
Color enhancer	–	0.001	0.001	0.001
Squid oil	–	9.0	9.3	3.4
Wheat gluten	–	10.0	15.0	15.0
Krill meal	–	2.5	2.5	2.5
Chilean fishmeal 67% CP	–	27.0	28.5	35.3
Squid meal	–	2.5	2.5	2.5
Fresh sardines, *Sardinops sagax*	100.0	47.1	25.2	27.1
Water	–	0.0	15.0	12.2
Total	100.0	100.0	100.0	100.0
Composition				
Dry matter	25.1	32.3	32.4	63.3
Crude protein	17.8	38.9	38.9	43.9
Crude lipid	10.5	14.7	14.7	9.7

which had been tested in a cage system at the Nagasaki Prefecture with PBFT consuming the manufactured feed at levels similar to that of a control group fed mackerel.

12.5.9 Brood Fish Diets

Among fish nutrition topics, brood-stock nutrition is perhaps the area that has been less explored and that is certainly the case in scombrid research. There are some studies, however, that have demonstrated a positive effect of targeted nutrition on the reproductive performance of marine fish. The quality of eggs, sperm quality, and seed production are visibly improved when fish are fed high-quality nutrients essential for reproduction. It can be safely assumed that this is also the case for scombrid fish, but studies on broodstock nutrition for cultured scombrids are certainly needed to corroborate these assumptions.

As a starting point, seminal studies on broodstock nutrition can be used to establish a baseline for tuna broodstock nutrition. Among them, the works of Izquierdo et al. (2001), Watanabe and Kiron (1995), Watanabe et al. (1984a,b,c, 1991a,b), Fernandez-Palacios et al. (1995, 1997), Blom and Dabrowski (1995), Harris (1984), and Craik (1985) were used to summarize the effects of applied nutrition on reproductive outcomes (Table 12.7).

Several food items shown to be valuable for broodstock diets of other marine species can potentially be valuable for tuna as well. Among these, squid protein is reported to have beneficial effects on egg quality (Fernandez-Palacios et al., 1997). Broodstock fed squid protein-based diets develop eggs with higher protein levels and squid-fed female fish produce ~40% more eggs/kg than those fed FM-based diets. Also, the inclusion of paprika in brood fish diets has been reported to increase carotenoid levels in gonadal tissue which in turn is believed to improve the reproductive performance of broodstock. Within the REPRODOTT framework the addition of 2% paprika to a squid-mackerel diet improved all reproductive indices measured on male and female ABFT (Partridge, 2013).

In addition, valuable information can be drawn from efforts oriented at "closing the cycle" of several scombrid species around the world. For example in Japan the close-cycle cultivation of PBFT involves feeding laboratory produce animals with sand lance *Ammodytes personatus*. As juveniles grow diets are modified to incorporate anchovy *Engraulis japonicus*, sardine *Sardinops melanosticus*, jack mackerel *Trachurus japonicus*, mackerel scad *Decapterus tabl*, chub mackerel *Scomber japonicus*, spotted chub mackerel *S. australasicus*, and Japanese common squid *Todarodes pacificus*. Typically brood fish are fed a ration equivalent to 1–11% of their body weight (Masuma et al., 1991; Kumai, 1997; Miyashita et al., 2001; Sawada et al., 2005; Masuma, 2006; Masuma et al., 2010).

Working with YFT broodstock in Panama, Margulies et al. (2010) documented the use of squid and sardines at rates ranging from 1% to 5% depending on fish size and water temperature. Although there is a marked preference for sardines by YFT, the squid rations help improve egg quality in spawning females. A diet consisting of 50% squid and 50% fish seems to provide adequate nutrition for broodstock YFT (Margulies et al., 2010) that spawns almost continuously as long as water temperature remains above 23.3°C.

12.5.10 Nutritional Manipulation of Flesh Color

Tuna meat purchasing decisions are influenced by color more than any other quality factor because consumers use it as an indicator of freshness/wholesomeness. Although sardine/squid diets provide acceptable nutrition for fattening tuna, prolonged feeding with these items results in tuna meat that is whiter than market preference. Additionally, traditional sushi/sashimi

TABLE 12.7 Effects of Brood-Stock Nutrition Which May Apply to Scombrids

Affecting & Affected Factors	Izquierdo et al. (2001)	Watanabe and Kiron (1995), Watanabe et al. (1984a,b,c, 1991a,b)	Fernandez-Palacios et al. (1995, 1997)	Blom and Dabrowski (1995)	Harris (1984), Craik (1985)
Food restriction	Inhibition of gonadal maturation				
Fecundity	Determines egg quality. Reduced by nutrient imbalance on the brain-pituitary-gonad endocrine system. Importance of the fatty acid DHA in embryo and larvae development.	Significantly affected by dietary essential fatty acid content in diets.	Significantly increased with an increase in dietary *n*-3 HUFA levels up to 1.6% (present at high levels in tuna diets, comprised of pelagic fish and squid). Affected by vitamin E.	Affected by ascorbic acid.	
Fertilization	Affected by dietary eicosapentaenoic (EPA) and arachidonic (AA) levels.	Vitamins E and C provide protective role for sperm cells by reducing risk of lipid peroxidation.	Vitamin E, important nutrient.		Carotenoids, important nutrients.
Embryo development	Vitamin A, important requirement during gonadal maturation and spawning, due to important role in bone development, retina formation and differentiation of immune cells. Research needed on the dietary requirements for thiamin (vitamin B1) and pyridoxine (vitamin B6) (No information exists on these requirements for tuna).	Affected by level of protein intake (low protein-calorie diet, low reproductive performance).	Number of normal eggs increase with increase in n-3 HUFA levels in brood-stock diets and increase in the level of dietary a-tocopherol (from 22 to 125 mg/kg).		

Note: There is no specific information in the literature on the subjects of dietary nutrients, vitamins A, E, and C and requirements of protein and energy for good reproductive performance in tuna broodstock.

standards are not clearly defined and the muscle coloration of tuna can be scored subjectively by master chefs in Japan or elsewhere.

Being cognizant of the subjectivity in grading the quality of fish and of conditions that may adversely affect flesh color, growers eagerly await for clearly defined tuna grading standards. To attain this objective a research group at Texas A&M University sampled PBFT fetching top value at both the Tsukiji and Nagoya tuna auctions in Japan (Buentello, 2010, unpublished data). Color was assessed using a Konica Minolta Chroma Meter CR-410 which converts a subjective hue into an objective X-Y-Z tri-stimulus (red, green, and blue) chromaticity value, and conducted proximate composition analysis using established methodologies for moisture, lipid, crude protein, and ash (Webb and Gatlin, 2003). This objective assessment now allows for the exploration of the effects of vitamins E and C on meat color as well as iron and other dietary factors. In tuna, as in beef, the oxidative state of myoglobin is responsible for the bright cherry-red preferred color. Myoglobin concentrations increase with the animals' age; thus, the muscle from older fish tends to be of a darker red in appearance (Ochiai et al., 2010). Highly active muscles also have more myoglobin, for example, tuna loins. Other color-imparting substances such as carotenes, astaxantin, and canthaxantin that not only can prevent oxidation also emit light in the 650–700 nm visible spectrum thus affecting color.

Finally, because not only myoglobin concentration but its oxidative state dictates in a significant way the expressed color of fresh tuna flesh, the type of processing—e.g., carbon monoxide spraying—also influences the meat's color and general appeal to the customer.

12.6 YAKE NIKU: SPONTANEOUS BURN-FLESH SYNDROME, STRESS, AND NUTRITION MANAGEMENT

One extraordinary attribute of scombrid fish is their ability to maintain muscle temperature above ambient temperature—up to 11°C higher. The mechanism by which tuna conserve heat is by retention of metabolic heat through vascular countercurrent heat exchangers or retia (Watson et al., 1988). However, this mechanism, which places tuna in a unique category among teleosts, may be detrimental for fish kept in cages during periods of peak summer temperatures in subtropical waters (reaching >30°C).

Although tuna species that occur naturally in subtropical waters and water temperatures between 24°C and 28°C favor larval development of several scombrid species (Mather et al., 1995; Wexler et al., 2003; Sawada et al., 2005; Kitagawa et al., 2010), larger "endothermic" captive individuals are unable to regulate body temperature by seeking deep, cool water. Also, when a vigorously contracting skeletal muscle must function anaerobically (as during harvest/slaughter), pyruvate cannot be oxidized further due to lack of oxygen. Under these conditions, pyruvate is reduced to lactate or lactic acid (Hochachka and Brill, 1987). The combined action of high temperatures and anaerobic exercise place extraordinary demands on acid-base regulatory systems of tuna causing a condition referred to as "Yake Niku" by the Japanese, or burnt-flesh syndrome. This condition accounts for internal muscle bleeding, soft and pale flesh, thus rendering a product bitter to taste and of no market value.

In 2009, our research team collected PBFT and YFT tissues (off the coast of Ensenada and La Paz, Mexico, respectively) during the slaughter process and this took place over extended grow-out cycles. We recorded fish weight, fork length, and roundup time before slaughter and also collected logbooks with environmental data. We analyzed plasma lactate, cortisol and glucose concentrations, plasma osmolarity and plasma chloride sodium and potassium, muscle glycogen and creatinine, as well as both plasma and muscle-protein bound and free amino acids. Our unpublished data (Buentello et al., 2008) indicates that the combination of high temperature in tropical/subtropical settings together with an increased anaerobic metabolism during harvest leads to accelerated lactate production. In excessive amounts, lactate overwhelms the consumption capacity of liver, kidney, and myocardium. This upsets the buffering capacity of tuna, resulting in acidosis which in turn leads to an increased protein catabolism.

To solve this particular problem we have designed specific harvesting strategies and nutritional manipulations (supplementation of vitamin D, thiamine, iron, coenzyme Q, L-carnitine, and riboflavin) to counteract sharp decreases of bicarbonate in cultured tuna thus preventing the occurrence of Yake Niku.

12.6.1 Dilution of Mercury Footprint via Applied Nutrition

One chief limitation that currently affects the tuna industry is the public's perception that all tuna are equal; thus, the fundamental concern over eating fish contaminated with mercury (Hg)—a heavy metal that affects both the human brain and nervous system and can cause serious health problems—applies the same for cultured and wild-caught tuna.

Most Hg pollution comes from the burning of fossil fuels in coal-fired power plants, disposal of Hg-containing products in incinerators and landfills, mining operations, and industrial uses like chlorine production. Levels of Hg in the environment have increased 3- to 5-fold in the past century as a result of human activities and are reaching threshold levels that threaten human health and environmental security, as well as the future of the global fishing industry (Storelli et al., 2005). Methylmercury (MHg)—the organic form Hg assumes in fish—is a potent neurotoxin that poses the greatest risk to the developing fetus and young children and has the potential to accumulate to toxic levels in fish tissues, an important source of human Hg exposure (Tchounwou et al., 2003). The cumulative nature of MHg results in those fish that are older and of higher trophic level being typically found to have the highest Hg content, and potentially pose the greatest health threat (Balshaw et al., 2007). Among those species recognized as potentially accumulating elevated Hg levels, tuna are one of the most frequently consumed and commercially available groups of fish worldwide (Burger et al., 2005).

It is our contention that the uptake risks are fundamentally different for captive versus free-ranging fish. Most of the capture-based tuna aquaculture operations in the world use fresh/frozen, locally caught sardine as primary feed. Specific concession areas are purposely sought away from cities, tourist developments or industrialized premises. In the wild, bioaccumulation of MHg results in apex-predator fish and older specimens typically exhibiting the highest Hg concentrations. In contrast, tuna in cages are fed clupeid fish for the most part, or fish and squid located low in the food chain and thus have low Hg levels, often undetectable. Therefore, a double effect results in both dilution of Hg via tuna growth (muscle and fat deposition) and reduction in Hg via the excretion of protein metabolites due to normal protein turnover. A study conducted in 2008 in our laboratory (Buentello et al., 2009) collected muscle samples, viscera, environmental, and morphometric data of PBFT held in net pens off the coast of Ensenada, Mexico. After inductively coupled mass spectroscopy analysis, results indicated that Hg levels (ppm) in cultured tuna muscle were of significantly lesser magnitude as other apex predators including wild PBFT of similar size, sharks, swordfish, and tilefish. The average Hg content in muscle from ranched tuna was 0.2 ppm. Taking into account the average Hg footprint in free ranging tuna, bioaccumulation, and dilution, a $\sim$30 kg PBFT fattened in cages for 4 months should not result in Hg levels surpassing the recommended maximum intake (FAO/WHO, 2005) of 1 ppm even when consumed frequently (4 times/week).

12.7 FUTURE PERSPECTIVES AND GLOBAL CONSIDERATIONS

Tuna farming has transcended its infancy and is now recognized as one of the most profitable forms of aquaculture worldwide, and is economically relevant to both developed and developing countries. However, in keeping with the increased interest in the responsible cultivation of these species, several viability issues are currently being addressed. Among them, research on nutrition of tuna species plays a pivotal role in fostering the success of this industry.

Beyond feed representing one of the largest operating expenses in tuna ventures, it is now evident that nutrition research is also essential to maximize profit, optimize product quality, and positively affect the disease resistance of scombrid species. For the foreseeable future, tuna nutrition will follow the path paved already for other marine fish such as the Atlantic salmon in the determination of nutrient requirements, utilization of alternative feed ingredients and supplements, and optimization of weaning and grow-out diets, taking into account the unique scombrid physiology and metabolic needs. The achievement of these nutritional objectives will resolve some of the most critical issues currently limiting the success and permanence of the tuna aquaculture industry.

REFERENCES

Ackman, R.G., 1980. Fish lipids. Part 1. In: Connell, J.J. (Ed.), Advances in Fish Science and Technology. Fishing News (Books) Ltd., Farnham, Surrey, pp. 86–103.

Aguado-Gimenez, F., Garcia-Garcia, B., 2005. Growth, food intake and feed conversion rates in captive Atlantic bluefin tuna (*Thunnus thynnus* Linnaeus, 1758) under fattening conditions. Aquaculture Res. 5 (36), 610–614.

Aguado-Gimenez, F., Martinez, F.J., Garcia-Garcia, B., 2004. In vivo total nitrogen and total phosphorous digestibility in Atlantic Bluefin Tuna (*Thunnus thynnus thynnus* Linnaeus, 1758) under industrially intensive fattening conditions in Southeast Spain Mediterranean coastal waters. Aquaculture Nutr. 10, 413–419.

Aguado-Gimenez, F., Garcia-Garcia, B., Hernández-Lorente, M.D., Cerezo-Valverde, J., 2006. Gross metabolic waste output estimates using a nutritional approach in Atlantic bluefin tuna (*Thunnus thynnus*) under intensive fattening conditions in western Mediterranean Sea. Aquaculture Res. 37, 1254–1258.

Allain, V., 2005. What do tuna eat? A tuna diet study. SPC Fisheries Newsletter #112.

Alvarez-Gonzalez, C.A., Cervantes-Trujano, M., Tovar-Ramırez, D., Conklin, D.E., Nolasco, H., Gisbert, E., et al., 2006. Development of digestive enzymes in California halibut *Paralichthys californicus* larvae. Fish Physiol. Biochem. 31, 83–93.

Alverson, F.G., 1963. The food of yellowfin and skipjack tunas in the eastern tropical Pacific Ocean. Inter-Am. Trop. Tuna Comm. Bull. 7, 293–396.

Baeverfjord, G., Krogdahl, Å., 1996. Development and regression of soybean meal induced enteritis in Atlantic salmon, *Salmo salar* L., distal intestine: a comparison with the intestines of fasted fish. J. Fish Dis. 19, 375–387.

Balshaw, S., Edwards, J.W., Ross, K.E., Daughtry, B.J., 2007. Mercury in seafood: mechanisms of accumulation and consequences for consumer health. Rev. Environ. Health 22, 91–113.

Barrows, F.T., Bellis, D., Krogdahl, A., Silverstein, J.T., Herman, E.M., Sealey, W.M., et al., 2008. Report of the Plant Products in Aquafeed Strategic Planning Workshop: An integrated, interdisciplinary research roadmap for increasing utilization of plant feedstuffs in diets for carnivorous fish. Rev. Fish. Sci. 16, 449–455.

Biswas, B.K., 2010. Establishment of formulated diet for rearing juvenile bluefin tuna, *Thunnus orientalis*. Bull. Fish. Lab. Kinki Univ. 12, 71–29.

Biswas, A.K., Nozaki, J., Kurata, M., Takii, K., Kumai, H., Seoka, M., 2006. Effect of *Artemia* enrichment on the growth and survival of Pacific bluefin tuna *Thunnus orientalis* (Temminck & Schlegel) larvae. Aquaculture Res. 37, 1662–1670.

Biswas, B.K., Ji, S.-C., Biswas, A.K., Seoka, M., Kim, Y.-S., Kawasaki, K.-I., et al., 2009a. Dietary protein and lipid requirements for the Pacific bluefin tuna *Thunnus orientalis* juvenile. Aquaculture 288, 114–119.

Biswas, B.K., Ji, S.C., Biswas, A.K., Seoka, M., Kim, Y.S., Takii, K., 2009b. A suitable dietary sugar level for juvenile Pacific bluefin tuna, *Thunnus orientalis*. Aquaculture Sci. 57, 99–108.

Biswas, A., Biswas, B.K., Ito, J., Takaoka, O., Yagi, N., Itoh, S., et al., 2011. Soybean meal can partially replace enzyme-treated fish meal in the diet of juvenile Pacific bluefin tuna, *Thunnus orientalis*. Fish. Sci. 77, 615–621.

Biswas, B.K., Ji, S.-C., Biswas, A.K., Seoka, M., Kim, Y.-S., Kawasaki, K.-I., et al., 2013. The optimal dietary level of ascorbic acid for juvenile Pacific bluefin tuna, *Thunnus orientalis*. Aquaculture Int. 21, 327–336.

Blank, J.M., Morrissette, J.M., Landeira-Fernandez, A.M., Blackwell, S.B., Williams, T.D., Block, B.A., 2004. In situ cardiac performance of Pacific bluefin tuna hearts in response to acute temperature change. J. Exp. Biol. 207, 881–890.

Blank, J.M., Farwell, C.J., Morrissette, J.M., Schallert, R.J., Block, B.A., 2007a. Influence of swimming speed on metabolic rates of juvenile Pacific bluefin tuna and yellowfin tuna. Physiol. Biochem. Zool. 80 (2), 167–177.

Blank, J.M., Morrissette, J.M., Farwell, C.J., Price, M., Schallert, R.J., Block, B.A., 2007b. Temperature effects on metabolic rate of juvenile Pacific bluefin tuna *Thunnus orientalis*. J. Exp. Biol. 210, 4254–4261.

Blom, J.H., Dabrowski, K., 1995. Reproductive success of female rainbow trout (*Oncorhynchus mykiss*) in response to graded dietary ascorbyl monophosphate levels. Biol. Reprod. 52, 1073–1080.

Booth, M.A., Moses, M.D., Allan, G.L., 2013. Utilization of carbohydrate by yellowtail kingfish *Seriola lalandi*. Aquaculture 376–379, 151–161.

Booth, M.A., Anderson, A.J., Allan, G.L., 2006. Investigation of the nutritional requirements of Australian snapper *Pagrus auratus* (Bloch & Schneider 1801): digestibility of gelatinized wheat starch and clearance of an intra-peritoneal injection of D-glucose. Aquaculture Res. 37, 975–985.

Bou, M., Todorčević, M., Fontanillas, R., Capilla, E., Gutiérrez, J., Navarro, I., 2014. Adipose tissue and livermetabolic responses to different levels of dietary carbohydrates in gilthead sea bream (*Sparus aurata*). Comp. Biochem. Physiol. Part A 175, 72–81.

Brauge, C., Medale, F., Corraze, G., 1994. Effect of dietary carbohydrate levels on growth, body composition and glycaemia in rainbow trout, *Oncorhynchus mykiss*, reared in seawater. Aquaculture 123, 109–120.

Brill, J.R., 1987. On the standard metabolic rate of tropical tunas, including the effect of body size and acute temperature change. Fish. Bull. 85, 25–35.

Bruce, B.P., 1997. A feasibility study of methods to assess and manage waste dispersal and deposition from the southern Bluefin tuna (*Thunnus maccoyii*) farms of Boston Bay, Port Lincoln, South Australia. Master of Science (Ecological Management) Thesis. University of Adelaide, South Australia.

Buckley, T.W., Miller, B.S., 1994. Feeding habits of yellowfin tuna associated with fish aggregation devices in American Samoa. Bull. Mar. Sci. 55 (2-3), 445–459.

Budge, S.M., Iverson, S.J., Koopman, H.N., 2006. Studying trophic ecology in marine ecosystems using fatty acids: a primer on analysis and interpretation. Mar. Mamm. Sci. 22, 759–801.

Buentello, J.A., Pohlenz, C., Neill, W.H., Gatlin III, D.M., Ascencio, F.J., 2008. Physiological indicators for tuna cultured in sea cages: a preliminary approach to prevention of the burnt-flesh syndrome. World Aquaculture 2008, Busan, South Korea.

Buentello, J.A., Pohlenz, C., Neill, W.H., Taylor, R., Ascencio, F.J., Gatlin III, D.M., 2009. An industry-oriented sampling strategy to detect any bioaccumulation of mercury in cultured pacific bluefin tuna (*Thunnus thynnus orientalis*). World Aquaculture 2009, Veracruz, Mexico.

Buentello, J.A., Pohlenz, C., Margulies, D., Scholey, V.P., Wexler, J.B., Tovar, D., et al., 2011. A preliminary study of digestive enzyme activities and amino acid composition of early juvenile yellowfin tuna (*Thunnus albacares*). Aquaculture 312, 205–211.

Bureau, D.P., Harris, A.M., Cho, C.Y., 1999. Apparent digestibility of rendered animal protein ingredients for rainbow trout (*Oncorhynchus mykiss*). Aquaculture 180, 345–358.

Burger, J., Stern, A.H., Gochfeld, M., 2005. Mercury in commercial fish: optimizing individual choices to reduce risk. Environ. Health Perspect. 113, 1–6.

Cahu, C.L., Zambonino Infante, J.L., Barbosa, V., 2003. Effect of dietary phospholipid level and phospholipid:neutral lipid value on the development of sea bass (*Dicentrarchus labrax*) larvae fed a compound diet. Br. J. Nutr. 90, 21–29.

Cahu, C.L., Gisbert, E., Villeneuve, L.A.N., Morais, S., Hamza, N., Wold, P.-A., et al., 2009. Influence of dietary phospholipids on early ontogenesis of fish. Aquaculture Res. 40, 989–999.

Carey, F.G., Kanwisher, J.W., Don Stenens, E., 1984. Bluefin tuna warm their viscera during digestion. J. Exp. Biol. 109, 1–20.

Carter, C.G., Seeto, G.S., Smart, A., Clarke, S., van Barneveld, R.J., 1998. Correlates of growth in farmed juvenile southern Bluefin tuna *Thunnus maccoyii* (Castelnau). Aquaculture 161, 107–119.

Carter, C.G., Brandsen, M.P., van Barneveld, R.J., Clarke, S.M., 1999. Alternative methods for nutrition research on the southern Bluefin tuna, *Thunnus maccoyii* (Castenau): In vitro digestibility. Aquaculture 179, 57–70.

Catacutan, M.R., Coloso, R.M., 1995. Effect of dietary protein to energy ratios on growth, survival, and body composition of juvenile Asian seabass, *Lates calcarifer*. Aquaculture 131, 125–133.

Catalan, I.A., Tejedor, A., Alemany, F., Reglero, P., 2011. Trophic ecology of Atlantic bluefin tuna *Thunnus thynnus* larvae. J. Fish Biol. 78, 1545–1560.

Chen, B.N., Qin, J.G., Kumar, M.S., Hutchinson, W.G., Clarke, S.M., 2006. Ontogenetic development of digestive enzymes in yellowtail kingfish *Seriola lalandi* larvae. Aquaculture 260, 264–271.

Cho, C.Y., Bureau, D.P., 1995. Determination of the energy requirements of fish with particular reference to salmonids. J. Appl. Ichthyol. 11, 141–163.

Clarkson, P.M., 1996. Nutrition for improved sports performance: current issues on ergogenic aids. Sports Med. 21 (6), 393–401.

Clay, D., 1988. Fat, water, protein and ash of bluefin tuna collected in the Gulf of St. Lawrence. ICCAT Coll. Vol. Sci. Pap. 28, 196–202.

Cowey, C.B., 1979. Protein and amino acid requirement of finfish. In: Halver, J.E., Tiews, K. (Eds.), Finfish Nutrition and Fishfeed Technology, vol. 1. Heenemann, Berlin, pp. 3–16.

Craik, J.C.A., 1985. Egg quality and egg pigment content in salmonid fishes. Aquaculture 47, 61–88.

Date, K., Yamamoto, Y., 1988. Seasonal variations with growth in nutritive components in meat of cultured yellowtail *Seriola quinqueradiata*. Nippon Suisan Gakkaishi 54, 1041–1047.

De la Gandara, F., 2012. From capture to self-sustained aquaculture and domestication of bluefin tuna *Thunnus thynnus*. SELFDOTT project periodic report. 488 pp. http://www.repositorio.ieo.es/e-ieo/bitstream/handle/10508/1119/SELFDOTT%20FINAL%20REPORT%20%28ABSTRACT%29.pdf?sequence=1.

Dickson, K.A., 1995. Unique adaptations of the metabolic biochemistry of tunas and billfishes for life in the pelagic environment. Environ. Biol. Fishes 42, 65–97.

Dymowska, A.K., Manfredi, T., Rosenthal, J.J.C., Seibel, B.A., 2012. Temperature compensation of aerobic capacity and performance in the Antarctic pteropod, *Clione antarctica*, compared with its northern congener, *C. limacine*. J. Exp. Biol. 215, 3370–3378.

Ege, R., Krogh, A., 1914. On the relation between the temperature and the respiratory exchange in fishes. Int. Rev. Gesamten Hydrobiol. Hydrogr. 1, 48–55.
Einen, O., Roem, A.J., 1997. Dietary protein/energy ratios for Atlantic salmon in relation to fish size: growth, feed utilization and slaughter quality. Aquaculture Nutr. 3 (2), 115–126.
Essed, Z., Fernandez, I., Alarcon, F.J., Diaz, M., Moyano, J.J., 2003. In vitro protein hydrolys is by diges tive enzymes of *Thunnus thynnus*. Cah. Options Méditerr. 60, 57–59.
Estess, E.E., Coffey, D.M., Shimose, T., Seitz, A.C., Rodriguez, L., Norton, A., et al., 2014. Bioenergetics of captive Pacific bluefin tuna (*Thunnus orientalis*). Aquaculture 434, 137–144.
FAO, Food and Agriculture Organization of the United Nations/World Health Organization, 2005. Codex Alimentarius Commission. FAO, Rome, Italy.
Fernandez-Palacios, H., Izquierdo, M.S., Robaina, L., Valencia, A., Salhi, M., Vergara, J., 1995. Effect of n-3 HUFA level in broodstock diets on egg quality of gilthead seabream (*Sparus aurata* L). Aquaculture 132, 325–337.
Fernandez-Palacios, H., Izquierdo, M., Robaina, L., Valencia, A., Salhi, M., Montero, D., 1997. The effect of dietary protein and lipid from squid and fish meals on egg quality of broodstock for Gilthead seabream (*Sparus aurata*). Aquaculture 148, 233–246.
Fiedler, P.C., Bernard, H.J., 1987. Tuna aggregation and feeding near fronts observed in satellite imagery. Cont. Shelf Res. 7 (X), 871–881.
Fitzgibbon, Q.P., Seymour, R.S., Ellis, D., Buchanan, J., 2007. The energetic consequence of specific dynamic action in southern bluefin tuna *Thunnus maccoyii*. J. Exp. Biol. 210, 290–298.
Fitzgibbon, Q.P., Baudinette, R.V., Musgrove, R.J., Seymour, R.S., 2008. Routine metabolic rate of southern bluefin tuna (*Thunnus maccoyii*). Comp. Biochem. Physiol. 152 (2), 231–238.
Francis, G., Makkar, H.P.S., Becker, K., 2001. Antinutritional factors present in plant-derived alternate fish feed ingredients and their effects in fish. Aquaculture 199, 197–227.
Gagne, F., Blaise, C., 2004. Shell protein characteristics and vitellogenin-like proteins in brine shrimp *Artemia franciscana* exposed to municipal effluent and 20-hydroxyecdysone. Comp. Biochem. Physiol. C Toxicol. Pharmacol. 138 (4), 515–522.
Gatlin III, D.M., Barrows, F.T., Brown, P., Dabrowski, K., Gaylord, T.G., Hardy, R.W., et al., 2007. Expanding the utilization of sustainable plant products in aquafeeds – a review. Aquaculture Res. 38, 551–579.
Glencross, B.D., van Barneveld, R.J., Carter, C.G., Clarke, S.M., 1999a. On the path to a manufactured feed for farmed bluefin tuna. World Aquaculture Mag. 30 (3), 42–46.
Glencross, B.D., van Barneveld, R.J., Carter, C.G., Clarke, S.M., 1999b. Factors influencing feed intake and feed conversion in farmed southern bluefin tuna (*Thunnus maccoyii*). World Aquaculture Society. Sydney, NSW, Australia, April/May 1999.
Glencross, B.D., Carter, C., Gunn, J., Van Barneveld, R., Rough, K., Clarke, S., 2002a. Southern bluefin tuna *Thunnus maccoyii*. In: Webster, C.D., Lim, C.E. (Eds.), Nutrient Requirements and Feeding of Finfish for Aquaculture. CABI Publishing, pp. 159–171.
Glencross, B.D., Clarke, S.M., Buchanan, J.G., Carter, C.G., van Barneveld, R.J., 2002b. Temporal growth patterns of farmed juvenile southern bluefin tuna, *Thunnus maccoyii* (Castelnau) fed moist pellets. J. World Aquaculture Soc. 33 (2), 138–145.
Graham, J.B., Dickson, K.A., 2001. Anatomical and physiological specializations for endothermy. In: Block, B.A., Stevens, E.D. (Eds.), Tuna: Physiology, Ecology, and Evolution. Academic Press, San Diego, pp. 121–168.
Graham, J.B., Laurs, R.M., 1982. Metabolic rate of the albacore tuna *Thunnus alalunga*. Mar. Biol. 72, 1–6.

Gunn, J., Hartog, J., Rough, K., 2001. The relationship between food intake and visceral warming in southern bluefin tuna (*Thunnus maccoyii*). In: Sibert, J.R., Nielsen, J.L. (Eds.), Electronic Tagging and Tracking in Marine Fisheries, vol. 1. Kluwer Academic Publishers, Dordrecht, pp. 109−130.

Guppy, M., Hochachka, P.W., 1978. Controlling the highest lactate dehydrogenase activity known in nature. Am. J. Physiol. (RICP) 3 (2), R136−R140.

Haga, Y., Naiki, T., Tazaki, Y., Shirai, T., Takaki, Y., Tanaka, Y., et al., 2011. Improvement in the feeding activity, early growth and survival of Pacific bluefin tuna *Thunnus orientalis* larvae fed a casein peptide-based microdiet supplemented with inosine monophosphate. Fish. Sci. 77, 245−253.

Halver, J.E., 1957. Nutrition of salmonid fishes. J. Nutr. 62, 225−243.

Harris, L.E., 1984. Effects of a broodfish diet fortified with canthaxanthin on female fecundity and egg color. Aquaculture 43, 179−183.

Hemre, G.I., Mommsen, T.P., Krogdahl, Å., 2002. Carbohydrates in fish nutrition: effects on growth, glucose metabolism and hepatic enzymes. Aquaculture Nutr. 8, 175−194.

Hirano, T., Nakamura, H., Suyama, M., 1980. Quality of wild and cultured ayu − II Seasonal variation of proximate composition. Nippon Suisan Gakkaishi 46, 75−78.

Hochachka, P.W., Brill, R.W., 1987. Autocatalitic pathways to cell death: a new analysis of the tuna burn problem. Fish Physiol. Biochem. 4 (2), 81−87.

Hosokawa, H., Takii, K., Shimeno, S., Takeda, M., 2001. Identification of feeding stimulants for yellowtail in muscle extract of jack mackerel. Suisanzoshoku 49, 225−229 (in Japanese with English abstract).

Hua, K., Bureau, D.P., 2012. Exploring the possibility of quantifying the effects of plant protein ingredients in fish feeds using meta-analysis and nutritional model simulation-based approaches. Aquaculture 356−357, 284−301.

Ikeda, I., Hosokawa, H., Shimeno, S., Takeda, M., 1988. Identification of feeding stimulant for jack mackerel in its muscle extract. Nippon Suisan Gakkaishi 54, 229−233 (in Japanese with English abstract).

Ingham, L., Arme, C., 1977. Intestinal absorption of amino acids by rainbow trout, *Salmo gairdneri* (Richardson). J. Comp. Physiol. 117, 323−334.

Ishihara, K., Saito, H., 1996. The docosahexaenoic acid content in the lipid of juvenile bluefin tuna *Thunnus thynnus* caught in the sea of the Japanese coast. Fish. Sci. 62, 840−841.

Izquierdo, M., Fernandez-Palacios, H., Tacon, A., 2001. Effect of broodstock nutrition on reproductive performance of fish. Aquaculture 197, 25−42.

Ji, S.-C., Takaoka, O., Seoka, M., Kohbara, J., Hosokawa, H., Shimeno, S., et al., 2007. Identification of feeding stimulants for juvenile Pacific bluefin tuna, *Thunnus orientalis* in muscle extract of horse mackerel, *Trachurus japonicus*. J. Korean Aquaculture Soc. 20, 19−25 (in Korean with English abstract).

Ji, S.-C., Takaoka, O., Biswas, A., Seoka, M., Ozaki, K., Kohbara, J., et al., 2008. Dietary utility of enzyme-treated fish meal for juvenile Pacific bluefin tuna *Thunnus orientalis*. Fish. Sci. 74, 54−61.

Kaji, T., 2000. Studies on the Early Development of Bluefin and Yellowfin Tuna. Ph.D. Thesis. Kyoto University, Kyoto, Japan.

Kaji, T., Tanaka, M., Takahashi, Y., Oka, M., Ishibashi, N., 1996. Preliminary observations on development of Pacific bluefin tuna *Thunnus thynnus* (Scombridae) larvae reared in the laboratory, with special reference to the digestive system. Mar. Freshwater Res. 47, 261−269.

Kaji, T., Tanaka, M., Oka, M., Takeuchi, H., Ohsumi, S., Teruya, K., et al., 1999. Growth and morphological development of laboratory-reared yellowfin tuna *Thunnus albacares* larvae and juveniles, with special emphasis on the digestive system. Fish. Sci. 65, 700−707.

Kitagawa, T., Kato, Y., Miller, M.J., Sasai, Y., Sasaki, H., Kimura, S., 2010. The restricted spawning area and season of Pacific bluefin tuna facilitate use of nursery areas: a modeling approach to larval and juvenile dispersal processes. J. Exp. Mar. Biol. Ecol. 393, 23–31.

Kohbara, J., Miyazaki, T., Takii, K., Hosokawa, H., Ukawa, M., Kumai, H., 2006. Gustatory responses in pacific bluefin tuna *Thunnus orientalis* (Temminck and Schlegel). Aquaculture Res. 37, 847–854.

Korsmeyer, K.E., Dewar, H., 2001. Tuna metabolism and energetics. Fish Physiol. 19, 35–78.

Korsmeyer, K.E., Dewar, H., Lai, N.C., Graham, J.B., 1996. The aerobic capacity of tunas: adaptation for multiple metabolic demands. Comp. Biochem. Physiol. 113A, 17–24.

Kubo, T., 2013. Study of swimming behavior and digestion of cultured bluefin tuna, *Thunnus thynnus orientalis*, Based on the Body Temperature. Bull. Fish. Lab. Kinki Univ. 13, (in press).

Kumai, H., 1997. Present state of bluefin tuna aquaculture in Japan. Suisanzoshoku 45, 293–297.

Lee, S.M., Jeon, I.G., Lee, J.Y., 2002. Effects of digestible protein and lipid levels in practical diets on growth, protein utilization and body composition of juvenile rockfish (*Sebastes schlegeli*). Aquaculture 211, 227–239.

Leef, M.J., Carter, C.G., Nowak, B.F., 2012. Assessment of nutritional status and digestive physiology in southern bluefin tuna *Thunnus maccoyii* fed a modified baitfish diet. Aquaculture 350–353, 162–168.

Logan, J.M., Rodríguez-Marín, E., Goñi, N., Barreiro, S., Arrizabalaga, H., Golet, W., et al., 2011. Diet of young Atlantic bluefin tuna (*Thunnus thynnus*) in eastern and western Atlantic foraging grounds. Mar. Biol. 158, 73–85.

Margulies, D., 1993. Assessment of the nutritional condition of larval and early juvenile tuna and Spanish mackerel (Pisces: Scombridae) in the Panama Bight. Mar. Biol. 115, 317–330.

Margulies, D., Scholey, V.P., Wexler, J.B., Santiago, M.C., 2010. Research on the reproductive biology and early life history of yellowfin tuna (*Thunnus albacares*) as the IATTC's Achotines Laboratory. Inter-American Tropical Tuna Commission. Joint International Symposium of Kinki University and Setouchi Town on The 40th Anniversary of Pacific Bluefin Tuna Aquaculture. In: Miyashita, S., Takii, K., Sakamoto, W. (Eds.), Towards the Sustainable Aquaculture of Bluefin Tuna, vol. 3153. Amal Biswas Fisheries Laboratories, Kinki University, Shirahama, Wakayama, Japan, pp. 649–2211.

Martínez-Montaño, E., Peña, E., Focken, U., Viana, M.T., 2010. Intestinal absorption of amino acids in the Pacific bluefin tuna (*Thunnus orientalis*): In vitro uptake of amino acids using hydrolyzed sardine muscle at three different concentrations. Aquaculture 299, 134–139.

Martínez-Montaño, E., Peña, E., Viana, M.T., 2013. Intestinal absorption of amino acids in the Pacific bluefin tuna (*Thunnus orientalis*): In vitro lysine-arginine interaction using the everted intestine system. Fish Physiol. Biochem. 39, 325–334.

Masuma, S., 2006. Studies on Broodstock Management and Spawning Ecology of Bluefin and Yellowfin Tuna in Captivity. Ph.D. Thesis. Kyushu University, Fukuoka, Japan (in Japanese with English summary).

Masuma, S., Oka, M., Kanematu, M., Tezuka, N., Teruya, K., Fushimi, H., et al., 1991. Feeding and growth of captive bluefin in Yaeyama. Saibaigiken 20, 35–40 (in Japanese).

Masuma, S., Miyashita, S., Yamamoto, H., Kumai, H., 2008. Status of bluefin tuna farming, broodstock management, breeding and fingerling production in Japan. Rev. Fish. Sci. 16, 385–390.

Masuma, S., Takebe, T., Sakakura, Y., 2010. Review of the broodstock management and larviculture of the Pacific northern bluefin tuna in Japan. Aquaculture 315, 2–8.

Masumoto, T., 2002. Yellowtail, *Seriola quinqueradiata*. In: Webster, C.D., Lim, C. (Eds.), Nutrient Requirements and Feeding of Finfish for Aquaculture. CABI Publishing, New York, pp. 131–146.

Mather, F.J., Mason Jr., J.M., Jones, A., 1995. Historical document: life history and fisheries of Atlantic bluefin tuna. NOAA Tech. Mem. No. NMFS-SEFSC-370, US Department of Commerce, NOAA, NMFS, Southeast Fisheries Science Center. http://ia341024.us.archive.org/0/items/historicaldocume00math/historicaldocume00math.pdf.

Matus de la Parra, A., Rosas, A., Lazo, J.P., Viana, M.T., 2007. Partial characterization of the digestive enzymes of Pacific bluefin tuna *Thunnus orientalis* under culture conditions. Fish Physiol. Biochem. 33, 223–231.

Medina, I., Aubourg, S.P., Martin, R.P., 1995. Composition of phospholipids of white muscle of six tuna species. Lipids 30, 1127–1135.

Miyashita, S., 2002. Studies on the seedling production of the Pacific bluefin tuna *Thunnus thynnus orientalis*. Bull. Fish. Lab. Kinki Univ. 8, 1–171.

Miyashita, S., Kato, K., Sawada, Y., Murata, O., Ishitani, Y., Shimizu, K., et al., 1998. Development of digestive system and enzyme activities of larval and juvenile bluefin tuna, *Thunnus thynnus*, reared in the laboratory. Suisanzoshoku 46, 111–120.

Miyashita, S., Sawada, Y., Okada, T., Murata, O., Kumai, H., 2001. Morphological development and growth of laboratory-reared larval and juvenile *Thunnus thynnus*. Fish. Bull. 99, 601–616.

Moon, T.W., 2001. Glucose intolerance in teleost fish: fact or fiction. Comp. Biochem. Physiol. B Biochem. Mol. Biol. 129, 243–249.

Morais, S., Mourente, G., Ortega, A., Tocher, J.A., Tocher, D.R., 2011. Expression of fatty acyl desaturase and elongase genes, and evolution of DHA:EPA ratio during development of unfed larvae of Atlantic bluefin tuna (*Thunnus thynnus* L.). Aquaculture 313, 129–139.

Morishita, T., Uno, K., Imura, N., Takahashi, T., 1987. Variation with growth in the proximate compositions of cultured red sea bream. Nippon Suisan Gakkaishi 53, 1601–1607.

Morote, E., Olivar, M.P., Pankhurst, P.M., Villate, F., Uriarte, M., 2008. Trophic ecology of bullet tuna *Auxis rochei* larvae and ontogeny of feeding-related organs. Mar. Ecol. Prog. Ser. 353, 243–254.

Mourente, G., Tocher, D.R., 2009. Tuna nutrition and feeds: current status and future perspectives. Rev. Fish. Sci. 17, 374–391.

Mourente, G., Megina, C., Díaz-Salvago, E., 2002. Lipids in female northern bluefin tuna (*Thunnus thynnus thynnus* L.) during sexual maturation. Fish Physiol. Biochem. 24, 351–363.

Mourente, G., Quintero, O., Canavate, J.P., 2015. Trophic links of Atlantic Bluefin tuna (*Thunnus thynnus* L.) inferred by fatty acid signatures. J. Exp. Mar. Biol. Ecol. 463, 49–56.

Moyes, C.D., 1996. Cardiac metabolism in high performance fish. Comp. Biochem. Physiol. 113A (1), 69–75.

Nakamura, Y.-N., Ando, M., Seoka, M., Kawasaki, K., Tsukamasa, Y., 2005. Comparison of the proximate compositions, breaking strength and histological structure by the muscle positions of the full-cycle cultured Pacific bluefin tuna *Thunnus orientalis*. Fish. Sci. 71, 605–611.

Nakamura, Y., Mori, K., Saitoh, K., Oshima, K., Mekuchi, M., Sugaya, T., et al., 2013. Evolutionary changes of multiple visual pigment genes in the complete genome of Pacific bluefin tuna. PNAS 110, 11061–11066.

NRC, Nutrient Requirements of Fish and Shirmp, 2011. Animal Nutrition Council, Washington, DC, 376 pp.

Ochiai, Y., Watanabe, Y., Ozawa, H., Ikegami, S., Uchida, N., Watabe, S., 2010. Thermal denaturation profiles of tuna myoglobin. Biosci. Biotechnol. Biochem. 74 (8), 1673–1679.

Ohkubo, N., Matsubara, T., 2002. Sequential utilization of free amino acids, yolk proteins and lipids in developing eggs and yolk-sac larvae of barfin flounder *Verasper moseri*. Mar. Biol. 140, 187–196.

Ohnishi, T., Biswas, A., Kaminaka, K., Nakao, T., Nakajima, M., Sakakibara, N., et al., 2014. Energy partitioning in cultured juvenile Pacific bluefin tuna, *Thunnus orientalis* (Temminck & Schlegel, 1844). Aquaculture Res., 1–10.

Ottolenghi, F., Silvestri, C., Giordano, P., Lovatelli, A., New, M.B., 2004. Capture-based aquaculture: the fattening of eels, groupers, tunas and yellowtails. Food and Agriculture Organization of the United Nations. FAO, Rome.

Papatryphon, E., Soares, J.H., 2000. The effect of dietary feeding stimulanys on growth performance of striped bass, Morone saxatilis, fed-a-plant-feedstuff-based diet. Aquaculture 185, 329–338.

Partridge, G.J., 2013. Closed-cycle hatchery production of tuna. In: Allan, G., Burnell, G. (Eds.), Advances in Aquaculture Hatchery Technology. Woodhead publishing, pp. 457–497.

Patterson, T.A., Evans, K., Carter, T.I., Gunn, J.S., 2008. Movement and behavior of large southern bluefin tuna (*Thunnus macoyii*) in the Australian region determined using pop-u satellite archival tags. Fish. Oceanogr. 17 (5), 352–367.

Perrin, W.F., Warner, R., Fiscus, H., Holts, D.B., 1973. Stomach contents of porpoise, *Stenella spp.* and yellowfin tuna, *Thunnus albacares*, in mixed-species aggregations. Fishery Bull. 71 (4), 1077–1092.

Phillips, A.M., 1962. Effect of diet and water temperature on the blood phosphorous of brook trout. Prog. Fish Culturist 24, 22–25.

Richards, W.J., Dove, G.R., 1971. Internal development of young tunas of the genera *Katsuwonus*, *Euthynnus*, *Auxis*, and *Thunnus* (Pisces, Scombridae). Copeia 1971, 72–78.

Roncarati, A., Dees, A., Forlini, L., Melotti, P., 2008. Status of Atlantic and Mediterranean bluefin tuna (*Thunnus thynnus*, L.) populations following the recent European conservation prescribing rules. World Aquaculture 2008, Busan, South Korea.

Rønnestad, I., Thorsen, A., Finn, R.N., 1999. Fish larval nutrition: a review of recent advances in the roles of amino acids. Aquaculture 177, 201–216.

Rønnestad, I., Tonheim, S.K., Fyhn, H.J., Rojas-Garcia, C.R., Kamisaka, Y., Koven, W., et al., 2003. The supply of amino acids during early feeding stages of marine fish larvae: a review of recent findings. Aquaculture 227, 147–164.

Rooker, J.R., Alvarado, J.R., Block, B.A., Dewar, H., de Metrio, G., Corriero, A., et al., 2007. Life history and stock structure of Atlantic Bluefin Tuna (*Thunnus thynnus*). Rev. Fish. Sci. 15, 265–310.

Roy, B.C., Miyake, Y., Ando, M., Kawasaki, K.-I., 2010. Proximate and fatty acid compositions in different flesh cuts of cultured, cultured fasted, and wild Pacific bluefin tuna (*Thunnus orientalis*). J. Aquat. Food Prod. Technol. 19, 284–297.

Ruohonen, K., Vielma, J., Grove, D.J., 1999. Low-protein supplement increases protein retention and reduces the amounts of nitrogen and phosphorous wasted by rainbow trout fed on low-fat herring. Aquaculture Nutr. 5, 83–91.

Saeki, K., Kumagai, H., 1984. Chemical components in ten kinds of wild and cultured fishes. Nippon Suisan Gakkaishi 50, 1551–1554.

Sahlmann, C., Gu, J., Kortner, T.M., Lein, I., Krogdahl, Å., Bakke, A.M., 2015. Ontogeny of the digestive system of Atlantic salmon (Salmo salar L.) and effects of soybean meal from start-feeding. PLOS one, 1–23.

Saito, H., Mrata, M., 1998. Origin of the monoene fats in the lipid of midwater fishes: relationship between the lipids of myctophids and those of their prey. Mar. Ecol. Prog. Ser. 168, 21–33.

Saito, H., Watanabe, T., Murase, T., 1995. The fatty acid composition characteristic of a highly migratory fish, with seasonal variation of docosahexaenoic acid content in lipid of bonito (*Euthynnus pelamis*). Biosci. Biotechnol. Biochem. 59, 2186–2188.

Saito, H., Ishihara, K., Murase, T., 1996. Effect of prey lipids on the docosahexaenoic acid content of total fatty acids in the lipids of *Thunnus albacares* yellowfin tuna. Biosci. Biotechnol. Biochem. 60, 962–965.

San Martin, M.F., Barbosa-Canovas, G.V., Swanson, B.G., 2002. Food processing by high hydrostatic pressure. Crit. Rev. Food Sci. Nutr. 42 (6), 627–645.

Satoh, K., Tanaka, T., Masujima, M., Okazaki, M., Kato, Y., Shono, H., et al., 2013. Relationship between the growth and survival of larval Pacific bluefin tuna, *Thunnus orientalis*. Mar. Biol. 160, 691–702.

Sawada, T., Takahashi, K., Hatano, M., 1993. Triglyceride composition of tuna and bonito orbital fats. Nippon Suisan Gakkaishi 59, 285–290.

Sawada, Y., Miyashita, S., Aoyama, M., Kurata, M., Mukai, Y., Okada, T., et al., 2000. Rotifer size selectivity and optimal feeding density of bluefin tuna, *Thunnus thynnus*, larvae. Suisanzoshoku 48, 169–177.

Sawada, Y., Okada, T., Miyashita, S., Murata, O., Kumai, H., 2005. Completion of the Pacific bluefin tuna *Thunnus orientalis* (Temminck et Schlegel) life cycle. Aquaculture Res. 36, 413–421.

Seoka, M., Takaoka, O., Takii, K., Kumai, H., 1998. Triacylglycerol and phospholipid contents in developing Japanese flounder eggs. Fish. Sci. 64, 654–655.

Seoka, M., Kurata, M., Hatanaka, Y., Biswas, A.K., Ji, S.C., Kumai, H., 2007a. Possible nutrients in *Artemia* affecting the larval growth of Pacific bluefin tuna *Thunnus orientalis*. Suisanzoshoku 55, 55–64.

Seoka, M., Kurata, M., Kumai, H., 2007b. Effect of docosahexaenoic acid enrichment in *Artemia* on growth of Pacific bluefin tuna *Thunnus orientalis* larvae. Aquaculture 270, 193–199.

Seoka, M., Kurata, M., Tamagawa, R., Biswas, A.K., Biswas, B.K., Yong, A.S.K., et al., 2008. Dietary supplementation of salmon roe phospholipid enhances the growth and survival of Pacific bluefin tuna *Thunnus orientalis* larvae and juveniles. Aquaculture 275, 225–234.

Seoka, M., Bandou, T., Suimon, M., Sawada, Y., 2010. Toyama squid *Watasenia scintillans* meal as a protein source in artificial diet for juvenile Pacific bluefin tuna *Thunnus orientalis*. Suisanzoshoku 58, 143–144.

Shimizu, Y., Tada, M., Endo, K., 1973. Seasonal variations in chemical constituents of yellowtail muscle – I. Water, lipid and crude protein. Nippon Suisan Gakkaishi 39, 993–999.

Smullen, R., 2009. A world first tuna formulated feed success for southern Bluefin tuna, Thunnus maccoyii. Sustainable Aquaculture of the Bluefin and Yellowfin tuna. Closing the life cycle for commercial production. The 2nd Global COE Program Symposium of Kinki. South Australia Research and Development Institute (SARDI), Australia, November 30–December 2, 2009.

Storelli, M.M., Giacominelli-Stuffler, R., Storelli, A., Marcotrigiano, G.O., 2005. Accumulation of mercury, cadmium, lead and arsenic in swordfish and bluefin tuna from the Mediterranean Sea: a comparative study. Mar. Pollut. Bull. 50, 993–1018.

Suyama, M., Hirano, T., Suzuki, T., 1986. Buffering capacity of free histidine and its related dipeptides in white and dark muscles of yellow tuna. Nippon Suisan Gakkaishi 52, 2171–2175.

Swimmer, Y., McNaughton, L., Moyes, C., Brill, R., 2004. Metabolic biochemistry of cardiac muscle in three tuna species (bigeye, *Thunnus obesus*; yellowfin, *T. albacares*; and skipjack, *Katsuwonus pelamis*) with divergent ambient temperature and oxygen tolerances. Fish Physiol. Biochem. 30, 27–35.

Takaoka, O., Takii, K., Nakamura, M., Kumai, H., Takeda, M., 1990. Identification of feeding stimulants for M\marbled rockfish. Nippon Suisan Gakkaishi 56, 345–351 (in Japanese with English abstract).

Takii, K., Ji, S.-C., 2005. Feeding stimulants for Pacific bluefin tuna *Thunnus orientalis*. In: The 21st Century COE program of Kinki University (Ed.), Kinki University 21st Century COE Program Intermediate Report, Kinki University, Osaka, pp. 143–146.

Takii, K., Miyashista, S., Seoka, M., Tanaka, Y., Kubo, Y., Kumai, H., 1997. Changes in chemical contents and enzyme activities during embryonic development of bluefin tuna. Fish. Sci. 63, 1014–1018.

Takii, K., Hosokawa, H., Shimeno, S., Ukawa, M., Kotani, A., Yamada, Y., 2005. Anesthesia, fasting tolerance, and nutrient requirement of juvenile northern bluefin tuna. Fish. Sci. 71, 499–503.

Tanabe, T., 2001. Feeding habits of skipjack tuna *Katsuwonus pelamis* and other tuna *Thunnus spp.* juveniles in the tropical western Pacific. Fish. Sci. 67, 563–570.

Tchounwou, P.B., Ayensu, W.K., Ninashvili, N., Sutton, D., 2003. Environmental exposure to mercury and its toxicopathologic implications for public health. Environ. Toxicol. 18, 149–175.

Tibbetts, S.M., Lall, S.P., Anderson, D.M., 2000. Dietary protein requirement of juvenile American eel (*Anguilla rostrata*) fed practical diets. Aquaculture 186, 145–155.

Tocher, D.R., 2003. Metabolism and function of lipids and fatty acids in teleost fish. Rev. Fish. Sci. 11, 107–184.

Tocher, D.R., Bendiksen, E.A., Campbell, P.J., Bell, J.G., 2008. The role of phospholipids in nutrition and metabolism of teleost fish. Aquaculture 280, 21–34.

Torres, A.P., Reglero, P., Balbin, R., Urtizberea, A., Alemany, F., 2011. Coexistence of larvae of tuna species and other fish in the surface mixed layer in the NW Mediterranean. J. Plankton Res. 33 (12), 1793–1812.

Tsukamasa, Y., Nakamura, Y.-N., Ando, M., Seoka, M., Kawasaki, K.-I., 2007. Changes in myoglobin content and proximate compositions of the dorsal ordinary muscles of full cycle cultured Pacific bluefin tuna *Thunnus orientalis* (Temminck et Schlegel) with body size. Aquaculture Res. 38, 201–205.

Uotani, I., Saito, T., Hiranuma, K., Nishikawa, Y., 1990. Feeding habit of bluefin tuna *Thunnus thynnus* larvae in the western North Pacific Ocean. Nippon Suisan Gakkaishi 56, 713–717.

Urán-Carmona, P.A., 2008. Etiology of Soybean-Induced Enteritis in Fish. Ph.D. thesis. Wageningen University, The Netherlands.

Van Barneveld, R.J., Davis, B.J., Smart, A., Clarke, S., Carter, C., Tivey, D., et al., 1997. The nutritional management of caged southern bluefin tuna. In: Rowe, J.B., Nolan, J.V. (Eds.), Recent Advances in Animal Nutrition in Australia. Department of Animal Science, University of New England, Armidale, New South Wales, pp. 88–97.

Ward, T.M., Hoedt, F., McLeay, L., Dimmlich, W.F., Kinloch, M., Jackson, G., et al., 2001. Effects of the 1995 and 1998 mass mortality events on the spawning biomass of sardine, *Sardinops sagax*, in South Australian waters. ICES J. Mar. Sci. 58, 865–875.

Watanabe, T., 2002. Strategies for further development for aquatic feeds. Fish. Sci. 68, 242–252.

Watanabe, T., Kiron, V., 1995. Broodstock management and nutritional approaches for quality offsprings in the red sea bream. In: Bromage, N.R., Roberts, R.J. (Eds.), Broodstock Management and Egg and Larval Quality. Cambridge Univ. Press, Cambridge, p. 424.

Watanabe, T., Arakawa, T., Kitajima, C., Fujita, S., 1984a. Effect of nutritional quality of broodstock diets on reproduction of red sea bream. Nippon Suisan Gakkaishi 50 (3), 495–501.

Watanabe, T., Ohhashi, S., Itoh, A., Kitajima, C., Fujita, S., 1984b. Effect of nutritional composition of diets on chemical components of red sea bream broodstock and eggs produced. Nippon Suisan Gakkaishi 50 (3), 503–515.

Watanabe, T., Itoh, A., Murakami, A., Tsukashima, Y., 1984c. Effect of nutritional quality of diets given to broodstocks on the verge of spawning on reproduction of red sea bream. Nippon Suisan Gakkaishi 50 (6), 1023–1028.

Watanabe, T., Lee, M., Mizutani, J., Yamada, T., Satoh, S., Takeuchi, T., et al., 1991a. Effective components in cuttlefish meal and raw krill for improvement of quality of red sea bream *Pagrus major* eggs. Nippon Suisan Gakkaishi 57 (4), 681–694.

Watanabe, T., Fujimura, T., Lee, M.J., Fukusho, K., Satoh, S., Takeuchi, T., 1991b. Effect of polar and nonpolar lipids from krill on quality of eggs of red seabream *Pagrus major*. Nippon Suisan Gakkaishi 57 (4), 695–698.

Watanabe, T., Viyakarn, V., Kimura, H., Ogawa, K., Okamoto, N., Iso, N., 1992. Utilization of soybean meal as a protein source in a newly developed soft-dry pellet for yellowtail. Nippon Suisan Gakkaishi 58, 1761–1773.

Watanabe, T., Viswanath, K., Shuichi, S., 1997. Trace minerals in fish nutrition. Aquaculture 151, 185–207.

Watson, C., Bourke, R.E., Brill, R.W., 1988. A comprehensive theory on the etiology of burnt tuna. Fish. Bull. 86 (2), 367–372.

Webb, K.A., Gatlin, D.M., 2003. Effects of dietary protein level and form on production characteristics and ammonia excretion of red drum *Sciaenops ocellatus*. Aquaculture 225, 17–26.

Wexler, J.B., Scholey, V.P., Olson, R.J., Margulies, D., Nakazawa, A., Suter, J.M., 2003. Tank culture of yellowfin tuna, *Thunnus albacares*: developing a spawning population for research purposes. Aquaculture 220, 327–353.

Wilson, R.P., 1989. Amino acids and proteins. In: Halver, J.E. (Ed.), Fish Nutrition, second ed. Academic Press, San Diego, CA, pp. 111–151.

Yufera, M., Darias, M.J., 2007. The onset of feeding in marine fish larvae. Aquaculture 268, 53–63.

Chapter 13

Genetics in Tuna Aquaculture

Yoshifumi Sawada and Yasuo Agawa
Fisheries Laboratories, Kindai University, Kushimoto, Wakayama, Japan

13.1 INTRODUCTION

Human beings have been breeding animals for more than 10,000 years (Shoda, 2010). During this period, new breeds and strains have been developed with desirable phenotypic traits such as rapid growth, higher tolerance to environmental conditions, enhanced disease resistance and improved meat quality. In aquaculture, selective breeding has a much shorter history but has been conducted for several species including Atlantic salmon (*Salmo salar*: Gjedrem, 1983), rainbow trout (*Oncorhynchus mykiss*: Donaldson and Olson, 1957), tilapia (*Oreochromis niloticus*: Eknath et al., 1993), and carp (*Cyprinus carpio*: Babouchkine, 1987; Gjedrem, 2005). Selective breeding programs for aquatic species provide better outcomes compared to terrestrial livestock. This higher response to selection of farmed aquatic species can be attributed to their high fecundity enabling higher selection intensity. In addition, large phenotypic and genetic variations exist in the genetically underdeveloped fish traits. However, in the tuna aquaculture industry, selection programs are not commonly used and production still relies heavily on catching wild fingerlings and broodstock. The completion of the life cycle of Pacific bluefin tuna (PBFT) has provided the potential to develop selective breeding programs for this species and this is now underway in Japan (Sawada et al., 2004). In this process, biotechnology will play an important role in the area of tuna reproduction, breeding, and harvested product quality management.

13.2 TARGET TRAITS IN TUNA BREED IMPROVEMENT

In any tuna breeding program the following biological traits are important targets for selection, fast growth rate, high survival rate during early development, high disease resistance, superior traits in reproduction, and high meat quality.

D.D. Benetti, G.J. Partridge, A. Buentello (Eds): Advances in Tuna Aquaculture.
DOI: http://dx.doi.org/10.1016/B978-0-12-411459-3.00013-8

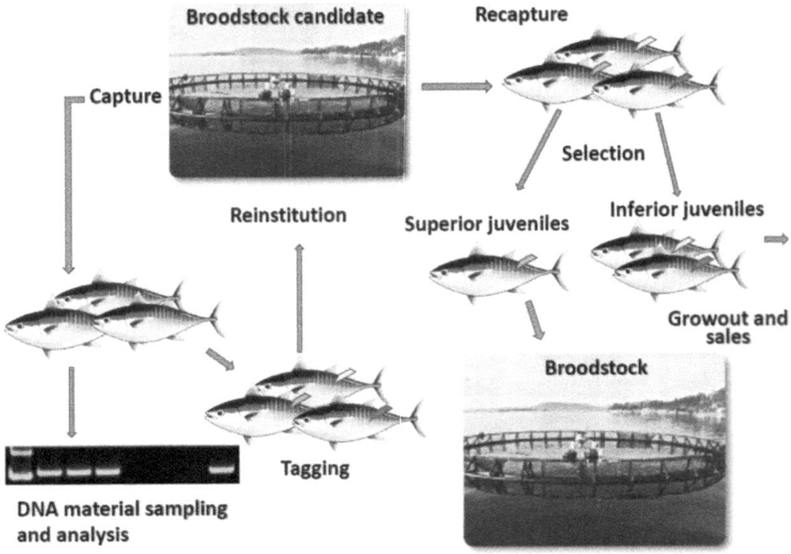

FIGURE 13.1 Diagram of tuna individual selection for breed improvement. The procedure of individuals having superior genetic traits encompasses capture, DNA material sampling and analysis, and tagging where multiple times of juvenile handling is inevitable.

Although tuna have extremely high growth rate compared to other fishes (Miyashita, 2002), it is desirable to further increase the speed of growth and development by selective breeding in order to improve production efficiency.

Developing breeds of tuna that are selectively optimized for commercial aquaculture is a dynamic and intensive process. Steps involve broodstock capture and maturation concurrent with genetic sampling and tagging. Records are kept to correlate larval success, growout metrics, and other significant biological traits to breeding individuals, and subsequent resampling for future brood fish. Figure 13.1 depicts an outline of the steps involved in the continuous improvement of tuna breeds.

The technology for tuna fingerling production is still underdeveloped and the survival rate during the hatchery phase is 10% at most. Improvements to survival rate will be attained by the improvement of larval and juvenile rearing techniques but selection of breeders whose offspring show high survival during early development can also be a goal of a selective breeding program. In the larval rearing of tuna, low survival rate due to an unsuitable rearing environment, such as water surface condition and current in the rearing water is the largest obstacle for their mass production (Miyashita, 2002; Sawada et al., 2005; Kurata et al., 2011). Therefore, it is important to create strains which have tolerance to environmental stress. Domestication and selection of families with high survival will be attained naturally in the early stage of tuna aquaculture technology development, as poor larval survival inevitably leads to selection against such breeds. However, as a more positive step, selection of domesticated and high survival breeds for specific environments should be implemented in tuna aquaculture.

PBFT do not tend to suffer from serious diseases from 1 year after hatching. However, PBFT juveniles experience viral, bacterial, and parasitic diseases that lead to high mortality (Munday et al., 2003), and development of disease-resistant strains is also important.

Under aquaculture conditions, female adult tuna with high reproductive activity are scarce (Masuma et al., 2010) compared to males, the majority of which reach maturity at a younger age than females (Sawada et al., 2008). Therefore, prolific breeders are required for stable reproduction especially for female tuna in tuna aquaculture.

One of the most important criteria in evaluating quality of the aquaculture-produced tuna is the muscle fat content. Fatty portions such as the belly flaps called "toro" are preferable as raw materials for the preparation of sushi and sashimi, also deriving a higher market value than other muscle sections. However, demand also exists for leaner meat called "akami". Selective breeding programs can be directed at controlling muscle fat contents is in PBFT muscle.

13.3 GENETIC TECHNOLOGIES IN TUNA BREEDING

13.3.1 Genomic DNA Analysis

Genomic information is applicable to the improvement of tuna aquaculture by using DNA marker-assisted selection. For this aim, it is necessary to establish a series of DNA microsatellite markers. Although tuna DNA microsatellite markers have been developed by various scientists (Takagi et al., 1999; McDowell et al., 2002; Nakadate and Chow 2008; Tagami et al., 2008; Morshima et al., 2009; Agawa et al., 2009; Clark et al., 2004; Takagi et al., 2001), the number of microsatellites is still insufficient to identify the desired traits or to make a high-density genetic map of tuna in the breeding program.

In the analysis of genes, such as base sequence determination, it is inefficient and difficult to use genomic DNA. As a substitute, genome libraries, which are composed of cloning vectors (inserted fragments of targeted genome DNA), are ordinarily utilized. To make a genome library, genome DNA is fractionated by the use of restricted enzymes, sonication or shearing, then the fragments are inserted into genetically engineered vectors, such as plasmids, bacteriophages (such as phage λ), bacterial artificial chromosomes (BACs), or yeast artificial chromosomes (YACs). For PBFT genome, a BAC library has been established (Yagishita et al., 2006).

Identification of sex-specific DNA markers of PBFT is a good example of genomic DNA analysis (Agawa et al., 2011). Under aquaculture conditions,

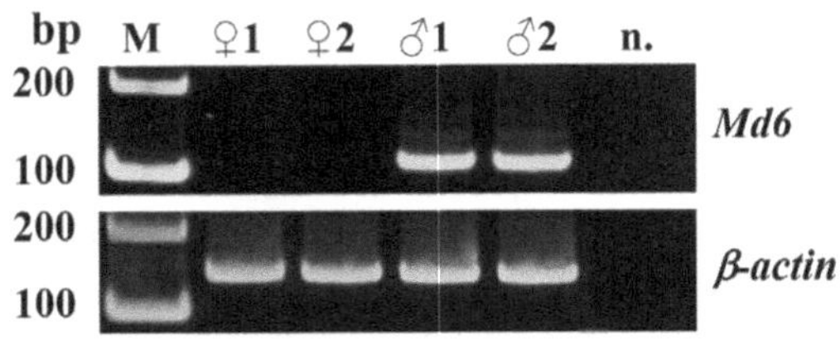

FIGURE 13.2 The sex-specific DNA PCR products were confirmed by gel electrophoresis. M; molecular size markers. The sizes (base pairs) were indicated left of the fragments. Two females and two males PBFTs were used. n.; negative control means PCR were conducted without template DNA. Md6; male delta 6 locus, male positive PCR (Agawa et al., 2015). β-actin; beta actin gene PCRs were used for positive control.

it has been reported that adult females have a low percentage of participation in reproduction (Masuma et al., 2010), and to make a broodstock group with a high percentage of females is necessary for the stable reproduction and for securement of appropriate genetic diversity in the laboratory hatched and raised PBFT strains. Because adult PBFT are large in size, vulnerable, and difficult to handle, sex determination should be conducted during early developmental stages of juveniles. Therefore, for PBFT juveniles, whose sexes cannot be distinguished phenotypically, sex identification by genetics using small biotic samples such as fin tips is very useful.

These biotic samples can then be screened using AFLP-selective DNA amplification including primer-specific secondary amplifications for increased fidelity and sequenced through high-throughput next-generation sequencing technologies (Agawa et al., 2015). Combined with common electrophoresis methodologies, male characteristic DNA fragments (Figure 13.2) can be observed and are attributable to specific breeding fish.

As a more comprehensive result for PBFT, draft analysis of the genomic DNA was performed by the Fisheries Research Agency (FRA) cooperatively with the University of Tokyo and Kyusyu University (Saito, 2010; Fisheries Research Agency, 2010). In this project, they reportedly found approximately 86,000 microsatellite DNA markers. These results of genomic analysis will contribute largely to the PBFT breeding programs.

13.3.2 Analysis of Mitochondrial DNA Polymorphism

The mitochondrial DNA D-loop region in tuna of the genus *Thunnus* is highly polymorphic as in other fishes (Bremer et al., 1998; Bremer et al., 1999; Chow et al., 2000), and this variation can be detected using conventional PCR-RFLP analysis in several tuna and billfish species (Chow et al., 1997; Chow et al., 2000; Chow and Takeyama, 2000; Niwa et al., 2003). The analysis of mitochondrial DNA polymorphism has been well used as the tool to identify tuna species (Chow and Kishino, 1995; Takeyama et al., 2001).

This kind of genetic tool to identify tuna species is useful to prevent mislabeling of tuna fishery and aquaculture products. To sustain consumers' confidence, such effort should be encouraged.

Mitochondrial DNA polymorphism has also been used to analyze the genetic structure of natural populations of yellowfin tuna (YFT) (Scoles and Graves, 1993; Ward et al., 1997; Wu et al., 2010), bigeye (Chow et al., 2000; Grewe and Hampton, 1998; Grewe et al., 2000; Appleyard et al., 2002; Durand et al., 2005; Chiang et al., 2006; Chiang et al., 2008), albacore (Chow and Ushiyama, 1995; Viñas et al., 1999), and Atlantic bluefin tuna (ABFT) (Ely et al., 2002.; Viñas et al., 2003; Boustany et al., 2008; Carlsson et al., 2004). To raise the precision of these analyses, genome DNA microsatellite markers are often combined (Nakadate, 2005). Such combination can detect genetic foreignness among pedigrees and even individuals with high sensitivity. Therefore, this combined analysis of mitochondrial and genome DNA polymorphism can be used in genetic monitoring for spawning ecology of captive tuna (Niwa et al., 2003), and it has the possibility to elucidate how individual variation among broodstock, degree of maturity, and nutritional state influence the egg quality and larval vitality. In addition, it can be a very powerful tool to register newly developed tuna strains in aquaculture and to identify artificially hatched and released stocks to the open sea for natural population enhancement in the near future.

13.3.3 Other Genetic Technologies

Chromosome manipulation in cultured fishes is effective to improve their performance. Inviable aneuploids such as triploids have been induced to improve growth (Arai et al., 1991; Hussain, 1996; Pandian and Koteeswaran, 1998; Arai 2001). Cloned fishes have been also produced in rainbow trout *Oncorhynchus mykiss* (Okada, 1985), common carp *Cyprinus carpio* (Komen et al., 1991), tilapia *Oreochromis niloticus* (Müller-Belecke and Hörstgen-Schwark, 1995; Müller-Belecke and Hörstgen-Schwark, 2000), olive flounder "hirame" *Paralichthys olivaceus* (Yamamoto, 1999), and red sea bream *Pagrus major* (Kato et al., 2002). Cloned strains have a possibility of being advantageous in the mass production by their homogeneity in their superior phenotypic traits such as the digestion ability of nutrients, sensitivity to medicinal agents, and environmental stress resistance. The useful characteristic of these technologies is to create strains without breeding potential. It will contribute to more efficient mass production of cultured species. At present, there is no example of application of these technologies in tuna aquaculture industry.

Genetic modifications should not be used by the tuna aquaculture industry until rigorous examination in terms of food safety is completed. In addition, genetically-modified tuna escapees could potentially interfere with the genetic material of wild tuna populations should they breach their enclosures in net pens. Although containment of other fin fish species is easily attained

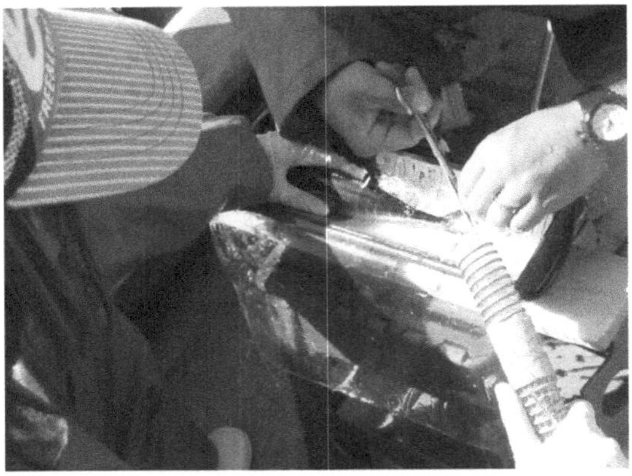

FIGURE 13.3 Sampling of DNA material and tagging of a PBFT juvenile. Tuna juveniles have low handling tolerance.

in land-based culture facilities, tuna species require significant space/water volume such as the one provided by massive net pens. Today, it is impractical to practice commercial tuna grow out in land-based facilities.

Finally, in practical tuna breed improvement, it is necessary to develop a new technology for the treatment of live individuals. To create a new variety having superior traits for mating, it is necessary to make a brood stock group by the selection of live individuals having targeted phenotypic characteristics. In the DNA marker-assisted selection to obtain such varieties, collecting DNA samples from each individual is necessary. These individuals should be tagged and kept alive during the genetic examination. However, tuna are very difficult to handle, even for the juveniles that are smaller and easier to handle than huge adult tuna (Ottolenghi, 2008). Handling stress using current technologies during genetic sampling (Figure 13.3) results in accumulation in plasma lactate, cortisol, and glucose (Addis et al., 2012), yet is currently necessary for a breed development.

For example, PBFT juveniles have a severe mortality by handling stress in the transfer from land-based rearing tanks to sea cages for further grow out culture. Therefore, it is necessary to develop the technologies for tagging of small-sized juveniles of 20–30 cm TL and for collection of DNA samples without injuring many small-sized juveniles. Therefore, it will be fundamental to elucidate the details of stress response in their handling which is unclear at present.

13.4 PROTECTION OF INTELLECTUAL PROPERTY OF TUNA AQUACULTURE PRODUCTS AND THEIR BREEDING METHODS

When new tuna breeds are established or breeding methods are newly developed, it is very important to protect the intellectual property of such breeds and breeding methods from the view point of the tuna industry. Securement of intellectual property rights over the improved breeds guarantees inventors' rights of preferential utilization both domestically and abroad (Fisheries Research Agency, 2010). Genetics and molecular biology provide scientific basis to protect intellectual property, for example, by pursuing patents of new breeds and breeding methods by providing complete traceability of the products and live stocks of fishes. However, care should be exercised in the necessary condition to protect intellectual property, for example, requirement of patentability. Generally, the nucleotide sequences of DNA fragments which are not eligible for patent are those which do not have the identified function or suggestion of usefulness of the synthesized protein (Japan Patent Office, 1999). For newly developed breeds of tuna, development of legal systems such as the reinforcement of the Plant Variety Protection and Seed Act in Japan will be very useful to protect their intellectual property. In addition, product branding of new breeds by trademark registration will be an important strategy to protect their intellectual property. These efforts to protect intellectual property should be exercised not only for the producers' benefit but also for that of consumers.

REFERENCES

Addis, P., Secci, M., Locci, I., Cau, A., 2012. Harvesting, handling practices and processing of bluefin tuna captured in the trap fishery: possible effects on the flesh quality. Collect. Vol. Sci. Pap. ICCAT 67, 390–398.

Agawa, Y., Honroyo, T., Kurata, M., Okada, T., Sawada, Y., 2009. Introduction of selective breeding of Pacific bluefin tuna *Thunnus orientalis* and its molecular biological approaches. In: Proceedings of the 2nd Global COE Program Symposium of Kinki University, Adelaide, Australia, Dec. 1–3 2009, pp. 8–11.

Agawa, Y., Iwaki, M., Komiya, T., Honryo, T., Tamura, K., Kouhei, T., et al., 2015. Identification of male sex-linked DNA sequence of the cultured Pacific bluefin tuna *Thunnus orientalis*. Fish. Sci. 81 (1), 113–121.

Agawa, Y., Komiya, T., Honroyo, T., Kurata, M., Okada, T., Murata, O., et al., 2011. Screening of the male characteristic DNA fragment using AFLP analysis in Pacific Bluefin tuna, *Thunnus orientalis* (in Japanese with English abstract). Nippon Suisan Gakkaishi 77, 639–646.

Appleyard, S.A., Ward, R.D., Grewe, P.M., 2002. Genetic stock structure of bigeye tuna in the Indian Ocean using mitochondrial DNA and microsatellites. J. Fish. Biol. 60, 767–770.

Arai, K., 2001. Genetic improvement of aquaculture finfish species by chromosome manipulation techniques in Japan. Aquaculture 197, 205–228.

Arai, K., Matsubara, K., Suzuki, R., 1991. Karyotype and erythrocyte size of spontaneous tetraploidy and triploidy in the loach *Misgurnus anguillicaudatus*. Nippon Suisan Gakkaishi 57, 2157–2172.

Babouchkine, Y.P., 1987. La sélection d'une carpe résistant à l'hiver. In: Tiews, K. (Ed.), Proceedings of World Symposium on Selection, Hybridization, and Genetic Engineering in Aquaculture, Bordeaux, 27–30 May 1986, vol. 1. HeenemannVerlagsgesellschaft mbH, Berlin, pp. 447–454.

Boustany, A.M., Reeb, C.A., Block, B.A., 2008. Mitochondrial DNA and electronic tracking reveal population structure of Atlantic bluefin tuna (*Thunnus thynnus*). Mar. Biol. 156, 13–24.

Bremer, A.J.R., Naseli, I., Ely, B., 1999. Heterogeneity of northern bluefin tuna populations. ICCAT Coll. Vol. Sci. Pap. XLIX, 127–129.

Bremer, A.J.R., Stequert, B., Robertson, N.W., Ely, B., 1998. Genetic evidence for inter-oceanic subdivision of bigeye tuna (*Thunnus obesus*) populations. Mar. Biol. 132, 547–557.

Carlsson, J., Jan, R., McDowell, L., Diaz-Jaimes, P., Carlsson, J.E.L., Boles, S.B., et al., 2004. Microsatellite and mitochondrial DNA analyses of Atlantic bluefin tuna (*Thunnus thynnus thynnus*) population structure in the Mediterranean Sea. Mol. Ecol. 13, 3345–3356.

Chiang, H.C., Hsu, C.C., Lin, H.D., Ma, G.C., Chiang, T.Y., Yang, H.Y., 2006. Population structure of bigeye tuna (*Thunnus obesus*) in the South China Sea, Philippine Sea and western Pacific Ocean inferred from mitochondrial DNA. Fish. Res. 79, 219–225.

Chiang, H.C., Hsu, C.C., Wu, G.C.C., Chang, S.K., Yang, H.Y., 2008. Population structure of bigeye tuna (*Thunnus obesus*) in the Indian Ocean inferred from mitochondrial DNA. Fish. Res. 90, 305–312.

Chow, S., Kishino, H., 1995. Phylogenetic relationships between tuna species of the genus *Thunnus* (Scombridae: Teleostei): inconsistent implications from morphology, nuclear and mitochondrial genomes. J. Mol. Evol. 41, 741–748.

Chow, S., Okamoto, H., Miyabe, N., Hiramatsu, K., Barut, N., 2000. Genetic divergence between Atlantic and Indo-Pacific stocks of bigeye tuna (*Thunnus obesus*) and admixture around South Africa. Mol. Ecol. 9, 221–227.

Chow, S., Okamoto, H., Uozumi, Y., Takeuchi, Y., Takeyama, H., 1997. Genetic stock structure of the swordfish (*Xiphias gladius*) inferred by PCR-RFLP analysis of the mitochondrial control region. Mar. Biol. 127, 359–367.

Chow, S., Takeyama, H., 2000. Nuclear and mitochondrial DNA analyses reveal four genetically separated breeding units of the swordfish (*Xiphias gladius*). J. Fish. Biol. 56, 1087–1098.

Chow, S., Ushiyama, H., 1995. Global population structure of albacore (*Thunnus alalunga*) inferred by RFLP analysis of the mitochondrial ATPase gene. Mar. Biol. 123, 39–45.

Clark, T.B., Ma, L., Saillant, E., Gold, J.R., 2004. Microsatellite DNA markers for population-genetic studies of Atlantic bluefin tuna (*Thunnus thynnus thynnus*) and other species of genus *Thunnus*. Mol. Ecol. Notes 4, 70–73.

Durand, J.-D., Collet, A., Chow, S., Guinand, B., Borsa, P., 2005. Nuclear and mitochondrial DNA markers indicate unidirectional gene flow of Indo-Pacific to Atlantic bigeye tuna (*Thunnus obesus*) populations, and their admixture off southern Africa. Mar. Biol. 147, 313–322.

Eknath, A.E., Tayamen, M.M., Palada-de Vera, M.S., Danting, J.C., Reyes, R.A., Dionisio, E.E., et al., 1993. Genetic improvement of farmed tilapias: the growth performance of eight strains of *Oreochromis niloticus* tested in different farm environments. Aquaculture 111, 171–188.

Ely, B., Stoner, D.S., Bremer, A.J.R., Dean, J.M., Addis, P., Cau, A., et al., 2002. Analyses of nuclear IdhA gene and mtDNA control region sequences of Atlantic northern bluefin tuna populations. Mar. Biotechnol. 4, 583–588.

Fisheries Research Agency, 2010. Strategy for genomic research in fisheries science. http://www.fra.affrc.go.jp/pressrelease/pr21/220331/0331besshi1.pdf.

Gjedrem, T., 1983. Genetic variation in quantitative traits and selective breeding in fish and shellfish. Aquaculture 33, 51–72.

Gjedrem, T., 2005. Status and scope of aquaculture. In: Gjedrem (Ed.), Selection and Breeding Programs in Aquaculture. Springer, Dordrecht, pp. 1–8.

Grewe, P.M., Appleyard, S.A., Ward, R.D., 2000. Determining genetic stock structure of bigeye tuna in the Indian Ocean using mitochondrial DNA and DNA microsatellites. Fish. Res. Dev. Corp. Rep. 97/122, CSIRO Marine Research, Hobart.

Grewe, P.M., Hampton, J., 1998. An assessment of Bigeye (*Thunnus obesus*) Population Structure in the Pacific Ocean, Based on Mitochondrial DNA and DNA Microsatellite Analysis. CSIRO Marine Research, Hobart.

Hussain, M.G., 1996. Advances in chromosome engineering research in fish: review of methods, achievements and applications. Asian Fish. Sci. 9, 45–61.

Japan Patent Office, 1999. Comparative study on the patentability of DNA fragments in tripartite patent offices (in Japanese). http://www.jpo.go.jp/torikumi/kokusai/kokusai3/tizai3.htm.

Kato, K., Hayashi, R., Yuasa, D., Yamamoto, S., Miyashita, S., Murata, O., et al., 2002. Production of cloned red sea bream, *Pagrus major*, by chromosome manipulation. Aquaculture 207, 19–27.

Komen, J., Bongers, A.B.J., Richter, C.J.J., Van Muiswinkel, W.B., Huisman, E.A., 1991. Gynogenesis in common carp (Cyprinus carpio L.): II. The production of homozygous clones and F1 hybrids. Aquaculture 92, 127–142.

Kurata, M., Seoka, M., Nakagawa, Y., Ishibashi, Y., Kumai, H., Sawada, Y., 2011. Promotion of initial swimbladder inflation in Pacific bluefin tuna, *Thunnus orientalis* (Temminck and Schlegel), larvae. Aquaculture Res. 43, 1–10.

Masuma, S., Takebe, T., Sakakura, Y., 2010. A review of the broodstock management and larviculture of the Pacific northern bluefin tuna in Japan. Aquaculture 315, 2–8.

McDowell JR, Diaz-Jaimes, P., Graves, J.E., 2002. Isolation and characterization of seven tetranucleotide microsatellite loci from Atlantic northern bluefin tuna *Thunnus thynnus thynnus*. Mol. Ecol. Notes 2, 214–216.

Miyashita, S., 2002. Studies on the seedling production of the Pacific bluefin tuna, *Thunnus thynnus orientalis*. Bull. Fish. Lab. Kinki Univ. 8, 1–171.

Morshima, K., Yamamoto, H., Sawada, Y., Miyashita, S., Kato, K., 2009. Developing 23 new polymorphic microsatellite markers and simulating parentage assignment in the Pacific bluefin tuna, *Thunnus orientalis*. Mol. Ecol. Resour. 9, 790–792.

Müller-Belecke, A., Hörstgen-Schwark, G., 1995. Sex determination in tilapia (*Oreochromis niloticus*): sex ratios in homozygous gynogenetic progeny and their offspring. Aquaculture 137, 57–65.

Müller-Belecke, A., Hörstgen-Schwark, G., 2000. Performance testing of homozygous lines in *Oreochromis niloticus*. Aquaculture 184, 67–76.

Munday, B.L., Sawada, Y., Cribb, T., Hayward, C.J., 2003. Diseases of tunas, *Thunnus* spp. J. Fish. Dis. 26, 187–206.

Nakadate, M., 2005. Genetic isolation between Atlantic and Mediterranean albacore populations inferred from mitochondrial and nuclear DNA markers. J. Fish. Biol. 66, 1545–1557.

Nakadate, M., Chow, S., 2008. Isolation and characterization of single copy nuclear DNA markers in Atlantic bluefin tuna *Thunnus thynnus*. Fish. Sci. 74, 1333–1335.

Niwa, Y., Nakazawa, A., Margulies, D., Scholey, V.P., Wexler, J.B., Chow, S., 2003. Genetic monitoring for spawning ecology of captive yellowfin tuna (*Thunnus albacares*) using mitochondrial DNA variation. Aquaculture 218, 387–395.

Okada, H., 1985. Studies on the artificial sex control in rainbow trout, *Salmo gairdneri*. Scientific Report of the Hokkaido Fish Hatchery 40, 1–49. (In Japanese with English summary.) Hokkaido National Fisheries Research Institute, Japan.

Ottolenghi, F., 2008. Capture-based aquaculture of bluefin tuna. In: Lovatelli, A., Holthus P.F. (Eds.), Capture-based aquaculture. Global overview. FAO Fisheries Technical Paper. No. 508. Rome, FAO. pp. 169–182.

Pandian, T.J., Koteeswaran, R., 1998. Ploidy induction and sex control in fish. Hydrobiologia 384, 167–243.

Saito, K., 2010. A large-scale sequencing project of Pacific tuna genome (in Japanese). Nippon Suisan Gakkaishi 76, 1132–1134.

Sawada, Y., Miyashita, S., Murata, O., Kumai, H., 2004. Seedling production and generation succession of the Pacific bluefin tuna, *Thunnus orientalis*. Mar. Biotechnol. 6, S327–S331.

Sawada, Y., Okada, T., Miyashita, S., Murata, O., Kumai, H., 2005. Completion of the Pacific bluefin tuna *Thunnus orientalis* (Temminck et Schlegel) life cycle. Aquaculture Res. 36, 413–421.

Sawada, Y., Seoka, M., Kato, K., Tamura, T., Nakatani, M., Hayashi, S., et al., 2008. Testes maturation of reared Pacific bluefin tuna, *Thunnus orientalis* Temminck and Schlegel, at two plus years old. Fish. Sci. 73, 1070–1077.

Scoles, D.R., Graves, J.E., 1993. Population genetic structure of yellowfin tuna, *Thunnus albacares*, from the Pacific Ocean. Fish. Bull. 91, 690–698.

Shoda, K., 2010. History of animal breeding and progress of genetics (in Japanese). In: Shoda, Y. (Ed.), World History of Breeding. Yushukan, Tokyo, pp. 1–22.

Tagami, M., Arakawa, T., Sano, H., Fuji, K., Hasegawa, O., Sakamoto, T., et al., 2008. Ultra-high throughput microsatellite marker development for linkage map construction by the next generation DNA sequencers. 5th World Fisheries Congress Program & Abstracts, 5th World Fisheries Congress Committee, Yokohama, Japan, pp. 373.

Takagi, M., Okamura, T., Chow, S., Taniguchi, N., 2001. Preliminary study on genetic stock structure of albacore (*Thunnus alalunga*) inferred from microsatellite DNA analysis. Fish. Bull. 99, 697–701.

Takagi, M., Okmura, T., Chow, S., Taniguchi, N., 1999. PCR primers for microsatellite loci in tuna species of the genus *Thunnus* and its application for population genetics study. Fish. Sci. 65, 571–576.

Takeyama, H., Chow, S., Tsuzuki, H., Matsunaga, T., 2001. Mitochondrial DNA sequence variation within and between *Thunnus* tuna species and its application to species identification. J. Fish. Biol. 58, 1646–1657.

Viñas, J., Pla, C., Tawil, M.Y., Hattour, A., Farrugia, A.F., de la Serna, J.M., 2003. Mitochondrial genetic characterization of Bluefin tuna (*Thunnus thynnus*) from three Mediterranean (Libya, Malta, Tunisia); and one Atlantic locations (Gulf of Cadiz). Col. Vol. Sci. Pap. ICCAT 55 (3), 1282–1288.

Viñas, J., Santiago, J., Pla, C., 1999. Genetic characterization and Atlantic–Mediterranean stock structure of albacore, *Thunnus alalunga*. ICCAT Coll. Vol. Sci. Pap. 49, 188–191.

Ward, R.D., Elliott, N.G., Innes, B.H., Smolenski, A.J., Grewe, P.M., 1997. Global population structure of yellowfin tuna, *Thunnus albacares*, inferred from allozyme and mitochondrial DNA variation. Fish. Bull. 95, 566–575.

Wu, G.C.-C., Chianga, H.-C., Choua, Y.-W., Wong, Z.-R., Hsuc, C.-C., Chend, C.-Y., et al., 2010. Phylogeography of yellowfin tuna (*Thunnus albacares*) in the Western Pacific and the Western Indian Oceans inferred from mitochondrial DNA. Fish. Res. 105, 248–253.

Yagishita, N., Oohara, I., Kobayashi, T., 2006. Construction of a BAC Library for the Pacific Bluefin Tuna *Thunnus orientalis*. Ecology and Aquaculture of Bluefin Tuna 21st century COE program, Kinki University joint symposium on Bluefin Tuna, Kinki University, Japan, pp. 85–88.

Yamamoto, S., 1999. Studies on sex-manipulation and production of cloned populations in hirame, *Paralichthys olivaceus* (Temminck et Schlegel). Aquaculture 173, 235–246.

Chapter 14

Tuna Economics and Markets

Jose Fernandez-Polanco and Ignacio Llorente
Department of Business Management, University of Cantabria, Cantabria, Spain

14.1 INTRODUCTION

The present chapter focuses on the economics of tuna production and markets for tuna products. Whilst wild catches still represent the bulk of tuna supply, some of the most popular species are facing conservational constraints for growth as a consequence of long years of overexploitation (Ottolenghi, 2008). Tuna ranching is still constrained by the dependence on wild catches of seed stock and is therefore subject to the same conservational concerns associated with the wild fishery. The economic aspects and issues of tuna production vary significantly when shifting from the wild fishery to aquaculture. Although tuna farming is still strongly dependent on wild catches, the differences in cost structures, organizational procedures, production planning, and market orientation make it radically different to fisheries from an economic and managerial point of view (Shamshak and Anderson, 2009; Shamshak, 2011). Tuna aquaculture is focused today on bluefin tuna species, promoted by the incentives of the high prices in the Japanese market, and the expectation of a global growth in demand. Tuna ranching has expanded considerably in the last decade. However, the limitations for growth will maintain scarcity in supply in the medium term unless the tuna reproductive cycle is fully developed in captivity and implemented at a reasonable cost.

Tuna production is driven by two main markets in the global seafood trade. The more traditional canned tuna market and the newly developed sushi and sashimi market are radically different. White meat species are dominant in the canned tuna market whilst red meat is preferred in the sushi and sashimi market. The differentiation in terms of species also results in differentiation in terms of harvesting techniques according to the quality grades required in each case. The canned tuna industry is entirely supplied by the traditional wild fishery, and there is no expectation of change in the near

D.D. Benetti, G.J. Partridge, A. Buentello (Eds): Advances in Tuna Aquaculture.
DOI: http://dx.doi.org/10.1016/B978-0-12-411459-3.00014-X

future (Hamilton et al., 2011). Modern tuna aquaculture emerged in the last two decades as a relevant supplier to the Japanese market for sushi and sashimi (Longo, 2011; Kurokura et al., 2012). However, new emergent regions in canned tuna consumption and the globalization of the Japanese culinary style are opening new opportunities for growth for an industry which has faced important transformations in recent years.

The chapter is structured in three main sections. In the first section the sources of tuna supply for the world markets are described. Reviews and considerations regarding economic management of natural resources and environmental issues of tuna production are also presented in this section. The second section deals with the production stages and cost structure in tuna ranching. In the third section the two main markets and their key issues are described, concluding with a brief overview on public concerns regarding tuna consumption.

14.2 MANAGING THE WORLD SUPPLY OF TUNA

The signs of stagnation in the yield of the wild fishery, including overfishing issues with some stocks, have increased the interest toward the potential of aquaculture as a productive and cost-effective alternative. Bioeconomic models of the wild tuna fishery have been extensively studied with the aim of optimizing the economic return securing the sustainability of the stocks. Models of this kind have been recently applied to tuna ranching, in order to identify the volumes resulting in maximum profit according to economic and environmental constraints.

14.2.1 Tuna Supply from the Wild Fishery

From a commercial point of view, the most relevant tuna species are skipjack (*Katsuwonus pelamis*), albacore, or long fin tuna (*Thunnus alalunga*), yellowfin tuna (*Thunnus albacares*, YFT), bigeye tuna (*Thunnus obesus*), southern bluefin tuna (*Thunnus maccoyii*, SBFT), Pacific bluefin tuna (*Thunnus orientalis*, PBFT), and Atlantic bluefin tuna (*Thunnus thynnus*, ABFT). The destination, consumption, and status of conservation vary significantly across species. The bluefin tuna are the most appreciated commodity in the sashimi market and are the fisheries suffering the highest pressure at present. The catches of bluefin tuna are subjected to quotas, and their trade is monitored and certified.

According to the latest FAO fishery databases (FAO, 2015), world production of the main commercial tuna species surpassed 4 million tons in 1999 and reached 4.6 million tons in 2011. Wild catches represent 96.7% of total supply. These figures indicate an increase of 1,019.5% since the earliest statistics of 1950, but only 12.5% since 2000. Issues of overfishing and conservation measures are behind the stagnation in the growth of the tuna fishery production in the new century, and these may be better understood when

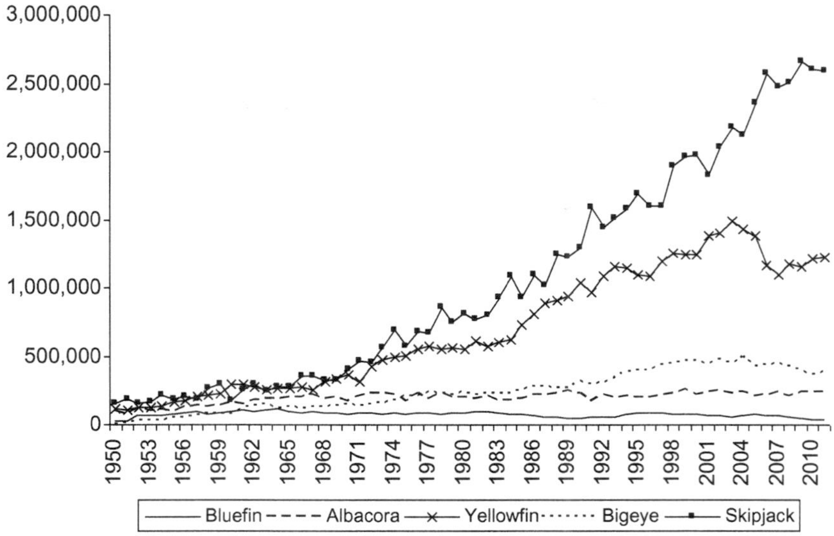

FIGURE 14.1 Evolution of tuna catches in tons. *Source: FAO, Fishstat plus. Capture production 1950–2011.*

analyzing the evolution in the catches of the different species (Figure 14.1). Bluefin species represent 1% of global catches but have a higher value and are the most severely overexploited of all tunas, with a decrease in catches of 49.2% since 2000. Other species with falling production in the first decade of the century were YFT (−2.1%) and bigeye (−16.9%). These species represent 27.2% and 8.5% of total tuna catches. Albacore catches, representing 5.2% of total fishery supply, increased in output by 8.4%. Finally, skipjack, which is the major species in terms of volume, accounting for 58% of the total supply of wild tuna, have increased in catches by 31.6% between 2000 and 2011.

Almost all fishing gears are suitable for fishing tunas, but the efficiency strongly depends on the characteristics of the different species. The two most common methods of industrial tuna fishing are purse seines and long lines. Both can indiscriminately target any of the main commercial species. Catches for the canning industry in the present times are around 2.5 million tons/year.

14.2.2 Tuna Supply from Aquaculture

Three different methods of bluefin tuna culture are being used at different levels of development today in the world. Fattening of adult tunas remains today the most common farming method at the commercial scale. In this method, adult tunas are caught and then fattened for several months.

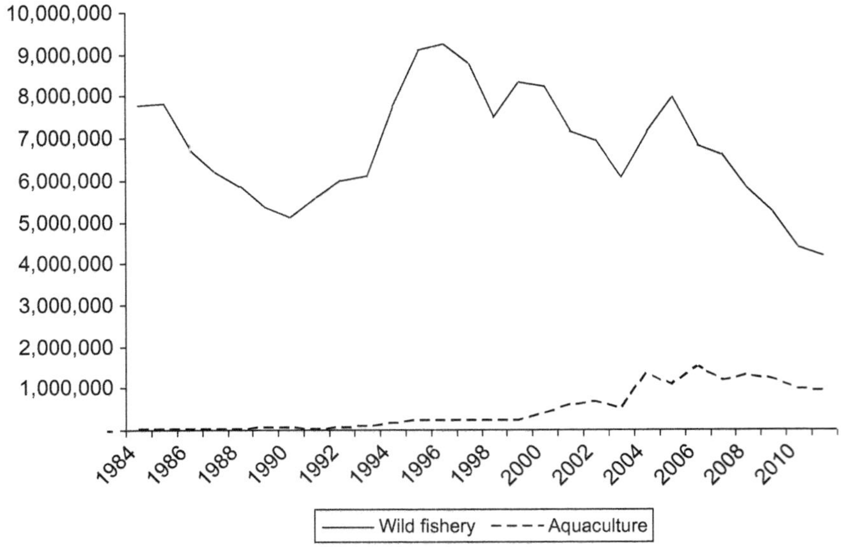

FIGURE 14.2 Compared production of bluefin species. Quantities in tons. *Source: FAO, Fishstat plus. Capture production 1950–2011.*

Culture of young PBFT started in Japan in 1969 using juvenile individuals less than 1-year-old (100–500 g) and is still in use today. The first official project to develop PBFT farming was undertaken in 1970 at Tokai University. This method requires 2 or 3 years prior to harvest and results in higher production costs than adult fattening. In 2002, Kinki University in Japan succeeded in producing fingerlings from eggs spawned from cultured tuna. Similar attempts have been made in Australia, Indonesia, and Spain where three spawning facilities have been built in recent years. Although the production cost is high and this system is still at an experimental stage, it is expected to decrease the dependency of wild tuna as a resource for aquaculture and, by extension, alleviate the pressure on the wild stocks.

Tuna ranching production reached only 0.2% of the total world tuna production in 2011 (FAO, 2015). However when bluefin tuna species are considered separately, aquaculture production represented 18% of the total world bluefin tuna production (Figure 14.2). According to FAO statistics, production dropped after 2006 as a consequence of the conservation measures to protect the overexploited wild stocks. Production was 8900 tons in 2011, equating to a fall of 39.2%, which is almost the same decrease recorded in wild catches (38.6%) over this time. This production does not include that from Japan, which was not listed in FAO production statistics in 2011.

Besides some production of YFT in Mexico and Oman between 2004 and 2008, tuna ranching is fully focused on the three different species of bluefin tuna (Figure 14.3). ABFT was the first species to be ranched at industrial

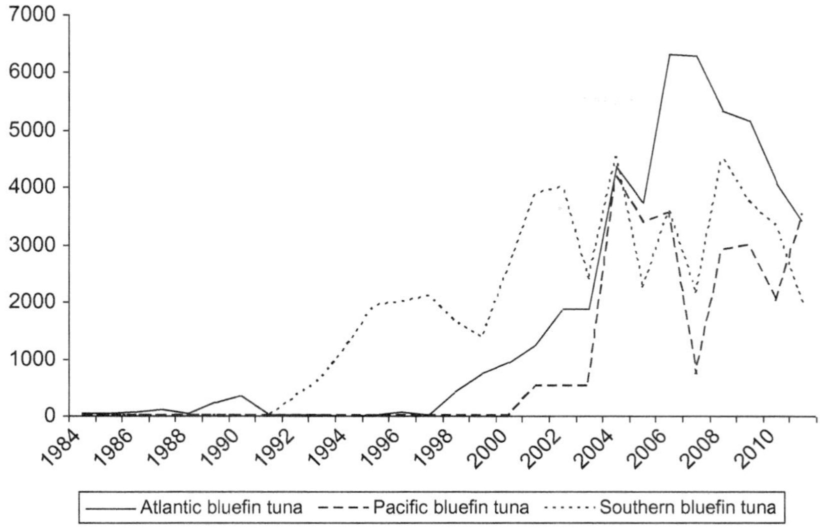

FIGURE 14.3 Aquaculture production of bluefin tuna. Quantities in tons. *Source: FAO, Fishstat plus. Capture production 1950–2011.*

scale in the mid-1980s. By 1996 the production of SBFT in Australia reached more than 2,500 tons and 4,500 tons in 2004. During the period 2000–2005, other Mediterranean countries started fattening ABFT at an industrial scale. As a result, in 2005, the production of ABFT exceeded the production of SBFT in terms of quantity (3,738 tons) and value (US$ 41,253,000). In 2006, the recovery plan for the former species started (ICCAT, 2006) and the availability of stock for the ranching industry in the Mediterranean declined. In 2001 Mexico emerged with the production of PBFT at a commercial scale, reaching an output of 4193 tons in 2004 (FAO, 2015). In the following years, production in Mexico fluctuated between 2000 and 3500 tons. With 3557 tons produced in 2011 (YFT, PBFT, ABFT, SBFT), Mexican PBFT was the main species of tuna aquaculture in terms of quantity (excluding Japan). However, in terms of value, the PBFT farmed in Mexico resulted in the lowest market value, below ABFT and the SBFT (FAO, 2015).

Farming YFT was first undertaken in Mexico. However, due to diminishing YFT wild stocks and some significant production obstacles like loss of flesh color (Morris, 2011), production was stopped in 2008.

According to the last Fishstat data (Table 14.1), Australia is the leading country in terms of value (US$44.26 million) with an output of 1,987 tons of SBFT in 2011 (Table 14.1). Malta and Croatia (ABFT) and Mexico (PBFT), are the next following countries generating US$29.32, US$26.80, and US $23.85 million respectively. Japan is not listed in the FAO tuna aquaculture statistics and thus it cannot be included in these groups due to homogeneity

TABLE 14.1 Production and Value of Farmed Bluefin Tuna in 2011 (Excluding Japan)

	Species	Value in Thousand US$	Quantity in Tons	US$/kg
Australia	SBFT	44,261	1987	22.3
Malta	Atlantic bluefin	29,320	960	30.5
Croatia	Atlantic bluefin	26,807	1610	16.6
Mexico	Pacific bluefin	23,851	3557	6.7
Spain	Atlantic bluefin	12,689	575	22.1
Turkey	Atlantic bluefin	1665	100	16.6
Greece	Atlantic bluefin	1582	95	16.6
Tunisia	Atlantic bluefin	770	70	11.0
Total		140,944	8954	15.7

Source: FAO, Fishstat plus. Capture production 1950–2011.

data issues. The value per kilo of Mexican farmed PBFT in 2011 was US$ 6.71 which is a significantly low price compared with SBFT which was (US $ 22.28) and ABFT from Malta (US$ 30.54).

The Australian tuna industry has led the development of tuna ranching activities in the world, both in terms of production and innovation. Production has been fluctuating with the evolution of the fishery and the management regulations of the stock. In 2011, production declined 40% due to a decrease in the quota of allowable catch (Patterson et al., 2012).

14.3 ECONOMICS OF TUNA RANCHING

Commercial scale tuna ranching is still a capture-based activity. The process begins with the catch of juveniles, in which the fishing gear is a key aspect. Juveniles have to be transported in optimal condition to the farming site, where they are to be provided with the appropriate feed. All these stages require additional equipment and specific raw materials which result in a cost structure substantially different than regular aquaculture operations. The main cost in this production system results from the acquisition and transport of juveniles, which varies depending on the days needed to arrive at the farming site. Feed cost can also change depending on the species used in the diet, whether they are local or imported and the status of the corresponding stocks. Tuna ranching is, therefore, strongly dependent on high market prices in order to balance the additional costs of a capture-based activity. To this extent, the future evolution of tuna ranching will be determined by the ability of wild supply stocks to recover and the potential impact on market prices.

Shamshak (2011), analyzed the economic feasibility of capture-based ABFT farming on the US east coast. The information was obtained from personal communications and consultation with experts in the field, peer-reviewed and gray literature and from data acquired during a site visit to a capture-based ABFT facility in Cartagena, Spain. In this work, the minimum equipment required consisted of towing cages, grow-out cages, a harvesting and feeding vessel, dive equipment, anchors, weights, and other mooring equipment. The costs associated with these assets was obtained from a supplier of ABFT farming equipment, and are detailed in Table 14.2.

The initial investment in the facility described by Shamshak and Anderson (2009) was $1,131,642. This amount results from 1 unit of each towing cage items, 2 units of each grow-out cage items, two diving vessels, and 1 unit of each of the other three vessel types. An additional amount of $10,000 was also considered for licenses and other administrative expenses. All the items acquired in the initial investment have a useful life and a salvage value, which implies that every year all the assets are depreciated and have depreciation cost depending on the initial value, the useful life, the salvage value, and the straight-line method applied. The financial cost associated with the different possible funding options also has to be considered. The interest on the principle of the loan stemming from the initial capital expenditures was assumed to be 7% in Shamshak and Anderson (2009). Every year the activity generates costs associated with docking and insuring vessels, for a value of $50,000, and repair and maintenance costs derived from cages, equipments, and vessels. Shamshak and Anderson (2009) assumed these costs to be $50,000 and $60,000 respectively, according to the data provided in the study that assessed the economics of the offshore aquaculture production of Pacific threadfin in Hawaii (Kam et al., 2003).

Shamshak and Anderson (2009) also estimated the variable costs from the data obtained for an ABFT farm in Cartagena, Spain (Table 14.3). An important source of the variable costs is the acquisition of wild captured tunas. In the Spanish case, the average daily cost of tugboats for transport of the towing cages to the farms was $6000. In that case, a period of 45 days to capture all the tuna available under the quota was estimated. This period elevated the cost to $270,000. Tuna farms have to pay for the purse seiners contracted to catch the fish, with an average cost of USD $9/kg. Another important cost is feed, which in the Spanish case was calculated to be

TABLE 14.2 Initial Investments for an Offshore Bluefin Tuna Farm

Item	Unit Price
Towing Cage	
Towing cage (50 × 25 m) supplied, constructed on-site and launched	$82,000
Towing ropes, bridles, weights etc. (per set)	$26,000
Towing net	$ 41,000
Grow-out Cage	
Mooring system per cage based on a grid system	$20,000
Farm cage triple ring collar (holding/on-growing)	$117,000
Farm site holding net	$35,000
Sea freight per complete unit	$23,000
Vessels	
Diving vessels	$103,000
Diving and sundry equipment	$75,000
Spares based on six cages	$51,000
Aquaculture support vessel	$250,000

Source: Shamshak and Anderson (2009).

TABLE 14.3 Components of Variable Costs

Cost	Value	Units
Feed cost	0.50	USD/kg
Acquisition cost	9	USD/kg
Towing cost	6000	USD/day
Fuel cost	3	USD/gallon
Per fish harvesting cost	1	USD/fish
Managerial labor	40	USD/h
Skilled diver labor	30	USD/h
General labor	20	USD/h

Source: Shamshak and Anderson (2009).

USD $0.50/kg. The fuel used by the vessels is a volatile cost that varies significantly between the different production areas. The farming activities in Cartagena assume a cost of USD $3 per gallon.

Labor costs depend on the level of qualification, the number of workers, and the number of hours of work which vary along the different production areas, the different facilities in the same area, and also along the different production seasons in the same facility. Normally there are three tiers of workers: manager, diver, and general labor. The minimum number of workers needed to operate a farm is four divers, four general workers and, one manager. The cost of the different tiers in Shamshak and Anderson (2009) was established from the labor cost of other aquaculture operations cited in the literature (Kam et al., 2003), and from a website that offers jobs for aquaculture technicians. When considering the harvesting cost which were not described in the paper, it was obtained a cost per unit as a result of dividing the total weekly harvesting cost by the number of fish harvested.

The main production cost in the simulations developed for ABFT production on the US east coast is the acquisition cost per kg of wild ABFT caught by purse seiners, which ranges from 77.3% to 91.2% of the total production cost. Other costs that can contribute heavily to the percentage of the total cost are feeding, which ranged from 2.2% to 17.1% and labor around 5%, which was similar between the different locations. Other minor costs were vessel trips, harvesting, maintenance, and lease, representing less than 1% of the total production cost.

14.4 THE MARKETS FOR FARMED TUNA

The market for tuna products is well differentiated between two broad kinds of products: raw and canned tuna. On one side, the most traditional market is for canned tuna, widely developed and globalized during the twentieth century. On the other hand, the sushi and sashimi market, which became a global trend in the last decade and is expanding outside Japan in other countries and regions. The differences in terms of the most common species used, quality requirements, and demand preferences result in two radically different markets. While bluefin tuna is the reference for the sashimi market, skipjack and YFT are the dominant species in the canned tuna market. As a result, tuna ranching is strongly dependent on the sashimi market. On the contrary, the canning industry is exclusively supplied by the wild fishery. The demand for canned tuna is distributed worldwide, led by large, vertically integrated processing companies. It is organized into multinational networks in which the multiple stages of tuna harvesting and processing can be undertaken in different countries before arriving to the final consumers. While integration is also becoming a trend in the sushi and sashimi markets, almost all trade is centralized in Japan, which concentrates almost 90% of the global demand of fresh and frozen bluefin tuna. The traditional markets for tuna products in developed countries have stagnated in the last decade. However, new markets have emerged in different regions of the world which have contributed to maintain the growth in the volume and incomes from world tuna trade. Traded volumes and prices keep on increasing; however, tuna markets are not immune to changes in public concerns. Increasing awareness regarding safety and sustainability of tuna products has affected a number of industry decisions and it is expected to grow in importance. Potential health risks for consumers and environmental concerns need to be assessed in order to avoid adverse consequences in the future.

14.4.1 The Global Sushi and Sashimi Market

Besides the ancient origins of the bluefin tuna fishery and the social impact of this activity on the livelihoods of coastal communities focused on this species, the high value of bluefin tuna is quite a recent phenomenon. Despite having important cultural and economic linkages with the fishing communities dependent on tuna catches, bluefin tuna was a species of little commercial interest in the western world besides recreational and other minor markets such as pet food (Longo, 2011). In many commercial fisheries, big bluefin tunas were even considered a nuisance predator of targeted species (Safina and Klinger, 2008). The growth and developments in production of this species of tuna have been driven by changes in seafood consumption in Japan, where bluefin tuna meat became an essential raw material for expensive sushi and sashimi preparations with the economic recovery after the war (Martínez-Garmendia and Anderson, 2005; Kurokura et al., 2012).

As previously discussed, the Japanese market far exceeds any other market in consumption and imports of bluefin tuna. In 2011, Japan accounted for 76% of the world imports of bluefin species of any kind, including fresh and frozen commodities. The share of the Japanese demand in global bluefin trade varies across commodities. Japan is the major market for all bluefin tuna species, with market shares in global imports exceeding 50% in ABFT and PBFT and over 80% in the case of SBFT (Table 14.4). Although much less significant than Japan, other countries such as the United States, Spain, and Italy have increased their consumption of bluefin tuna, affecting the Japanese share of the global demand. The increase of bluefin tuna imports into Japan was not a result of a decrease of domestic production but rather changes in the consumption structure in the market. Domestic production of PBFT in Japan has been relatively stable or has slightly increased after 1995, and imports from other countries have increased since 2000. As a result, the total supply of bluefin tuna in Japan has increased (Kurokura et al., 2012).

Tuna prices fell in Japan during the economic recession in the early 1990s. In the following years prices were also affected by the increasing imports of farmed ABFT from the Mediterranean. Sashimi-grade tuna prices

TABLE 14.4 Japanese Share of Global Bluefin Imports

	Atlantic and Pacific Bluefin Tuna		Southern Bluefin Tuna	
	Fresh	Frozen	Fresh	Frozen
2000	51%	55%	100%	100%
2006	48%	81%	91%	93%
2011	50%	70%	88%	95%

Source: FAO (2015). FishComm. Global Commodities Production and Trade (1976–2011).

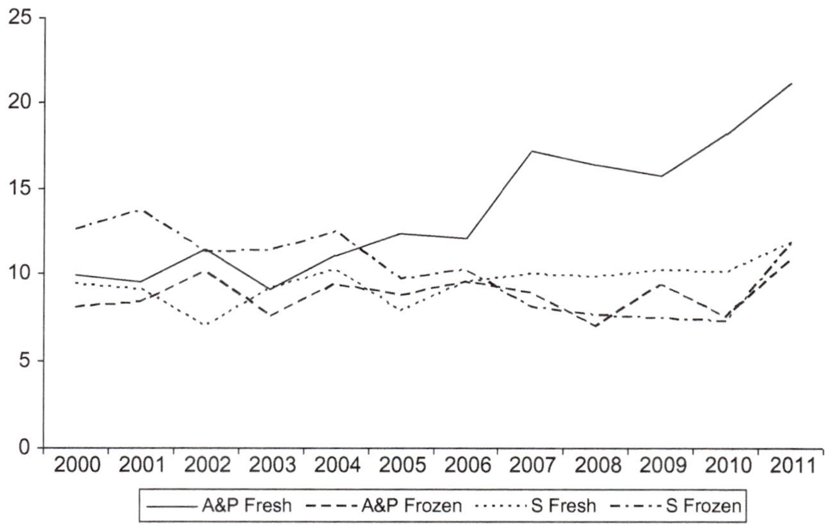

FIGURE 14.4 Evolution of the export price of bluefin tuna commodities. A&P = Atlantic and Pacific bluefin. S = SBFT. *Source: FAO (2015). FishComm. Global Commodities Production and Trade (1976–2011).*

have continued to remain stagnant. In the past decade, supermarket and restaurant chains have gained relevance in sales volume in front of the traditional auction system. Fresh tunas are normally marketed whole through auctions, while around 70–80% of the frozen products are sold to other agents outside the auction system. This shift in the distribution of tuna in Japan contributed to increase in Japanese consumers' preference toward lower-cost foods (Hamilton et al., 2011). The export price of bluefin tuna has been stable in the last decade in almost all commodities (Figure 14.4). Fresh ABFT and PBFT show the greatest increase in price, mainly motivated by a decrease in supply from the Mediterranean due to the implementation of conservation measures on the stocks. Fresh ABFT have the best possible quality according to Japanese consumers' preferences, which results in the highest market prices. Frozen bluefin and imported farmed tuna are not regarded as having the highest quality characteristics by sushi graders. These sources of supply opened up a medium-quality sushi market.

The difference between quantity and value of Japanese imports is an indicator of the high value and quality of bluefin tuna that is exported into Japan. The price of bluefin tuna is set according to a range of different attributes. Every fish is evaluated before being auctioned in Japan and it is priced according to the grade of each attribute. The grades of the attributes are strongly dependent of the harvesting procedures resulting in a price premium for specific techniques. Carroll et al. (2001), observed that differences in freshness, fat content, color and shape, along with weight, resulted in different price levels in the Japanese auctions. Similar results were found in a multispecies study in Hawaii (McConnell and Strand, 2000), where the species, fat content, size, and handling method were found to be causes of variation in the auction price.

The high price of bluefin tuna allowed supporting the costs of air transportation, improving the freshness of the fish auctioned in Japan. Improvements in food storage technology allowed for conservation of the most appreciated attributes of bluefin tuna. Finally, the expansion of middle-low range markets allowed the Japanese middle class to access products previously reserved for higher income classes, resulting in a boom in consumption (Longo, 2011).

Sushi has been consumed by the Japanese for centuries and is the most famous element in the traditional Japanese cuisine. However, the ways of preparation and the main raw materials and species have significantly changed over the years. The notion the world has today about sushi, in the form of raw or slightly macerated fish with rice, is actually a relatively new trend (Kurokura et al., 2012; Kamoey, 2015). Sushi evolved to become an expensive food after the Meiji restoration in 1868, and was eaten at special events. White fish species were generally preferred for sushi than red flesh. It was in the 1970s and 1980s, boosted by new interest from consumers on species with high fat content, when bluefin tuna became a luxury good (Longo, 2011). In the years of economic prosperity, maguro toro (bluefin tuna belly) became a symbol of status for the Japanese elites, and restaurants and sushi bars started including this special dish in their menus on a regular basis. The development of conveyor belt sushi bars made sushi available to the middle-class consumers in Japan, and contributed to the expansion of the Japanese seafood cuisine to other regions in the world. Sushi consumption extended to other nations in the 1990s. By the new century, bluefin became the most appreciated and expensive species of tunas in the world (Ottolenghi, 2008).

Conveyor belt sushi bars started in 1958 in Osaka and the system received a utility model patent in 1962. The new system allowed dealing with a large number of customers with very low labor requirements. The major restaurant chains entered into this business by the end of the 1970s, significantly increasing the number of restaurants. These restaurants deal in relatively low priced sushi, making the experience accessible to a wider range of income segments. Sushi bars could expand their business with a low margin policy with high turnover. The high price of the fatty meat of wild bluefin tuna makes it difficult to use under the strategy of these sushi chains. The major companies started importing farmed bluefin tuna for fatty meat after 2000, including it in the everyday menu by 2004, strengthening the linkage between the Japanese market and the overseas farmed bluefin tuna industry (Kurokura et al., 2012). The growth of sushi bar chain businesses in Japan is slowing down indicating some saturation. Preferences of the new generations, for whom salmon is more popular than tuna, can also affect the future evolution of the Japanese demand of farmed bluefin tuna (Kamoey, 2015). Japan's consumption of imported cultured bluefin tuna is not expected to increase in the future. However, sushi bars are expanding worldwide, which may compensate for a potential stagnation of the Japanese market.

Fatty meat of other tuna species is being also used in the modern sushi and sashimi market. Although the taste is not as good as bluefin, it increases the trade margins. Bigeye was considered a substitute of bluefin in the mid-1990s, and several studies were undertaken in order to analyze substitution across tuna species in the Japanese auctions (Bose and McIlgorm, 1996). Bigeye was found to be a substitute of YFT and YFT a substitute of albacore. However, substitution between bluefin and bigeye has not been proven. In recent years bigeye catches from the Japanese longline fleet have declined due to a number of factors including sustainability, political and economic issues. Import quantities of fresh and frozen bigeye also declined after 2003. Bigeye prices have increased significantly in the last decade, in a higher proportion than frozen bluefin. However, it is still competitive for medium grade sashimi since the price is lower and more stable than bluefin (Figure 14.5).

Estimating the consumption of sashimi outside Japan is almost impossible due to the levels of aggregation of the trade data. Fresh or frozen raw tuna can be consumed cooked in many different ways in the other nations, and it is difficult to know which percentage would end in the sushi market of every country. The second largest market in volume for sushi and sashimi is the United States, with an estimated share of 8 to 10% of global sashimi consumption. In the last decade the popularity of Japanese food increased across all European countries. Important markets have also grown in other Asian nations such as Taiwan, Korea and China, as well as in Australia, New Zealand, and Latin America. The future of the bluefin tuna market, and by extension of the tuna ranching industry, appears to be linked to the success of the Japanese cuisine in becoming an essential part of a global multicultural diet.

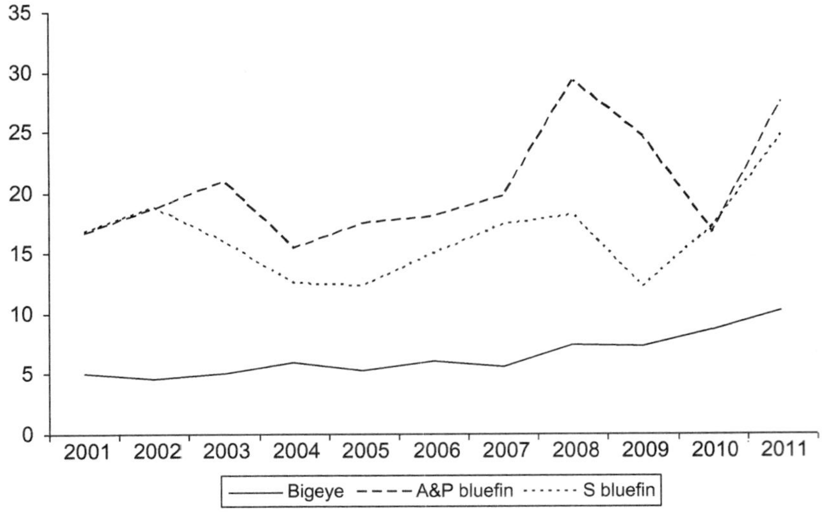

FIGURE 14.5 Evolution of the import price of frozen bluefin and bigeye tuna commodities. A&P = Atlantic and Pacific bluefin. S = SBFT. *Source: FAO (2015). FishComm. Global Commodities Production and Trade (1976–2011).*

14.4.2 Barriers for Tuna Consumption

Tuna consumption faces threats of discredit coming from information provided to consumers both from the mass media and governmental advisory decisions in some countries. Tuna species present two objective problems affecting public opinion and, subsequently, consumer's perceptions. The first relates to concerns about methyl-mercury and human health. The other relates to the sustainability of the tuna fishery. Safety and sustainability are two key concepts affecting consumers' decisions on food, in general, and on seafood in particular (Fernández-Polanco and Luna, 2012).

Methyl-mercury is an organic form of mercury which presents high levels in long-lived, predatory fish, such as tuna, shark, and swordfish. Despite the benefits of seafood consumption (FAO/WHO, 2011), the issues of methyl-mercury in tuna circulated across consumers, media, and governmental institutions which issued recommendations on limiting consumption in populations under risk, such as pregnant women and young children, and may have a negative effect on other species and market segments.

The issue of mercury in seafood was recently revitalized by the media and in 2004 the US Food and Drug Administration (FDA) issued an advisory recommending some parts of the population in special risk to avoid consumption of fish with higher mercury levels including several tuna species. The impact of these decisions and further negative information on US consumers' preferences and purchasing behavior was studied using samples of the populations in and out of these risk categories (Shaw Hughner et al., 2009).

The findings suggest that, in general, the expected benefits of fish consumption exceed the potential negative effects of harmful consequences in both groups. However, in the group at risk, knowledge of tuna's harmful attributes negatively impact purchasing behavior. For an out of risk individual, consumption of tuna products is much more influenced by other factors such as culture, tradition, education, or frequency of seafood consumption (Fernandes da Costa et al., 2011). However, such kinds of alerts may not only affect the consumption of the involved species but all seafood in general (Mallison, 2007), and it is a matter of concern for the whole fishing industry.

The other important concern which may affect consumers' perception of tuna products relates to harvest sustainability. Consumers' perception of the sustainability of tuna production is compromised by the status of overfishing in the bluefin tuna fishery (Fromentin, 2009). Fisheries' sustainability is a growing concern for consumers and institutions in developed countries. Different actions and tools like eco-labels and red lists have been put in the market arena in order to drive consumers to certain sustainable certified species and avoid those with higher environmental impacts. The Dolphin-Safe label appears to have been successful in increasing demand and perceived value for canned tuna in the US market (Teisl et al., 2002). However, one cannot expect that consumers and retailers will act in an environmentally friendly manner just by being aware of overfishing and other issues related with tuna fisheries. Such behavior starts with consumers and retailers' concerns, but will not result in a willingness to behave sustainably without the concurrency of other personal and social factors (Abdul-Muhmin, 2007). Such factors may result in different responses to sustainability issues when changing across countries and markets. Consumers in Japan are less aware of these kinds of issues than in the United States and Western Europe, and thus, the impact of eco-labels and other market-oriented tools regarding fisheries' sustainability is much lower (Uchida et al., 2014).

The success of eco-labels not only depends on consumers' awareness or concerns, but also on the implication of the influential stakeholders. The involved agents should comprise governmental or nongovernmental agencies, and should work in a perfect interaction with all stakeholders in the value chain of fishery products (Roheim, 2008). Demand for sustainable certified seafood is increasing in the large retail chains, in order to avoid losses derived from supply breakages and critics from the environmental groups. In the near future a certification scheme of any kind will be a requirement for operating in the mass markets through the retail chains. Several tuna fisheries have adopted certification schemes or are in the process of adoption. The main constraint is in the supply side, in particular in the case of canned tuna species. Canned tuna is generally sold in low margin markets and it is not clear that price premiums may be possible away from very selective segments. The certification processes result in important costs which could not be recovered if premium prices from final consumers are not secured. The potential negative market consequences from failing in the process or withdrawing from a certification program also prevents some tuna fisheries from attempting entering into certification schemes (Hamilton et al., 2011).

14.5 SUMMARY

Tuna species are some of the most popular fish for human consumption. The market has been growing in the past and it is still expanding and profitable in general terms. Tuna harvesting has faced important challenges in the past years, adapting the production processes in the search for improved efficiency and satisfaction of demand. Important changes have been undertaken in the production, organization, and marketing fields. Some of these changes are fully completed and others are still ongoing.

One of these challenges is in regards to the conservation of the wild stocks and the sustainability of the harvesting techniques. Production has increased exponentially since 1950, but signs of stagnation in the stocks of some species started appearing by the end of the 1990s. Although aquaculture production reached relevant levels of development in the last two decades, future growth is limited by the availability of wild stocks. The completion of the biological cycle with reproduction in captivity will set a turning point in the future of tuna harvesting. However, this is not yet a real alternative at an industrial scale, since there is still place for improvement in the technical aspects, and the costs are still too high for being considered a competitive option. Future development in this field will also be conditioned by the status of the wild stocks, which set the market price in the last instance. The conservation measures implemented in the Mediterranean and other bluefin fisheries could result in the recovery of the stocks and an increase in the supply to both the final consumption market and in the availability of juveniles for tuna ranching. The immediate consequence would be a decrease in the price which may widen the cost gap between full cycle aquaculture, ranching, and captures.

The markets for tuna products continue increasing based on the growth in consumption in new regions of the world and the dissemination of sushi as a global dietary trend. The traditional markets show signs of maturation but still represent a huge and profitable volume for business. Growing interest is focused on developing new value-added products which may help increasing the margins in a scenario of increasing raw materials price. Consumer concerns regarding safety and sustainability issues with tuna consumption and harvesting may rise as a potential threat for future market development. The industry will have to deal with these concerns which are growing across consumers. Besides the different marketing tools available, transparency and fluid communication between the industry and the public will have to be strengthened in the future.

REFERENCES

Abdul-Muhmin, A.G., 2007. Explaining consumers' willingness to be environmentally friendly. Int. J. Consum. Stud. 31, 237–247.

Bose, S., McIlgorm, A., 1996. Substitutability among species in the Japanese tuna market: a cointegration analysis. Mar. Resour. Econ. 11, 143–155.

Carroll, M.T., Anderson, J.L., Martínez-Garmendia, J., 2001. Pricing US North Atlantic bluefin tuna and implications for management. Agribusiness 17, 243–254.

FAO, 2015. FishstatJ. Available at: <http://www.fao.org/fishery/statistics/software/fishstat/en>.

FAO/WHO, 2011. Report of the Joint FAO/WHO Expert Consultation on the Risks and Benefits of Fish Consumption, January 25–29, 2010. FAO Fisheries and Aquaculture Report No. 978 FIPM/R978(En), Rome, Italy.

Fernandes da Costa, P.M., Hu, W., Pan, M., 2011. Ahi poke (raw tuna salad) consumption and consumer characteristics in Hawaii. Aquaculture Econ. Manag. 15 (4), 302–315.

Fernández-Polanco, J., Luna, L., 2012. Factors affecting consumers' beliefs about aquaculture. Aquaculture Econ. Manag. 16 (1), 22–39.

Fromentin, J.-M., 2009. Lessons from the past: investigating historical data from bluefin tuna fisheries. Fish and Fish. 10 (2), 197–216.

Hamilton, A., Lewis, A., McCoy, M.A., Havice, E., Campling, L., 2011. Market and Industry Dynamics in the Global Tuna Supply Chain. FFA.

ICCAT, 2006. Recommendation by ICCAT to establish a multi-annual recovery plan for bluefin tuna in the eastern Atlantic and Mediterranean. Available at: <https://www.iccat.int/Documents/Recs/compendiopdf-e/2006-05-e.pdf>.

Kam, L.E., Leung, P., Ostrowski, A.C., 2003. Economics of offshore aquaculture of Pacific threadfin (*Polydactylus sexfilis*) in Hawaii. Aquaculture 223 (1–4), 63–87.

Kamoey, A., 2015. *The Japanese Market for Seafood.* Globefish Research Programme, vol. 117. FAO, Rome.

Kurokura, H., Takagi, A., Sakai, Y., Yagi, N., 2012. Tuna goes around the world on sushi. Aquaculture Econ. Manag. 16 (2), 155–166.

Longo, S.B., 2011. Global sushi: the political economy of the Mediterranean bluefin tuna fishery in the modern era. J. World-Syst. Res. 17 (2), 403–427.

Mallison, A., 2007. Aquaculture – what retailers expect from producers. In: Arthur, R. Nierentz, J. (Eds.), Global Trade Conference on Aquaculture, May 29–31, 2007, Qingdao, China. FAO Fisheries Proceedings 9, Rome.

Martínez-Garmendia, J., Anderson, J.L., 2005. Conservation, markets, and fisheries policy: the North Atlantic bluefin tuna and the Japanese sashimi market. Agribusiness 21 (1), 17–36.

McConnell, K.E., Strand, I.E., 2000. Hedonic prices for fish: tuna prices in Hawaii. Am. J. Agricult. Econ. 82, 133–144.

Morris, T., 2011. Marine fish culture in Mexico. Amberjack, yellowtail farming developing. Global Aquaculture Advocate 14 (1), 76–77.

Ottolenghi, F., 2008. Capture based aquaculture of bluefin tuna. Capture-Based Aquaculture: Global Overview. FAO Fisheries Technical Paper No. 508, Rome.

Patterson, H., Stobutzki, I., Stephan, M., 2012. Southern Bluefin Tuna Fishery, in Fishery Status Reports 2012. Australian Bureau of Agricultural and Resource Economics and Sciences.

Roheim, C.A., 2008. The economics of ecolabeling. In: Ward, T., Phillips, B. (Eds.), Seafood Ecolabelling Principles and Practice. Wiley-Blackwell, Oxford. UK.

Safina, C., Klinger, D.H., 2008. Collapse of bluefin tuna in the western Atlantic. Conserv. Biol. 22, 243–246.

Shamshak, G.L., 2011. Economic evaluation of capture-based bluefin tuna aquaculture on the US east coast. Mar. Resour. Econ. 26 (4), 309–328.
Shamshak, G.L., Anderson, J.L., 2009. Dynamic stochastic adaptive bioeconomic model of off-shore bluefin tuna aquaculture. Aquaculture Econ. Manag. 13 (2), 155–175.
Shaw Hughner, R., Kurp Maher, J., Childs, N.M., Nganje, W.E., 2009. Fish: friend or foe? Food policy and subpopulation warnings for consumers. Food Policy 34, 185–197.
Teisl, M.F., Roe, B., Hicks, R.L., 2002. Can eco-labels tune a market? Evidence from dolphin-safe labeling. J. Environ. Econ. Manag. 43 (3), 339–359.
Uchida, H., Roheim, C.A., Wakamatsu, H., Anderson, C.M., 2014. Do Japanese consumers care about sustainable fisheries? Evidence from an auction of ecolabelled seafood. Aust. J. Agric. Resour. Econ. 58, 263–280.

Index

Note: Page numbers followed by "*f*" and "*t*" refer to figures and tables, respectively.

B

C

H

I

J

K

L

Kolibri 360 GmbH
Ferdinand-Jühlke-Straße 7
99095 Erfurt, Deutschland
produktsicherheit@kolibri360.de